Neugebauer

SOLAR COMPOSITION AND ITS EVOLUTION – FROM CORE TO CORONA

Cover illustration: Image of the solar corona as recorded by the outer LASCO coronagraph (C3) on 23 December 1996. The Sun is located in the constellation Sagittarius. The center of the Milky Way is visible, as well as the dark interstellar dust rift, which stretches from the south to the north. The cloudy solar wind can be seen along the ecliptic plane, both over the east and the west limb of the Sun, stronger over the latter. The image also shows Comet SOHO-6, one of 7 sungrazers discovered so far by LASCO, as its head enters the equatorial solar wind region. It eventually plunged into the Sun. (Courtesy of SOHO/LASCO consortium. SOHO is a project of international cooperation between ESA and NASA.)

Space Sciences Series of ISSI

Volume 5

The International Space Science Institute is organized as a foundation under Swiss law. It is funded through recurrent contributions from the European Space Agency, the Swiss Confederation, the Swiss National Science Foundation, and the Canton of Bern. For more information, see the homepage at http://vega.unibe.ch/

The titles in this series are listed at the end of this volume.

SOLAR COMPOSITION AND ITS EVOLUTION – FROM CORE TO CORONA

Proceedings of an ISSI Workshop
26–30 January 1998, Bern, Switzerland

Edited by

C. FRÖHLICH
Physikalisch-meteorologisches Observatorium, World Radiation Center, Davos, Switzerland

M. C. E. HUBER
Space Science Department of ESA, ESTEC, Noordwijk, The Netherlands

S. K. SOLANKI
Institute of Astronomy, ETH, Zürich, Switzerland

R. VON STEIGER
International Space Science Institute, Bern, Switzerland

KLUWER ACADEMIC PUBLISHERS
DORDRECHT / BOSTON / LONDON

Space Sciences Series of ISSI

A C.I.P. Catalogue record for this book is available from the Library of Congress.

ISBN 0-7923-5496-6

Published by Kluwer Academic Publishers,
P.O. Box 17, 3300 AA Dordrecht, The Netherlands.

Sold and distributed in North, Central and South America
by Kluwer Academic Publishers,
101 Philip Drive, Norwell, MA 02061, U.S.A.

In all other countries, sold and distributed
by Kluwer Academic Publishers,
P.O. Box 322, 3300 AH Dordrecht, The Netherlands.

Printed on acid-free paper

All Rights Reserved
©1998 Kluwer Academic Publishers
No part of the material protected by this copyright notice may be reproduced or
utilized in any form or by any means, electronic or mechanical,
including photocopying, recording or by any information storage and
retrieval system, without written permission from the copyright owner

Printed in the Netherlands.

TABLE OF CONTENTS

Foreword   xi

Unveiling the Secrets of the Sun
R. M. Bonnet   1

I: SOLAR INTERIOR

The 'Standard' Sun
J. Christensen-Dalsgaard   19

Shortcomings of the Standard Solar Model
W. A. Dziembowski   37

Microphysics: Equation of State
W. Däppen   49

Opacity of Stellar Matter
F. J. Rogers and C. A. Iglesias   61

Element Settling in the Solar Interior
S. Vauclair   71

Macroscopic Transport
J.-P. Zahn   79

Solar Neutrinos
Y. Suzuki   91

Lithium Depletion in the Sun: A Study of Mixing Based on Hydrodynamical Simulations
T. Blöcker, H. Holweger, B. Freytag, F. Herwig, H.-G. Ludwig, and M. Steffen   105

On the Velocity and Intensity Asymmetries of Solar p-mode Lines
M. Gabriel   113

Sensitivity of Low-frequency Oscillations to Updated Solar Models
J. Provost, G. Berthomieu, and P. Morel   117

Composition and Opacity in the Solar Interior
S. Turck-Chièze   125

Solar Models with Non-Standard Chemical Composition
S. Turcotte and J. Christensen-Dalsgaard   133

On the Composition of the Solar Interior
D. Gough   141

II: LOWER SOLAR ATMOSPHERE

Standard Solar Composition
N. Grevesse and A. J. Sauval   161

Structure of the Solar Photosphere
S. K. Solanki   175

The Structure of the Chromosphere
P. G. Judge and H. Peter — 187

The Solar Quiet Chromosphere–Corona Transition Region
L. S. Anderson-Huang — 203

FIP Fractionation: Theory
J.-C. Hénoux — 215

FIP Effect in the Solar Upper Atmosphere: Spectroscopic Results
U. Feldman — 227

Constraints on the FIP Mechanisms from Solar Wind Abundance Data
J. Geiss — 241

Element Separation in the Chromosphere
H. Peter — 253

Temporal Evolution of Artificial Solar Granules
S. R. O. Ploner, S. K. Solanki, A. S. Gadun, and A. Hanslmeier — 261

The Lower Solar Atmosphere
R. J. Rutten — 269

III: UPPER SOLAR ATMOSPHERE AND SOLAR WIND

Elemental Abundances in Coronal Structures
J. Raymond, R. Suleiman, J. L. Kohl, and G. Noci — 283

Structure of the Solar Wind and Compositional Variations
P. Bochsler — 291

The Solar Noble Gas Record in Lunar Samples and Meteorites
R. Wieler — 303

Atomic Physics for Atmospheric Composition Measurements
P. R. Young and H. E. Mason — 315

Solar Energetic Particles: Sampling Coronal Abundances
D. V. Reames — 327

UVCS/SOHO: The First Two Years
S. R. Cranmer, J. L. Kohl, and G. Noci — 341

The Expansion of Coronal Plumes in the Fast Solar Wind
L. Del Zanna, R. von Steiger, and M. Velli — 349

Fractionation of Si, Ne, and Mg Isotopes in the Solar Wind as Measured by SOHO/CELIAS/MTOF
R. Kallenbach, F. M. Ipavich, H. Kucharek, P. Bochsler, A. B. Galvin, J. Geiss, F. Gliem, G. Gloeckler, H. Grünwaldt, S. Hefti, M. Hilchenbach, and D. Hovestadt — 357

Solar EUV and UV Emission Line Observations Above a Polar Coronal Hole
K. Wilhelm and R. Bodmer — 371

Solar Energetic Particle Isotopic Composition
D. L. Williams, R. A. Leske, R. A. Mewaldt, and E. C. Stone — 379

O^{5+} in High Speed Solar Wind Streams: SWICS/Ulysses Results
R. F. Wimmer-Schweingruber, R. von Steiger, J. Geiss, G. Gloeckler,
F. M. Ipavich, and B. Wilken   387

Element and Isotopic Fractionation in Closed Magnetic Structures
T. H. Zurbuchen, L. A. Fisk, G. Gloeckler, and N. A. Schwadron   397

Composition Aspects of the Upper Solar Atmosphere
R. von Steiger   407

Is the Sun a Sun-like Star?
B. Gustafsson   419

Author Index   429

List of Participants   430

viii

ISSI Workshop
Solar Composition and Its Evolution – From Core to Corona
26–30 January 1998, Bern, Switzerland
Group Photograph

1.	R. von Steiger	13.	S. Turck-Chièze	25.	R. Rutten
2.	R. Mewaldt	14.	M. Asplund	26.	S. Turcotte
3.	C. Fröhlich	15.	M. Huber	27.	L. Anderson
4.	T. Zurbuchen	16.	R. Wimmer	28.	D. Gough
5.	R. Kallenbach	17.	C. Charbonnel	29.	W. Dziembowski
6.	S. Vauclair	18.	S. Solanki	30.	J. Raymond
7.	F. Rogers	19.	H. Peter	31.	S. Nusser Jiang
8.	J. Christensen-Dalsgaard	20.	Ph. Judge	32.	M. Gabriel
9.	J.-P. Zahn	21.	J.-C. Hénoux	33.	B. Gustafsson
10.	H. Holweger	22.	J. Provost	34.	J. Geiss
11.	J. Kohl	23.	P. Young	35.	K. Wilhelm
12.	N. Grevesse	24.	D. Reames	36.	U. Feldman

Not on this picture: P. Bochsler, R. Bonnet, L. Del Zanna, Y. Suzuki, K. Widing
Photograph by D. Taylor

FOREWORD

The discovery of chemical elements in celestial bodies and the first estimates of the chemical composition of the solar atmosphere were early results of Astrophysics – the subdiscipline of Astronomy that was originally concerned with the general laws of radiation and with spectroscopy.

Following the initial quantitative abundance studies by Henry Norris Russell and by Cecilia Payne-Gaposchkin, a tremendous amount of theoretical, observational, laboratory and computational work led to a steadily improving body of knowledge of photospheric abundances – a body of knowledge that served to guide the theory of stellar evolution. Solar abundances determined from photospheric spectra, together with the very similar abundances determined from carbonaceous chondrites (where extensive information on isotopic composition is available as well), are nowadays the reference for all cosmic composition measures.

Early astrophysical studies of the solar photospheric composition made use of atmosphere models and atomic data. Consistent abundances derived from different atmospheric layers and from lines of different strength helped to confirm and establish both models and atomic data, and eventually led to the now accepted, so-called "absolute" abundance values – which, for practical reasons, however, are usually given relative to the number of hydrogen nuclei.

Over the past decades, with solar-physics investigations reaching well beyond the photosphere, a more complex view of the Sun's composition has emerged. This development was stimulated to a large extent by use of space techniques, be it to gain access to solar radiation in the vacuum-ultraviolet and X-ray domains, or to study the solar wind and solar energetic particles by direct detection. Helioseismology investigations – in part also conducted from space – provided a further dimension to our knowledge of solar structure and composition. Specifically, space research has provided detailed insights

- into the structural, dynamic and compositional subtleties of the solar interior,
- into solar atmospheric structures, particularly those of the outer solar atmosphere, and their different composition, as well as
- into the varied chemical and isotopic composition and charge-state distribution of the solar wind and of solar energetic particles.

In addition, studies of the solar neutrino flux have revealed a deficit of electron neutrinos. This finding might be explained by an unexpected behaviour of the solar core, or by an adaption of the theory of elementary particle physics – i.e., by a change of assumptions either in astronomy or in physics. Given the current state of knowledge of the solar core from helioseismology studies, it seems more likely that the neutrino deficit will be explained (or, one might say, removed) by modifications of the physics of elementary particles rather than by a change of parameters describing the solar core.

The Sun now studied is far removed from its earlier models, a homogeneous spherical body described by temperature and density that varied with radial distance: chemical fractionation as well as considerable fine-structure in, and above the solar "surface" (as, e.g., magnetic loops), and internal mixing must now be taken into account. And, accordingly, attention in composition studies has shifted from "absolute solar" abundances toward differences in the chemical (and sometimes also isotopic) composition of the different parts of the Sun – more precisely: in its interior, in the structures that make up its atmosphere and in solar-wind and other, more energetic, particle streams. As was the case with the earlier photospheric composition studies, the investigation of abundance differences results in further clues and input for understanding, in particular, mass-transport, separation and heating mechanisms.

As the findings that led to the current view of the Sun and its composition stem to a large extent from space-based measurements, obtained with different techniques and methods, the convenors – C. Fröhlich, D. Gough, N. Grevesse, M. C. E. Huber, and S. Solanki – proposed to the International Space Science Institute (ISSI) in Bern, an institute which facilitates and supports such multidisciplinary events, to organise a Workshop on "Solar Composition and its Evolution – from Core to Corona". ISSI's management accepted the proposal and invited leading representatives of all scientific fields concerned in order to achieve an in-depth synopsis of the subject.

The Workshop was held from 26 to 30 January 1998. This volume contains the reviews and contributions presented, and summarises the resulting discussions. The written versions of all papers were refereed and some of the papers were considerably modified in the process. It is hoped that this has led to the coherent overview of the subject that was intended by the convenors of the Workshop.

We wish to express our sincere thanks to all those who have made this volume possible. First of all, we should like to thank the authors for writing original articles, for keeping to the various deadlines, and for producing unusually neat camera-ready versions. We also thank the reviewers for their critical and timely reports, which have significantly contributed to the quality of this volume. Finally, it is our pleasure to thank ISSI, in particular its directors J. Geiss and B. Hultqvist, for hosting and supporting this workshop, and the ISSI staff, R. Kallenbach, V. Manno, G. Nusser Jiang, U. Pfander, X. Schneider, D. Taylor, and S. Wenger, for the local organization and for assistance in the preparation of this volume. ISSI's perfectly smooth organisation permitted the Workshop participants to fully concentrate on the topics and to advance the subject by providing this up-to-date 'compte rendu' of the current knowledge of solar composition!

August 1998
C. Fröhlich, M. C. E. Huber, S. K. Solanki, R. von Steiger

UNVEILING THE SECRETS OF THE SUN

R.M. BONNET
European Space Agency
8-10 rue Mario Nikis
75015 Paris, France

1. Introduction

The Sun is our star. The source of light and energy which makes our life possible. A source of inspiration for scientists and for artists. However, until a few decades ago, we only knew that star through the observations of its yellow/orange visual disk and through the occurrence of a few rare eclipses, when the Moon covers nearly exactly the Sun's disk and reveals the corona, the spectacular extension of the solar atmosphere into interplanetary space. Everything we knew, in particular of the interior of the Sun, was derived from models built from the "observable" quantities: geometrical dimension, mass, brightness, surface temperature, ratio of abundance of helium over hydrogen, etc....

It should be no surprise that understanding the Sun, a star only eight light-minutes away from us, has been and is still today a task of major interest and scope. For astrophysicists, who have a model star at hand. For physicists, who want to check their theories on particles, cosmic rays, plasma phenomena and magnetic fields. For geophysicists and climatologists who want to understand the influence it exerts on the Earth. For all, the Sun is a unique, close-by, immense laboratory. After centuries of solar observations, only now do we have the necessary tools to start understanding this fantastic and complex machine.

2. Observing the Sun

Observing the Sun is not as easy as it might seem. Indeed, it is the only star close enough to easily resolve its surface features through a telescope. However, building a solar telescope requires special precautions, because the Sun is bright and its radiation heats up every part of the instruments. While it is easy to observe stars with the naked eye, it is very dangerous to do so in the case of the Sun, except under very special conditions, like when the sky is foggy, in the desert under sandstorm conditions, or sometimes during sunrise or sunset. This is the reason why we may assume that the existence of sunspots was noticed long before the invention of the telescope: they can be observed relatively easily on the disk under the conditions just mentioned. Similarly, the appearance of the corona can be seen, in principle, during the regularly occurring solar eclipses. Nevertheless, in the description of the

Babylonian eclipse of 1063 BC, the phenomenon of the corona seems to have been mentioned for the first time (Golub and Pasachoff, 1997), and it was only during the second half of the 19th century that the association of the corona with the Sun became firmly established. The first to describe sunspots and even to publish some drawings of their appearance on the disk was Galileo in 1612. Following Galileo, progress in the qualities of telescopes and the use of high-altitude observatories made it possible to observe more and more refined details on the surface. Solar granulation had been observed already at the end of the 19th century.

In spite of the Sun's proximity, several of its properties and characteristics have remained in the realm of speculation for a long time. Recurrent questions have dominated the life and the work of astrophysicists. What are the mechanisms that modulate the apparition of sunspots and faculae every 11 years or so? What are sunspots? Why is the temperature of the corona hotter than the Sun's "surface" below it? What is (are) the source(s) of the solar wind and what "pushes" it? Why are solar neutrinos missing? Is the solar constant a genuine constant? Just to mention a few.

The most spectacular progress in our knowledge of the Sun came just before the Second World War and has continued ever since due to major advances in instrumentation and observing techniques. One of them, was the invention of the coronagraph by Lyot (1930). Instead of having to wait for two years or so in the average for a suitable solar eclipse to occur, daily observations of the corona could become nearly routine, weather permitting. Similarly, the advent of radioastronomy and of space techniques after the war led to a genuine revolution in the understanding of our star. For the first time, the corona could be observed not only outside eclipses, but also without the help of a coronagraph, and in front of the disk.

Space observations above the Earth's atmosphere eliminate two major disadvantages of ground based observations: weather dependence and atmospheric turbulence on the one hand, and, on the other, atmospheric absorption which prevents spectral coverage at wavelengths shorter than 300 nm and longer than 2 μm. The combination of the increased opacity of the solar atmosphere at short (UV, EUV, X-rays) and long (IR) wavelengths, together with the raise in temperature of the outer solar layers makes it possible to observe the chromosphere and the corona outside eclipses and in front of the disk. Balloons, high-altitude aircraft, sounding rockets or artificial satellites bring within reach the whole electromagnetic spectrum. Above the atmosphere, the "seeing" is in theory perfect and limited only by diffraction, the mechanical stability of the instrument structure and the combined pointing capability of the internal optics and of the platforms which carry the telescopes.

Solar observations from space were pioneered at the US Naval Research Laboratory (Tousey, 1953) in 1946 when they successfully launched, and recovered, an ultra-violet spectrometer on board one of the V-2 rockets which were brought in pieces to the US by Wernher von Braun and his co-workers after the war. For the first time in history the very existence of ultraviolet radiation from the Sun could be confirmed, as well as the existence of an absorbing terrestrial ozone layer at altitudes up to 25 km.

Following this early success, a large number of rocket and satellite-borne spectrometers were launched, starting a very successful period for solar physics. Although that discipline was for a long time already exclusively European, the most spectacular space success came from the United States, with NASA's series of Orbiting Solar Observatories launched in the 60's and early 70's and with the Apollo Telescope Mount which in 1973 made an efficient use of the upper stage of the Saturn 5 rocket and contributed crucial observations of the upper solar atmosphere in the far ultraviolet and in X-rays as well as by use of a coronagraph. For the first time, coronal structures could be seen in detail in front of the disk, among other features, coronal holes, coronal loops and X-ray bright points. The phenomenon of coronal mass ejections was discovered as well. Coronal holes were originally named as such by Waldmeier in the fifties, later seen in front of the disk by OSO-4, bright points were discovered by an AS&E rocket flight.

In parallel, the first interplanetary probes could confirm the existence of the solar wind, inferred by Biermann (1951) in Germany from the shape of comet tails, and theoretically treated by Parker (1958) in the United States. These early measurements were remarkably precise and revealed the existence of a two-velocity regime solar-wind, as well as the sector structure of the interplanetary magnetic field and the recurrence of high velocity jets with the periodicity of 27 days imposed by solar rotation.

The great majority of these observations concerned the measurements of the ultraviolet radiation, which were crucial for linking solar and terrestrial phenomena as well as for studying the extension of the solar corona and of the solar wind to the limit of the Earth's magnetosphere. In this context, the measurement of the solar constant, the total solar radiative output, has long been considered essential in relation with our climate on Earth. Unfortunately, due to instrumental limitations, it could not be done with the required degree of accuracy of a few parts per million, and had to wait until the late 70's to be performed properly.

In the late sixties and early seventies, two new measurements were at the origin of another revolution and of a complete rejuvenation of the field: the detection of neutrinos and the observations of solar oscillations, allowing for the first time to sound the deep layers of the Sun underneath the photosphere.

The early neutrino measurements by Davis in the United States (Bahcall and Davis, 1976) revealed an intriguing phenomenon: only one third of the neutrinos emitted by the Sun during the fusion of hydrogen into helium in the very hot core of the Sun, could be captured. The mystery of the missing neutrinos has remained unsolved ever since. This surprising result has been confirmed by new and more accurate measurements from different experimenters all around the world. The "neutrino deficit" could be interpreted either by an imperfect representativity of the standard model of the Sun's interior, or by an improper knowledge of the nuclear cross sections, or by the behaviour of the neutrinos themselves.

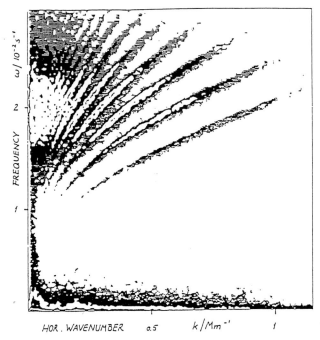

Figure 1. The best κ, ω diagram of high degree solar p-modes obtained by Deubner (1975), illustrating the state of the art in 1976.

In parallel, the discovery of the 5-minute oscillations by Leighton *et al.* (1962) and the confirmation by Deubner (1975) that they represented acoustic resonances in different cavities of the Sun's interior (Figure 1), opened a new prospect for investigating the physical properties of the solar interior. Helioseismology was born and, for the first time, it became possible to infer more precisely the solar model through the determination of the temperature, pressure and also of the chemical composition of the deep layers from accurate measurements of the frequencies of the oscillating modes, probing different acoustic resonance cavities at various depths in the solar interior.

Important progress was made when Grec, Fossat and Pommerantz (1980), taking advantage of the Antarctic summer, could observe the Sun without interruption for more than 4 consecutive days from the South Pole, thereby dramatically increasing the accuracy of the measurements. These observations proved the power of uninterrupted measurements, and have been at the origin of a genuine revolution in observational helioseismology. Following that success, two approaches were used to increase the continuity of observations. First, networks of ground stations installed all around the Earth (IRIS and GONG networks) would relay the collecting data from one station to the next. Second, space measurements, using eclipse-free trajectories, were undertaken on interplanetary probes (IPHIR on the Russian Phobos mission) or from SOHO in orbit around Lagrangian point L1.

I cannot end this historical sketch without mentioning the first ever attempt to observe the Sun above its poles. Indeed, all the observations mentioned above have been performed in the plane of the ecliptic, offering at best a grazing view of the solar poles which, as we guess, due to the radial structure of the polar magnetic field should present substantial differences from what is observed in the equatorial plane. The Europeans have developed a unique mission, Ulysses, orbiting the Sun perpendicularly to the ecliptic plane with a period of 6.3 years. Ulysses was launched in October 1990 by NASA with the Space Shuttle and passed successively over the South and North poles of the Sun successively in 1994 and 1995. A description of the mission and its results can be found in Wenzel *et al.* (1992) and in Marsden *et al.* (1996).

3. The Genesis of SOHO

Because of its importance to solar physics the SOHO spacecraft deserves particular attention. This is why the following two chapters are devoted to this unique observatory.

With its comprehensive set of instruments, SOHO makes a unique attempt at answering the not yet understood key questions in solar physics, in particular, origins and acceleration mechanisms of the solar wind, the heating of the corona and the physics of the deep interior of the Sun. It apparently succeeds in that task.

The history of SOHO has been described by Huber *et al.* (1996). The genesis of SOHO can be related to two missions studied by ESA in the late 70's and early 80's, which missed their final selection through the fierce competitive process which prevailed at that time in the European Space Agency science programme.

The first of these, the Grazing Incidence Solar Telescope, GRIST (Figure 2), was studied in the context of the utilisation of Spacelab and of the US space transportation System (STS). It was supposed to fill the gap left in the study of the far ultraviolet spectrum of the Sun by the Solar Maximum Mission (SMM) after NASA decided to drop the grazing-incidence spectrometer of that mission for budgetary and schedule reasons. GRIST was planned as a joint development between Europe and the US, and missed its selection at ESA partly for that very reason. Because of the unilateral decision of NASA in 1981 to eliminate their own satellite from the joint dual spacecraft ESA-NASA Out-of-Ecliptic mission, the ancestor of Ulysses, the relationship between ESA and NASA suffered a severe setback and the ESA selection committees gave a higher priority to ESA-only or to ESA-led missions.

The second aborted mission, the Dual Irradiance and Solar Constant Observatory, DISCO (Figure 3), had been proposed by a group of European scientists to study the solar output and its short and long term variability over at least half a

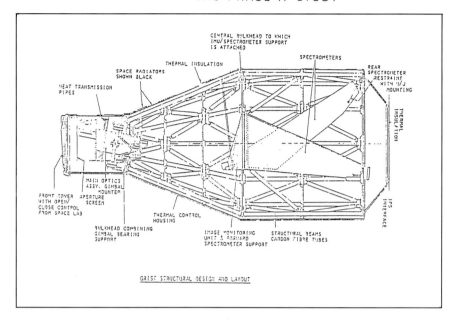

Figure 2. The cover of the GRIST Phase A study report in 1978.

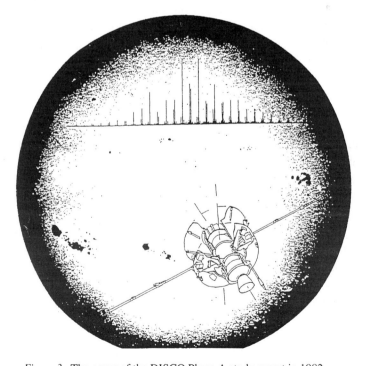

Figure 3. The cover of the DISCO Phase A study report in 1982.

solar cycle. DISCO was conceived as a small and inexpensive satellite, in a way a European precursor of the "smaller faster and cheaper" category that symbolises present-day NASA missions. DISCO was selected for an assessment and subsequently for a phase-A study. Its payload included a set of particle and field measurements to study the properties of the solar wind at 1 AU from the Sun, and a miniaturised version of the instrument used by G. Grec and E. Fossat at Nice Observatory to observe the solar oscillations from the South Pole. Such an instrument unhampered by atmospheric perturbations and operating 24 hours per day and 365 days a year, would be the best tool for detecting the global modes and hopefully the gravity oscillating modes of the Sun which, in theory, provide the best information on the physical conditions of the solar core.

The study team formed by ESA to assess the value and the feasibility of the DISCO mission included Ph. Delache, C. Fröhlich, D. Crommelynck, A. Balogh, C. Harvey and myself. Early in the study, Delache and myself proposed to assign DISCO to an orbit around L1. Such an orbit would offer the essential advantages of the required continuity. Indeed, SOHO is currently in such an orbit, which has abundantly proven its unique advantages. However, at completion of its phase-A in 1983, DISCO lost the final competition against the Infrared Space Observatory (ISO). This apparent misfortune in the end led ESA to introduce into its science programme two of its most prestigious missions, launched at only a 7 day interval at the end of 1995: ISO and SOHO.

SOHO was proposed to ESA in 1982 by the former GRIST team and a selection of the original DISCO scientists. After some evolution, it incorporated three sets of instruments, originally part of the GRIST and DISCO payloads, addressing helioseismology, chromosphere, coronal and solar wind observations as well as in-situ particle and field measurements. At L1, SOHO would therefore allow the most comprehensive study of the Sun from its central nuclear core to the remote end of the corona in the vicinity of the Earth.

SOHO started phase-A in November 1983, at the same time as Cluster, a fleet of 4 satellites in polar orbit, to study the Earth's magnetosphere, as it is influenced by the variability of the Sun's magnetic field. Both missions competed for selection against each other at that time. Fortunately, this competition never materialised, because the formulation of Horizon 2000 by ESA's Survey Committee in 1984 led to the combination of both SOHO and Cluster within the STSP cornerstone, the first to be implemented after the formal approval of ESA's Long Term plan by its Council at ministerial level in 1985 in Rome.

Contrary to DISCO, SOHO was studied with a large and early participation of US scientists, after the effects of the Ulysses crisis in 1981 had faded away. NASA had expressed a strong interest in participating in SOHO, thereby responding to the pressure of the US Solar Physics community which had little prospects for a future mission after SMM and which considered SOHO as the "only thing moving in town". However, NASA officials were not convinced of the interest of going to space for helioseismology measurements. They requested a convinc-

Figure 4. The cover of the SOHO Phase A study report in 1985.

ing demonstration that space was necessary. This was achieved through a series of dedicated meetings involving scientists on both sides of the Atlantic, Europeans being adamant that space was an absolute necessity if one wanted to eliminate atmospheric noise. The requirements for high spatial resolution to serve both the chromosphere-coronal instruments and one of the US seismology instruments (which was to become the Michelson Doppler Imager) led to defining SOHO as a 3-axis stabilised satellite with a one-arc-second stability (Figure 4), while its more modest precursor, DISCO, was a simple spinner.

4. The Challenge of Making SOHO

Following its inclusion in the Horizon 2000 programme, SOHO was soon confronted with difficult challenges. The first one was clearly its cost. All cornerstone missions of Horizon 2000 were assigned a budget capped at no more than 400 Million ECU (MECU) in 1984 economic conditions. Unfortunately, SOHO was only one of the two STSP missions, the four Cluster satellites formed the other part of the cornerstone. Both missions combined reached a budget nearly equal to about 750 MECU (590 without launch), i.e., twice the assigned limit. Three actions were taken to lower the cost: scientific descoping, search for international cooperation and use of the cheapest launch opportunities.

Scientific descoping affected both SOHO and Cluster. A committee chaired by D. Southwood managed to reduce the cost (without launches) from 590 MECU to 500 MECU, by limiting the number of instruments on SOHO and simplifying Cluster to a set of 4 identical satellites (originally Cluster had 3 identical satellites plus a central one). Unfortunately, that was not enough and a second descoping committee was set up, chaired this time by H. Balsiger. Through a better rationalisation of resources and by limiting the operations of Cluster to 2 years, the cost could be brought down to 484 MECU. This however included neither the launch costs of SOHO and Cluster, nor the cost of operating SOHO. It was assumed that international cooperation would give NASA the responsibility of launching and operating SOHO. NASA would also provide some hardware to both SOHO and Cluster, in particular tape recorders. It also assumed that the 4 Cluster spacecraft would be launched at marginal cost on board the second demonstration flight of the new Ariane 5 rocket. It was only when Artemis, an ESA telecommunications satellite which was supposed to fly on the first Ariane 5, suffered an important delay of several years that it was decided to swap Cluster on Ariane 501, a very unfortunate measure since the rocket failed, victim of unforgivable human mistakes (Cavallo, 1996). When phase B was started in 1991, the cost of SOHO and Cluster had gone down to 474 MECU, showing a remarkable saving of some 275 MECU!

The second main difficulty was of a technical and managerial nature and the consequence of the comprehensive character of the scientific objectives of SOHO which led ESA to define a payload including not less than 11 instruments plus all the associated electronics. The biggest instruments were provided by American principal investigators (LASCO, UVCS, MDI) and the true international character of the mission manifested itself very early on, evidencing difficulties due to substantial cultural differences between the two sides of the Atlantic. For the first time in the history of space cooperation between the US and Europe, US principal investigators were to fly their experiments on a European spacecraft led by a team of European engineers at ESTEC and they had to adhere to rules and methods to which they were not accustomed. NASA engineers themselves were not used to playing the role of minor partners and to not being in the driver's seat. Frictions inevitably developed at the beginning. Fortunately that period was followed by more fruitful relationships and the situation rapidly evolved to an era which has been characterised by an excellent working atmosphere between both teams.

Still on the technical side, the satellite had to be built and handled under very clean conditions from the point of view of dust as well as molecular contamination. This was required to avoid a situation where reflective and refractive optical parts in the experiments would soon have been blinded, a well-known problem for solar instruments which, without special precautions, cannot survive very long in orbit. As a consequence, it is not an exaggeration to say that SOHO is probably the cleanest spacecraft ever developed.

Last but not least, the requirement of continuity for the helioseismology observations made it imperative to have very reliable storage memories on board. Tape recorders (which in the agreement with NASA were provided by the US) did not seem to be reliable enough and new solid state memories had to be developed in Europe. SOHO carries in fact both types of memories and it is fair to say that none has failed in orbit.

Once in orbit, SOHO has indeed behaved flawlessly. Its technical performances are in conformity with, but generally far better than, the expectations and specifications. In addition, the more than perfect performance of the launcher and of the operations in orbit have saved a lot of fuel on board and result in a lifetime of more than 20 years, ten times more than the specifications. Today, SOHO is the most stable, the cleanest and the most powerful solar observatory ever operated in space.

Europe and the United States are clearly unveiling the secrets of our star, thanks to two European-led missions, SOHO and Ulysses. These two missions are at the core of a renewed era in solar physics which will most probably bring answers to many of the most fundamental questions which solar physicists have been working with for a long time.

The following chapter illustrates a few of the early results obtained with these two missions.

5. Early Results

5.1. THE SOLAR WIND

Figure 5 is a polar plot showing actual Ulysses measurements of the solar wind speed, of the magnetic field polarity, of the flux of energetic particles and of cosmic rays. Particularly striking is the uniform fast (750 km/s) solar wind filling a large fraction of the heliosphere above 30 degrees latitude in both hemispheres.

It has long been thought that the slow wind, being the type most commonly observed in the ecliptic, was the "true" solar wind, and that all high-speed streams represented disturbances. Ulysses has shown that the high-speed wind originates in the long-lived polar coronal holes and that the fast wind has a composition that is very similar to that of the photosphere, whereas the slow wind shows strong compositional biases compared with the photosphere. The clear implication is that it is the fast, rather than the slow wind, that represents an unbiased sample of solar material and is the true "quiet" state of the solar-wind phenomenon.

The mass, ionic charge and energy of the particles in the solar wind have been measured with an unprecedented accuracy by the CELIAS instrument on SOHO (Hovestadt et al., 1995). The CELIAS/MTOF spectrum shown in Figure 6 was accumulated over a three-day period optimized for observing species with masses above that of sulphur. This is why the peaks for calcium (mass 40) and iron (mass 56) are so dominant compared to, for example, oxygen.

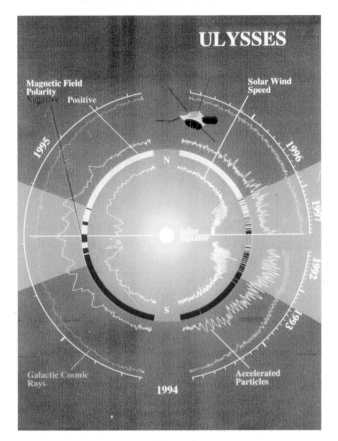

Figure 5. A polar plot showing the solar wind velocity, the magnetic field polarity and the flux of energetic particles and cosmic rays as measured by Ulysses over one complete orbit. As a result of its larger velocity at perihelion the measurements evidence less time resolution near the Sun's equator in 1994–1995.

The elements and isotopes for which SOHO has given the first observations include the silicon isotopes (masses 29 and 30), phosphorus (mass 31), a sulphur isotope (mass 34), chlorine (mass 35) and its isotope (mass 37), an argon isotope (mass 38), the calcium isotopes (masses 42 and 44), titanium (mass 48), chromium (mass 52) and its isotope (mass 53), the iron isotope (mass 54), manganese (mass 55), nickel (mass 58) and its isotopes (masses 60 and 62). For all but argon 38, these are the first *in-situ* solar-wind observations by any means. Other elements and isotopes are not observed routinely by conventional solar wind experiments. These include nitrogen (mass 14), a neon isotope (mass 22), sodium (mass 23), the magnesium isotopes (mass 26, 26) aluminium (mass 27), sulphur (mass 32), argon (mass 36), and calcium (mass 40).

SOHO is also addressing the key question of the mechanisms that accelerate the solar wind, by use of two different methods. The first one is based on the technique

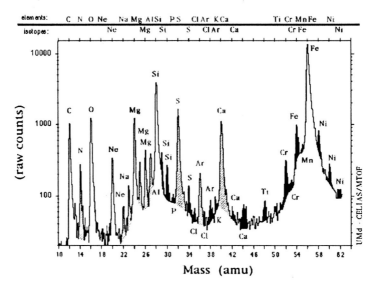

Figure 6. The Solar wind element and isotope composition as observed by the CELIAS/MTOF instrument onboard SOHO.

of "Doppler dimming" (Hyder and Lites, 1970) which consists in observing the profile of the coronal Lyman-alpha line (which is due to the fluorescence of neutral hydrogen atoms continuously created and destroyed by charge exchange in the solar corona). The width of the line is a measure of the random velocity of the protons in the corona. The intensity of the line, is a measure of the outflow velocity. The same process, applied to the O VI lines at 103.2 and 103.6 nm, can also be used for determining the outflow velocity of O VI ions. These measurements were efficiently performed by the UVCS instrument on SOHO, as illustrated in Figure 7 (Kohl *et al.*, 1998).

There the observed $1/e$ line-of-sight widths of the H I Ly-alpha and of the average O VI lines versus height are shown, together with the $1/e$ linewidths of the fluorescence that would be seen if thermal broadening were the only line-broadening mechanism, i.e. the line width corresponded to the electron temperature at the respective height. In fact, the linewidths vastly exceed those expected from the electron temperature. This has been interpreted as the consequence of ion-cyclotron resonant acceleration. It should be noted that everything seems to happen between 1.8 and 2 solar radii, and that – owing to its larger charge – the O ion is subject to a much stronger acceleration than the protons whose movements are reflected by the H-I Ly alpha profile. Because UVCS observations were made near the solar poles where the fast solar wind dominates, these results concern the acceleration of the fast wind.

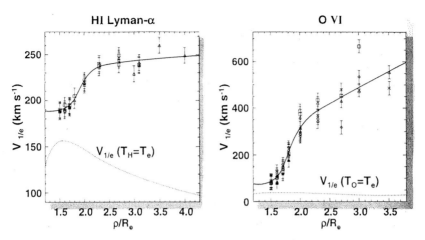

Figure 7. The line-of-sight linewidths of H-I Lyman alpha and O VI resonance lines observed above the polar coronal holes, evidencing the acceleration of the coronal plasma between 1.5 and 2.5 solar radii. The dotted curves indicate the line widths expected for the case where the random velocity of the H atoms and O^{5+} ions corresponds to the (modeled) electron temperature at the various heights.

By contrast, the second method most likely applies to the slow wind. It consists in detecting velocities and accelerations in the corona close to the Sun's limb through the continuous monitoring of coronal mass ejections (CME's) which are regularly observed at relatively low solar latitudes, by the 3 coronagraphs of the Large Angle Spectrometric Coronagraph (LASCO) which explores the corona from $1.3 R_\odot$ to $30 R_\odot$. (Brueckner *et al.*, 1995).

LASCO allows to follow the evolution and motions of CMEs and of all moving structures in the solar corona which are most probably passively tracing the outflow of the slow solar wind. Tracking 50–100 of the most prominent moving coronal features, Sheeley *et al.* (1997) have found that their radial speed typically doubles from 150 km/s near $5 R_\odot$ to 300 km/s near $25 R_\odot$, and their speed profiles cluster around a nearly parabolic path characterised by a constant acceleration of about 4 m/s^2 through most of the $30 R_\odot$ field of view of LASCO. This profile is consistent with an isothermal solar wind expansion at a temperature of about 1.1 MK and a sonic point near $5 R_\odot$.

5.2. HEATING OF THE CORONA

SOHO is now providing new and increasing evidence that the long-sought source of energy necessary to heat the corona might in fact be linked to the subphotospheric magnetic field. This conclusion comes out of the observations made by the Michelson Doppler Imager, MDI (Scherrer *et al.*, 1995), which measures both the solar magnetic field at the one-arc-sec scale and the plasma velocity on the surface. By use of 'time-distance' helioseismology (Duvall *et al.*, 1997), it is also possible to track the subsurface motions as they affect and shift the frequencies of the acoustic waves.

By relating surface and subsurface flows with the behaviour of the magnetic field, MDI has shown that the entire emerging magnetic flux is being replaced every 40 hours. It is most likely that coronal heating is due to the dissipation of this magnetic flux through reconnection. There is enough energy in the magnetic loops for that.

The MDI observations are corroborated by those made from the SOHO Coronal Diagnostic Spectrometer, CDS (Harrison *et al.*, 1995), which has observed "blinkers" (Harrison, 1997) occurring over the whole volume of the chromosphere at the impressive number of several thousands at any time, and most probably by the manifestation of reconnection events which are associated with high velocity flows, as observed by the German-French high resolution chromospheric spectrometer, SUMER (Wilhelm *et al.*, 1995; Innes *et al.*, 1997).

5.3. PROBING THE SOLAR INTERIOR

The high stability of SOHO, together with the remarkable performance of its helioseismology payload (MDI, GOLF and VIRGO), have yielded new results on the deep Sun with an astonishing accuracy, at distances to the centre as close as a few percent of the solar radius. In particular, the detailed analysis of the Doppler shifts in the wave frequencies as observed by MDI has revealed a completely new picture of the internal solar rotation, and the existence of a striking shear layer at the bottom of the convection zone. In mid-latitudes there is a rapid increase of the rotation velocity with depth, while below there appears to reign essentially rigid rotation. Figure 8 displays the variation with depth of the sound speed inside the Sun, or rather how it deviates from standard model predictions. These results are intimately connected with questions of solar composition and will most likely be discussed in the course of this workshop, and are therefore not described in detail here.

To get down to the very core of the Sun, we must have recourse to gravity oscillations. However, they are difficult to detect. A recent estimate of their amplitude close to the surface leads to a value of 0.06 mm/s. Although, the "Global Oscillations of Low Frequency" detector, GOLF (Gabriel *et al.*, 1995), should have sufficient sensitivity, these elusive waves have not been detected yet. More observing time is required to confirm or not their presence, as well as a substantial data reduction effort, using the data from all three SOHO helioseismology instruments to evaluate and quantify more precisely – and subtract – the solar noise measured by the GOLF instrument.

As for the solar constant, VIRGO (Fröhlich *et al.*, 1995) has shown clearly that it is far from being constant, showing strong variations related to the presence of faculae and sunspots (Fröhlich *et al.*, 1997) and a gradual increase since the beginning of 1997, probably corresponding to the rising phase of the new solar cycle (Figure 9, Fröhlich, 1998).

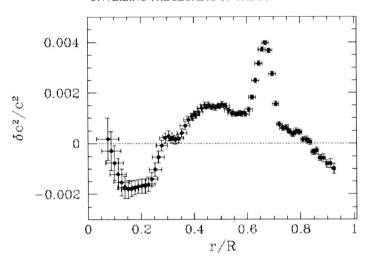

Figure 8. Deviation from the solar standard model of the actual sound velocity inside the Sun as deduced from the measurements performed by the MDI instrument onboard SOHO.

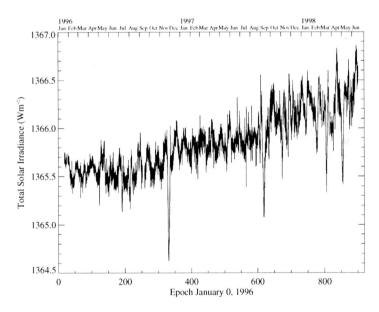

Figure 9. The total solar irradiance as measured by the VIRGO instrument onboard SOHO. Note the great accuracy of the measurements showing variations of a few parts per thousands.

These are only a few of the results achieved so far. Many others – more specifically relating to solar composition – will be described by the specialists in the course of this workshop showing the fantastic power of SOHO which in its two-year nominal mission has addressed and responded to the most pressing issues of solar physics today.

6. Conclusion

Little by little the secrets of our star are being unveiled thanks to the tremendous progress accomplished in the past thirty years in instrumentation and observing techniques. Space observatories, SOHO in particular, have dramatically improved our view of the Sun in its entirety, from the deep core through the corona to the heliosphere. SOHO is a time machine and as time passed during its nominal mission, new discoveries have accumulated, and more are expected due to the proximity of the forthcoming solar maximum.

SOHO and Ulysses, together with a fleet of other space missions and in coordination with the best ground based observatories, are in the process of sharply improving our knowledge of the Sun. The years to come are exciting and promise to be rich in new and fundamental results. In the end, the model of our Sun will be more firmly established, placing our knowledge of all stars on a much firmer basis. All those involved in the development of these stunning missions, engineers at ESTEC and in industry, scientists on both sides of the Atlantic, and all those who so successfully continuously operate SOHO and Ulysses must be congratulated for their work and efforts in contributing to such remarkable advances.

References

Bahcall, J. and Davis, R.: 1976, *Science* **191**, 264.
Biermann, L.F.: 1951, *Zeitschr. f. Astrophys.* **29**, 274.
Brueckner, G.E. *et al.*: 1995, *Solar Phys.* **162**, 357.
Cavallo, G.: 1996, *ESA Bulletin* **87**, 38.
Deubner, F.-L.: 1975, *Astron. Astrophys.* **44**, 371.
Duvall, T.L., Kosovichev, A.G., Scherrer, P.H. *et al.*: 1997, *Solar Phys.* **170**, 63.
Fröhlich, C. *et al.*: 1995, *Solar Phys.* **162**, 101.
Fröhlich, C. *et al.*: 1997, *Solar Phys.* **170**, 1.
Fröhlich, C.: 1998, personal communication.
Gabriel, A.H. *et al.*: 1995, *Solar Phys.* **162**, 61.
Golub, L. and Pasachoff, J.M.: 1997, *The Solar Corona*, Cambridge Univ. Press, p. 22.
Grec, G., Fossat, E. and Pommerantz, M.: 1980, *Nature* **288**, 541.
Harrison, R.A. *et al.*: 1995, *Solar Phys.* **162**, 233.
Harrison, R.A.: 1997, *Solar Phys.* **175**, 467.
Hovestadt, D. *et al.*: 1995, Solar Phys. 162, 441.
Huber, M.C.E., Bonnet, R.M., Dale, D.C., Arduini, M., Fröhlich, C., Domingo, V. and Whitcomb, G.: 1996, *ESA Bulletin* **86**, 25.
Hyder, C.L. and Lites, B.W.: 1970, *Solar Phys.* **14**, 147.
Kohl, J.L. *et al.*: 1998, *Astrophys. J.* **501**, L127.
Innes, D.E., Brekke, P., Germerott, D., and Wilhelm, K.: 1997, *Solar Phys.* **175**, 341.
Leighton, R.B., Noyes, R.W. and Simon, G.W.: 1962, *Astrophys. J.* **135**, 474.
Lyot, B.: 1930, *C. R. Acad. Sci. (Paris)* **99**, 580.
Marsden, R.G., Smith, E. J., Cooper, J.F. and Tranquille, C.: 1996, *Astron. Astrophys.* **316**, 279.
Parker, E.: 1958, *Astrophys. J.* **128**, 664.
Scherrer, P.H. *et al.*: 1995, *Solar Phys.* **162**, 129.
Sheeley, N.R. *et al.*: 1997, *Astrophys. J.* **484**, 472.
Tousey, R.: 1953, in *The Sun*, G. Kuiper (ed.), University of Chicago Press, 658.
Wenzel, K.P. *et al.*: 1992, *Astron. Astrophys. Suppl. Ser.* **92**, 207.
Wilhelm, K. *et al.*: 1995, *Solar Phys.* **162**, 189.

I: SOLAR INTERIOR

THE 'STANDARD' SUN
Modelling and Helioseismology

JØRGEN CHRISTENSEN-DALSGAARD
Teoretisk Astrofysik Center, Danmarks Grundforskningsfond, and Institut for Fysik og Astronomi, Aarhus Universitet, DK-8000 Aarhus C, Denmark

Abstract. The 'standard' solar model is based on a number of simplifying assumptions and depends on knowledge of the physical properties of matter in the Sun. Given these assumptions, the constraint that the model have the observed surface luminosity provides an estimate of the initial solar helium abundance. From helioseismic analyses further information can be obtained about the present composition, including a fairly precise measure of the envelope helium abundance and an estimate of the hydrogen profile in the radiative interior. It must be emphasized, however, that these inferences may suffer from systematic error arising from incomplete knowledge about the equation of state and opacity of the solar interior.

Key words: Sun, structure – Sun, composition – helioseismology

1. Introduction

Modelling the interiors of the Sun and other stars requires assumptions about their initial composition. Conversely, to the extent that observable properties of the stars depend on the initial or present composition, such observations can provide information about stellar composition, independent of direct spectroscopic measurement of the stellar surface abundances. This obviously assumes that other aspects of the physics of the stellar interiors are so well understood that the composition is the main uncertainty. (As is abundantly clear elsewhere in these proceedings, spectroscopic measurements share similar potential uncertainties.) An important example of such an inference of composition is the determination of the solar initial helium abundance through calibration of the present luminosity, discussed in section 2.3 below.

Helioseismology has placed stringent constraints on models of the structure of the solar interior. Although the oscillation frequencies are not directly sensitive to composition, it is possible, with additional information about the physics of the solar interior, to use the results of helioseismology to make inferences about aspects of the solar internal composition. Assuming only that the equation of state in the solar convection zone is known with sufficient accuracy, the convection-zone helium abundance may be inferred. With additional assumptions, the helium profile beneath the convection zone may also be constrained.

Here I summarize the physical assumptions underlying the model computations and discuss the influence of the composition on solar structure. I then present a few aspects of helioseismology and review the inferences on solar composition that have been made on the basis of helioseismic analyses.

2. Modelling the Solar Interior

I consider just what is normally termed 'standard' solar modeling. This involves a number of simplifications of what is in reality a complex physical system. The most serious is the assumption that there is no macroscopic motion, and hence mixing, outside the convectively unstable regions. This ignores a number of possible hydrodynamical instabilities associated, for example, with the establishment of the present rotation profile which exhibits a strong shear just beneath the convection zone. In addition, effects of rotation and magnetic fields are ignored, so that the model is spherically symmetric.

As a background for the following, I provide a brief review of the properties of 'standard' solar models. This also specifically defines the model used as reference, viz. Model S [see Christensen-Dalsgaard et al. (1996), where further details on solar modelling and on the specific model are given].

2.1. COMPUTATION OF SOLAR MODELS

Given the standard assumptions, the basic equations of stellar structure are quite simple:

$$\frac{dp}{dr} = -\frac{Gm\rho}{r^2} \quad \text{(hydrostatic equilibrium)}, \quad (1)$$

$$\frac{dm}{dr} = 4\pi\rho r^2 \quad \text{(mass)}, \quad (2)$$

$$L = -\frac{4\pi r^2 a\tilde{c}}{3\kappa\rho}\frac{dT^4}{dr} \quad \text{(radiative luminosity)}, \quad (3)$$

$$\frac{dL}{dr} = 4\pi\rho r^2 \epsilon \quad \text{(energy generation)}. \quad (4)$$

Here r is distance to the centre, p is pressure, ρ is density, m is the mass inside r, G is the gravitational constant, L is the luminosity at r, a is the radiation-density constant, $\tilde{c}$ is the speed of light, κ is the opacity, T is temperature, and ϵ is the rate of energy generation per unit mass. [In equation (4) I neglected small terms resulting from the change in internal energy and work done as a result of stellar evolution with time; these terms are of course included in the numerical calculations.] Equation (3) holds only in convectively stable regions; in convection zones energy is transported largely by convection leading, except very near the surface, to a temperature gradient that is essentially adiabatic,

$$\frac{d\ln T}{d\ln p} \simeq \left(\frac{\partial \ln T}{\partial \ln p}\right)_{ad} \equiv \nabla_{ad} . \quad (5)$$

3.2. Asymptotic Properties of Frequencies

The frequencies of acoustic modes approximately satisfy the *Duvall law*:

$$\int_{r_t}^{R} \left(1 - \frac{L^2 c^2}{\omega_{nl}^2 r^2}\right)^{1/2} \frac{\mathrm{d}r}{c} = \frac{[n + \alpha(\omega_{nl})]\pi}{\omega_{nl}}, \qquad (9)$$

where $L = l + 1/2$ (*e.g.* Christensen-Dalsgaard *et al.*, 1985); here r_t is the lower turning point, where the bracket in the integral vanishes, and $\omega_{nl} = 2\pi\nu_{nl}$ is the angular frequency. Also, α is a function of frequency which depends on the properties of the Sun relatively near the surface. That the observed frequencies of solar oscillation satisfy a relation of this general form was first found by Duvall (1982).

It is convenient to analyse the observations in terms of *differences* $\delta\omega_{nl} = \omega_{nl}^{(\mathrm{obs})} - \omega_{nl}^{(\mathrm{mod})}$ between the observed frequencies and those of a suitable reference solar model. From equation (9) we obtain (*e.g.* Christensen-Dalsgaard, Gough and Pérez Hernández, 1988) that

$$S_{nl} \frac{\delta\omega_{nl}}{\omega_{nl}} \simeq \mathcal{H}_1\left(\frac{\omega_{nl}}{L}\right) + \mathcal{H}_2(\omega_{nl}), \qquad (10)$$

where

$$S_{nl} = \int_{r_t}^{R} \left(1 - \frac{L^2 c^2}{r^2 \omega_{nl}^2}\right)^{-1/2} \frac{\mathrm{d}r}{c} - \pi \frac{\mathrm{d}\alpha}{\mathrm{d}\omega}, \qquad (11)$$

$$\mathcal{H}_1(w) = \int_{r_t}^{R} \left(1 - \frac{c^2}{r^2 w^2}\right)^{-1/2} \frac{\delta_r c}{c} \frac{\mathrm{d}r}{c}, \qquad \mathcal{H}_2(\omega) = \frac{\pi}{\omega}\delta\alpha(\omega). \qquad (12)$$

In equation (12), $\delta_r c$ is the sound-speed difference between Sun and model, at fixed r.

It is straightforward to carry out a separation of the frequency differences as in equation (10), *e.g.* by representing the two terms by suitable parametrizations as done, for example, by Christensen-Dalsgaard, Gough and Thompson (1989), who also demonstrated how to invert the expression for $\mathcal{H}_1(w)$ to infer the sound-speed difference. Furthermore, $\mathcal{H}_2(\omega)$ provides information about the near-surface layers, including the variation in γ_1 induced by helium ionization; hence it may be used to infer the helium abundance of the convection zone. We return to this in section 4.1 below.

To illustrate the behaviour expected from equation (10), Fig. 2 shows differences between observed frequencies and frequencies of Model S, suitably scaled. It is evident that these are dominated by the $\mathcal{H}_2$ contribution, which furthermore is a slowly varying function of frequency; this indicates that the main contribution to the errors arise in the superficial layers of the model (see also Christensen-Dalsgaard and Thompson, 1997). The residual after subtraction of $\mathcal{H}_2$ shows a

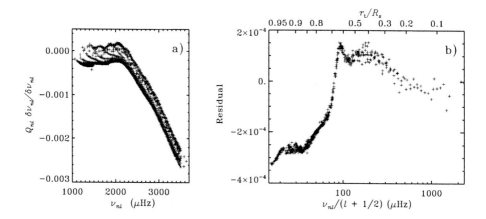

Figure 2. (a) Scaled frequency differences, in the sense (observation) − (model). The observations are a combination of BiSON whole-disk measurements (*e.g.* Elsworth *et al.*, 1994) and LOWL observations (Tomczyk *et al.* 1995), as described by Basu *et al.* (1997), while the computed frequencies are for Model S. The scale factor Q_{nl} corresponds essentially to the asymptotic factor S_n, normalized by its value for radial modes. (b) Scaled differences after subtraction of the fitted $\mathcal{H}_2(\omega)$, plotted against $\nu_{nl}/(l+1/2)$ which determines the lower turning point r_t, shown as the upper abscissa.

striking dependence on the depth of penetration of the modes, indicating a substantial source of error in the model just below the convection zone, at $r \simeq 0.7R$.

3.3. ANALYSIS OF FREQUENCY DIFFERENCES

Equation (10) was based on an asymptotic expression for the frequencies; thus, although it is fairly precise for the the observed solar modes, it suffers from systematic errors. More generally, one may linearize the relations between the adiabatic oscillation frequencies and the structure of the model; according to section 3.1 the latter can be characterized by two quantities, *e.g.* (c^2, ρ). However, account must also be taken of the fact that the observed oscillations are nonadiabatic, as well as of the uncertainties in the description of the near-surface properties of the model and oscillations. The result can be expressed as

$$\frac{\delta\omega_{nl}}{\omega_{nl}} = \int K_{c^2,\rho}^{nl}(r)\frac{\delta_r c^2(r)}{c^2(r)}\mathrm{d}r + \int K_{\rho,c^2}^{nl}(r)\frac{\delta_r \rho(r)}{\rho(r)}\mathrm{d}r + \frac{F_{\mathrm{surf}}(\omega_{nl})}{I_{nl}} + \epsilon_{nl} \ . \quad (13)$$

Here the kernels $K_{c^2,\rho}^{nl}(r)$, $K_{\rho,c^2}^{nl}(r)$ are functions computable from the reference model. The term in $F_{\mathrm{surf}}(\omega_{nl})$ results from the near-surface errors, I_{nl} being the mode inertia; the form of this term reflects the fact that the direct physical influence of the near-surface problems is a function of frequency, while the effects on the frequencies depend on the fraction of the Sun which takes part in the oscillations (*e.g.* Christensen-Dalsgaard and Berthomieu 1991). Finally, in equation (13) ϵ_{nl} are the errors in the observations. It should be noticed that there is a close similarity between equation (13) and the asymptotic equation (10): in particular, the term in F_{surf}/I_{nl} is approximately equivalent to $\mathcal{H}_2(\omega)$. As is evident from Fig. 2a, this term in fact dominates the differences between the observed and model frequencies.

Equation (13) forms the basis for inverse analyses of the frequency differences, aimed at inferring the structural differences between the Sun and the model. Here I consider techniques which proceed implicitly or explicitly by making linear combinations of the data. To be specific, the sound-speed difference $\delta_r c(r_0)$ in the vicinity of some location $r = r_0$ is inferred as a linear combination of the data (*i.e.*, the $\delta \omega_{nl}/\omega_{nl}$) with coefficients $c_{nl}(r_0)$, which according to equation (13) has the form

$$\sum_{nl} c_{nl}(r_0) \frac{\delta \omega_{nl}}{\omega_{nl}}$$
$$= \int \sum_{nl} c_{nl}(r_0) K^{nl}_{c^2,\rho}(r) \frac{\delta_r c^2(r)}{c^2(r)} dr + \int \sum_{nl} c_{nl}(r_0) K^{nl}_{\rho,c^2}(r) \frac{\delta_r \rho(r)}{\rho(r)} dr$$
$$+ \sum_{nl} c_{nl}(r_0) \frac{F_{\mathrm{surf}}(\omega_{nl})}{I_{nl}} + \sum_{nl} c_{nl}(r_0) \epsilon_{nl} \,. \qquad (14)$$

The properties of the inference depend on the *averaging kernel*

$$\mathcal{K}_{c^2,\rho}(r_0, r) = \sum_{nl} c_{nl}(r_0) K^{nl}_{c^2,\rho}(r) \,, \qquad (15)$$

which multiplies $\delta_r c^2/c^2$ in equation (14); if this is localized near $r = r_0$, and other terms in that equation are small, the combination of the data provides an average of $\delta_r c^2/c^2$ in the vicinity of r_0.

Here I concentrate on the so-called SOLA technique where the coefficients $c_{nl}(r_0)$ are chosen explicitly, to optimize the properties of $\mathcal{K}_{c^2,\rho}(r_0, r)$. It is based on the method of optimally localized averages of Backus and Gilbert (1968), as modified by Pijpers and Thompson (1992); the specific implementation for structure inversion was described by Basu *et al.* (1996). Here the goal is to choose the coefficients such that $\mathcal{K}_{c^2,\rho}(r_0, r)$ approximates a prescribed target $\mathcal{T}(r_0, r)$ which is suitably localized around $r = r_0$, while at the same time minimizing the influence of the remaining terms in equation (14). In particular, the coefficients are determined under the constraint that the effect of the near-surface term, for a slowly varying function $F_{\mathrm{surf}}(\omega)$, is suppressed. The width of the averaging kernel gives a measure of the resolution achieved in the inversion, while the last term in equation (14) provides a measure of the error in the inferred sound-speed difference.

Fig. 3 shows the results of inversion of the differences, illustrated in Fig. 2, between observed frequencies and those of Model S. Strikingly, the model has predicted the sound speed in the solar interior within an error (in c^2) of less than 0.5 %. Even so, the intrinsic precision of the determination, as obtained from the errors in the observations, is far higher; thus the differences between the true structure of the Sun and the model are highly significant. As already suggested by the residual

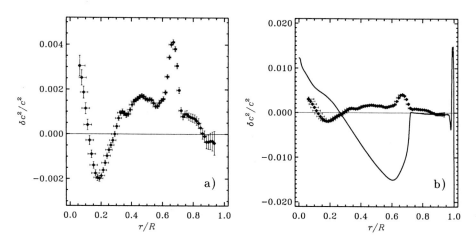

Figure 3. (a) Sound-speed difference, in the sense (Sun) − (model), inferred from inversion of the differences between the observed BiSON and LOWL frequencies and the frequencies of Model S. The vertical error bars are 1-σ errors on the inferred differences, while the horizontal bars provide a measure of the resolution of the inversion. (Adapted from Basu *et al.*, 1997.) (b) The same, on a different scale, compared with the sound-speed difference between models without and with settling of helium and heavy elements (solid line).

differences in Fig. 2b the dominant difference is the strong peak just below the convection zone, which might be indicative of partial mixing of the strong gradient of hydrogen abundance established by settling (*cf.* Fig. 1; see also section 4.2; and Dziembowski, 1998). For comparison, Fig. 3b shows the inferred $\delta c^2/c^2$ on a smaller scale, compared with the sound-speed difference between models without and with settling. It is clear that the agreement between Sun and model is greatly improved through the inclusion of settling and diffusion. Thus the inversion provides strong support for the importance of He settling in the Sun (*e.g.* Christensen-Dalsgaard, Proffitt and Thompson 1993).

3.4. Effects of Composition and Equation of State

To obtain information about composition from the inversions, additional constraints on the solar model must be imposed. Given that the frequencies depend on p, ρ and γ_1 which are all thermodynamical quantities, the most natural constraint is probably the equation of state. Specifically, we assume that $\gamma_1 = \gamma_1(p, \rho, Y, Z)$ may be obtained from the thermodynamical properties of the gas, although perhaps with an error from using an incorrect equation of state. For simplicity I also neglect the effect of Z, which in any case is constrained (at least in the convection zone) by the spectroscopic measurements. Then

$$\frac{\delta_r \gamma_1}{\gamma_1} = \left(\frac{\partial \ln \gamma_1}{\partial \ln p}\right)_{\rho,Y} \frac{\delta_r p}{p} + \left(\frac{\partial \ln \gamma_1}{\partial \ln \rho}\right)_{p,Y} \frac{\delta_r \rho}{\rho}$$

$$+ \left(\frac{\partial \ln \gamma_1}{\partial Y}\right)_{p,\rho} \delta_r Y + \left(\frac{\delta \gamma_1}{\gamma_1}\right)_{\text{int}}, \qquad (16)$$

where $(\delta\gamma_1/\gamma_1)_{\text{int}}$ is the intrinsic error in γ_1, *i.e.*, the difference between the values corresponding to the solar and the model equations of state, at given p, ρ and composition. Equation (16) and the equation for c may be used in equation (13). Introducing $u = p/\rho$ and using the linearized versions of equations (1) and (2) to eliminate $\delta_r p$ and $\delta_r \rho$, we finally obtain

$$\frac{\delta\omega_{nl}}{\omega_{nl}} = \int K^{nl}_{u,Y}(r) \frac{\delta_r u(r)}{u(r)} dr + \int K^{nl}_{Y,u}(r) \delta_r Y(r) dr$$
$$+ \int K^{nl}_{c^2,\rho}(r) \left(\frac{\delta\gamma_1}{\gamma_1}\right)_{\text{int}} dr + \frac{F_{\text{surf}}(\omega_{nl})}{I_{nl}} + \epsilon_{nl} \qquad (17)$$

(see also Basu and Christensen-Dalsgaard, 1997).

As discussed by Basu and Christensen-Dalsgaard, equation (17) may be used to infer the intrinsic error in γ_1, and hence to probe the equation of state of solar matter. Alternatively, if $(\delta\gamma_1/\gamma_1)_{\text{int}}$ is assumed to be negligible, the equation forms a convenient basis for inversion for $\delta_r u$. Indeed, since $K^{nl}_{Y,u}$ is proportional to $(\partial \ln \gamma_1/\partial Y)_{p,\rho}$ which is only substantial in the hydrogen and helium ionization zones, it is relatively easy to suppress the influence of the term in $\delta_r Y$ in equation (17), compared with the suppression of the corresponding term in $\delta_r \rho$ in equation (13) in inversion for $\delta_r c^2$ (*e.g.* Dziembowski, Pamyatnykh and Sienkiewicz, 1990). Finally, equation (17) may be inverted to determine $\delta_r Y$ in the helium ionization zones (*e.g.* Kosovichev *et al.*, 1992); since the convection zone is fully mixed, this provides a measure of the convection-zone value Y_e of the helium abundance.

4. Helioseismic Inferences of Solar Composition

4.1. The Helium Abundance in the Convection Zone

As noted in section 2.1, the structure of the convection zone depends essentially only on the equation of state and the composition, as well as on the value of the specific adiabat. It was noted by Gough (1984) that, as a result, the helium abundance Y_e in the convection zone could be obtained from a determination of the sound speed in the convection zone (see also Däppen and Gough 1986). This estimate was based on the dependence of γ_1 on Y_e in the second helium ionization zone; as an illustration, Fig. 4a shows γ_1 in models without and with helium settling, where Y_e is, respectively, 0.2635 and 0.2447. Obviously, any such determination would also be affected by errors in the assumed equation of state; Gough (1984) pointed out that with sufficiently detailed data it might be possible to separate the effects of uncertainties in the equation of state and differences in Y_e.

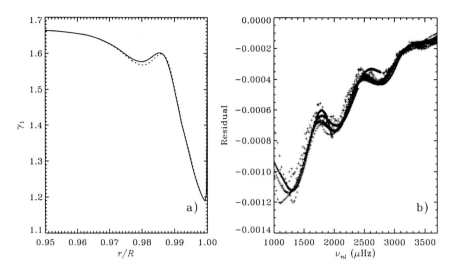

Figure 4. (a) Adiabatic exponent γ_1 in Model S, with helium and heavy-element settling (solid line) and in corresponding model without settling (dotted line). (b) Residual scaled frequency differences between these models, in the sense (no settling) - (settling), after subtraction of the term in $\mathcal{H}_1(\omega/L)$ (*cf.* eq. 10).

Much of the subsequent analysis has been based on the asymptotic representations in equation (9) and (10). By analysing the phase function $\alpha(\omega)$ in equation (9), Vorontsov, Baturin and Pamyatnykh (1991) estimated that $Y_e \simeq 0.25 \pm 0.01$; they noted that this value was substantially lower than the initial helium abundance Y_0 inferred from calibration of solar models (*cf.* eq. 8) and pointed out that this might be evidence for settling.

Pérez Hernández and Christensen-Dalsgaard (1994) based the analysis on the phase-function difference $\mathcal{H}_2(\omega)$ (*cf.* eq. 12). To illustrate this method, Fig. 4b shows the results of analysing frequency differences between models with and without helium settling according to equation (10). The difference in γ_1 between these two models (*cf.* Figure 4) causes an oscillatory behaviour in $\mathcal{H}_2$, the amplitude of which provides a measure of the difference in Y_e. In practice, errors in the immediate surface regions of the model make a substantially larger contribution to the frequency differences (*cf.* Fig. 2); however, as pointed out by Pérez Hernández and Christensen-Dalsgaard these contributions are slowly varying functions of frequency and hence can be filtered out. The resulting analysis indicated that the MHD equation of state provided a better representation of solar matter than did the simpler so-called CEFF equation of state (Christensen-Dalsgaard and Däppen 1992); using the MHD results, Pérez Hernández and Christensen-Dalsgaard inferred a value of Y_e of 0.242 ± 0.003, although recognizing that remaining errors in the equation of state could introduce larger errors in the inferred Y_e.

Basu and Antia (1995) considered both $\mathcal{H}_1(\omega/L)$ and $\mathcal{H}_2(\omega)$, for sets of reference models with varying Y_e, and computed with the MHD and the OPAL equations of state; they obtained $Y_e = 0.0246 \pm 0.001$ and $Y_e = 0.0249 \pm 0.001$ from the MHD and OPAL models, respectively. Basu (1998) applied a similar analysis to recent data from the SOI/MDI instrument on the SOHO spacecraft, obtaining $Y_e = 0.248 \pm 0.001$ from OPAL models and a slightly higher value from MHD models. Thus it appears that the determinations based on the asymptotic treatment are largely consistent, although with some remaining scatter resulting from the uncertainty in the equation of state. It is striking that the values obtained are in reasonable accord with the value of 0.2447 obtained in Model S.

In principle, inversion for $\delta_r Y$ on the basis of equation (17), which is independent of the asymptotic approximation, should provide a safer determination of Y_e, although evidently still subject to systematic errors due to the term $(\delta\gamma_1/\gamma_1)_{\text{int}}$ arising from possible errors in the equation of state. Dziembowski, Pamyatnykh and Sienkiewicz (1991) found $Y_e = 0.234 \pm 0.005$, with models using the MHD equation of state. A careful analysis of possible systematic errors in such determinations was presented by Kosovichev et al. (1992). More recently, Kosovichev (1997) obtained $Y_e = 0.232 \pm 0.006$ using an MHD reference model and $Y_e = 0.248 \pm 0.006$ with an OPAL model. It is interesting that for the MHD model this determination is consistent with that obtained by Dziembowski et al. while the OPAL model yields a result consistent with the asymptotic analyses. However, it appears that the numerical inversion is far more sensitive to differences in the equation of state than are the asymptotic procedures. The reason for this sensitivity is apparently not understood.

4.2. Composition of the Radiative Interior

The determination of the sound speed in the radiative interior evidently constrains the composition profile, although only in combination with further information about the physics of the interior. As a simple example, I note that, assuming the ideal gas law (eq. 6), $c^2 \simeq \gamma_1 k_B T/(\mu m_u)$; thus the determination of c^2 provides information about T and the composition only in the combination T/μ. To obtain quantitative information about the composition profile, the equations of stellar structure must be explicitly invoked. Two different procedures have been used: in one, the helioseismically inferred sound speed is imposed as a constraint on computations of models of the present Sun; in the second, the equations of stellar structure are used to formulate an inverse problem for corrections to the composition.

In the former case (Shibahashi and Takata 1996; Antia and Chitre 1998; Takata and Shibahashi 1998) equations (1) – (4) are solved, under the constraint that $\gamma_1 p/\rho$ match the helioseismically determined c^2, but with no prior assumptions about the hydrogen profile $X(r)$; however, obviously the equation of state, opacity and energy generation are assumed to be known. Furthermore, specification of the single function $c^2(r)$ does not allow determination also of the profile $Z(r)$ of heavy

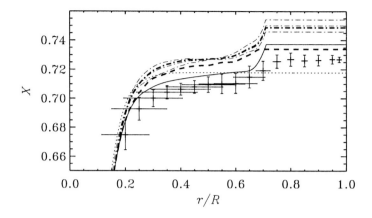

Figure 5. Results of inversions for the hydrogen profile. The dashed curve was obtained by Antia and Chitre (1998) from computation of a solar model under the constraint that the model sound speed agree with the helioseismically inferred sound speed. The dot-dashed curves were similarly obtained by Takata and Shibahashi (1998); the different curves illustrate the effects of variations in the assumed sound-speed profile for $r \leq 0.35R$. The symbols with error bars were obtained by inversion of equation (18) for $\delta_r Y$ (Kosovichev, 1997). The vertical error bars are 1-σ errors, as obtained from the errors in the observed frequencies, while the horizontal bars provide a measure of resolution. For comparison, the thin solid and dotted lines show model results, as in Fig. 1.

elements. Thus $Z(r)$ must be assumed; to estimate the resulting uncertainty Antia and Chitre used several profiles, obtained from models with settling of heavy elements and different values of Z_s. Computation of a model with the correct surface luminosity was found to require a modest change of the assumed nuclear reaction rates. Fig. 5 shows examples of the solutions obtained in this manner, compared with the model profiles already shown in Fig. 1a. Although there are considerable differences between the two results, they both show evidence for a gradient in X beneath the convection zone, which appears to be less steep than in the model with settling. This may in part by caused by the finite resolution of the inversion; however, any true reduction in the gradient could indicate that settling might be counteracted by the presence of weak mixing in this region (see also Dziembowski, 1998). It should also be noticed that the value obtained by Antia and Chitre in the convection zone agrees, perhaps fortuitously, with the value of Y_e obtained from asymptotic analysis. The difference between the results probably reflects the uncertainty caused by differences in the assumed $Z(r)$, as well as differences in the assumed physics of the solar interior and in the helioseismically inferred sound speed on which the analysis is based.

In the second approach, linearized versions of equations (1), (2) and (4) of hydrostatic equilibrium, mass and energy generation, as well as equations (3) or (5) for the temperature gradient, and of the expressions for the equation of state, opacity and energy-generation rate, are used to express $\delta_r c^2/c^2$ and $\delta\rho/\rho$ in equation (13) in terms of the differences $\delta_r Y$ and $\delta_r Z$ in $Y(r)$ and $Z(r)$ between the Sun and the model (*e.g.* Gough and Kosovichev 1990; Kosovichev 1997). Thus the inverse problem can be written as

$$\frac{\delta\omega_{nl}}{\omega_{nl}} = \int K^{nl}_{Y,Z}(r)\delta_r Y(r)\mathrm{d}r + \int K^{nl}_{Z,Y}(r)\delta_r Z(r)\mathrm{d}r + \frac{F_{\mathrm{surf}}(\omega_{nl})}{I_{nl}} + \epsilon_{nl} \ . \quad (18)$$

From this, the correction $\delta_r Y$ may be determined by inversion as before. Some results are shown in Fig. 5. They confirm the general trend obtained from the computations of constrained models by Antia and Chitre and by Takata and Shibahashi, including the composition gradient near the base of the convection zone. Even so, it is obvious that there are significant differences between the different inferences; it is likely that part of this difference is caused by the use of different assumptions about the physics. However, a more detailed comparison of these determinations, and an investigation of the systematic errors, is still required.

5. Conclusions

It is evident that any helioseismologically based inferences of the solar internal composition is strongly dependent on the assumptions of the physics of the solar interior. As an extreme example, the sound-speed difference between models without and with settling can probably be reproduced by a suitable, although quite likely unrealistic, change in the opacity, thus perhaps weakening the evidence for settling obtained in Fig. 3b. Even so, the results presented here show that substantial information has already been obtained. Using the asymptotic techniques the helium abundance in the convection zone has been determined as close to 0.25, with only modest sensitivity to the equation of state; compared with the initial helium abundance of 0.2713 required to calibrate the model, this provides additional evidence for settling. The rather greater corresponding sensitivity of the numerical inversions is still cause for some concern, however.

The general behaviour of the hydrogen abundance in the radiative interior is similar to that obtained in models with settling, with some evidence for mixing beneath the convection zone. Indeed, it has been found that models including mixing of this region are in better accord with the helioseismically inferred sound speed (*e.g.* Chaboyer, Demarque and Pinsonneault, 1995; Richard *et al.*, 1996). Elliott and Gough (1998) found that a simple extension of the fully mixed region could account almost perfectly for the sound-speed difference inferred in Fig. 3. As is well known additional evidence for mixing comes from the depletion, relative to the meteoritic values, of lithium and possibly beryllium in the solar photosphere. In fact, there is no doubt that mixing to some extent must take place below the convection zone, as a result of convective overshoot as well as of likely flows and instabilities caused by the strong shear in rotation in this region (e.g. Schou *et al.*

1998). Helioseismic inferences of the structure of this region will help constraining the radial extent and magnitude of the mixing, hence providing information on these dynamically very interesting processes.

Further improvements in the helioseismic inferences of solar composition are intimately linked to improved understanding of the microphysics. Given the uniform composition of the convection zone, it may be possible here to obtain some, although obviously never complete, separation between the uncertainties in the equation of state and errors in the composition. Beneath the convection zone, where possible errors in the opacity must also be taken into account and the composition varies with position, such a separation is not possible. A more complete understanding of the processes that might affect the composition in this region will require a combination of the helioseismic data with data on the surface composition. Studies of the composition of other stars will undoubtedly also be very valuable, in constraining models of mixing processes in stars.

Acknowledgements

I am very grateful to H. M. Antia, A. G. Kosovichev and H. Shibahashi for the results used in Fig. 5. I thank the anonymous referee for useful comments on an earlier version of the paper, and D. O. Gough for communicating these comments. The work reported here was supported in part by the Danish National Science Foundation through the establishment of the Theoretical Astrophysics Center.

References

Antia, H. M. and Chitre, S. M.: 1998, 'Determination of temperature and chemical composition profiles in the solar interior from seismic models', *Astron. Astrophys.*, submitted.
Backus, G. and Gilbert, F.: 1968, 'The resolving power of gross Earth data', *Geophys. J. R. astr. Soc.* **16**, 169–205.
Bahcall, J. N. and Pinsonneault, M. H.: 1995, 'Solar models with helium and heavy-element diffusion', *Rev. Mod. Phys.* **67**, 781–808.
Basu, S.: 1998, 'Effects of errors in the solar radius on helioseismic inferences', *Mon. Not. R. astr. Soc.* **298**, 719–728.
Basu, S. and Antia, H. M.: 1995, 'Helium abundance in the solar envelope', *Mon. Not. R. astr. Soc.* **276**, 1402–1408.
Basu, S. and Christensen-Dalsgaard, J.: 1997, 'Equation of state and helioseismic inversions', *Astron. Astrophys.* **322**, L5–L8.
Basu, S., Chaplin, W. J., Christensen-Dalsgaard, J., Elsworth, Y., Isaak, G. R., New, R., Schou, J., Thompson, M. J. and Tomczyk, S.: 1997, 'Solar internal sound speed as inferred from combined BiSON and LOWL oscillation frequencies', *Mon. Not. R. astr. Soc.* **291**, 243–251.
Basu, S., Christensen-Dalsgaard, J., Pérez Hernández, F. and Thompson, M. J.: 1996, 'Filtering out near-surface uncertainties from helioseismic inversions', *Mon. Not. R. astr. Soc.* **280**, 651–660.
Böhm-Vitense, E.: 1958, 'Über die Wasserstoffkonvektionszone in Sternen verschiedener Effektivtemperaturen und Leuchtkräfte', *Z. Astrophys.* **46**, 108–143.
Chaboyer, B., Demarque, P. and Pinsonneault, M. H.: 1995, 'Stellar models with microscopic diffusion and rotational mixing. I. Application to the Sun', *Astrophys. J.* **441**, 865–875.

Christensen-Dalsgaard, J. and Berthomieu, G.: 1991, 'Theory of solar oscillations', in Cox, A. N., Livingston, W. C. and Matthews, M. (eds), *Solar interior and atmosphere*, Space Science Series, University of Arizona Press, pp. 401–478.
Christensen-Dalsgaard, J. and Däppen, W.: 1992, 'Solar oscillations and the equation of state', *Astron. Astrophys. Rev.* **4**, 267–361.
Christensen-Dalsgaard, J. and Thompson, M. J.: 1997, 'On solar p-mode frequency shifts caused by near-surface model changes', *Mon. Not. R. astr. Soc.* **284**, 527–540.
Christensen-Dalsgaard, J., Däppen, W., Ajukov, S. V., *et al.*: 1996, 'The current state of solar modeling', *Science* **272**, 1286–1292.
Christensen-Dalsgaard, J., Duvall, T. L., Gough, D. O., Harvey, J. W. and Rhodes Jr, E. J.: 1985, 'Speed of sound in the solar interior', *Nature* **315**, 378–382.
Christensen-Dalsgaard, J., Gough, D. O. and Pérez Hernández, F.: 1988, 'Stellar disharmony', *Mon. Not. R. astr. Soc.* **235**, 875–880.
Christensen-Dalsgaard, J., Gough, D. O. and Thompson, M. J.: 1989, 'Differential asymptotic sound-speed inversions', *Mon. Not. R. astr. Soc.* **238**, 481–502.
Christensen-Dalsgaard, J., Proffitt, C. R. and Thompson, M. J.: 1993, 'Effects of diffusion on solar models and their oscillation frequencies', *Astrophys. J.* **403**, L75–L78.
Cox, A. N., Guzik, J. A. and Kidman, R. B.: 1989, 'Oscillations of solar models with internal element diffusion', *Astrophys. J.* **342**, 1187–1206.
Däppen, W. and Gough, D. O.: 1986, 'Progress report on helium abundance determination', in Gough, D. O. (ed.), *Seismology of the Sun and the distant Stars*, Reidel, Dordrecht, pp. 275–280.
Däppen, W.: 1998, 'Microphysics: equation of state', *Space Sci. Rev.*, this volume.
Duvall, T. L.: 1982, 'A dispersion law for solar oscillations', *Nature* **300**, 242–243.
Dziembowski, W. A., Pamyatnykh, A. A. and Sienkiewicz, R.: 1990, 'Solar model from the helioseismology and the neutrino flux problem', *Mon. Not. R. astr. Soc.* **244**, 542–550.
Dziembowski, W. A., Pamyatnykh, A. A. and Sienkiewicz, R.: 1991, 'Helium content in the solar convective envelope from helioseismology', *Mon. Not. R. astr. Soc.* **249**, 602–605.
Dziembowski, W. A.: 1998, 'Shortcomings of standard solar model', *Space Sci. Rev.*, this volume.
Elliott, J. R. and Gough, D. O.: 1998, 'Calibration of the thickness of the solar tachocline', *Astrophys. J.*, submitted.
Elsworth, Y., Howe, R., Isaak, G. R., McLeod, C. P., Miller, B. A., New, R., Speake, C. C. and Wheeler, S. J.: 1994, 'Solar p-mode frequencies and their dependence on solar activity: recent results from the BISON network', *Astrophys. J.* **434**, 801–806.
Gough, D. O.: 1984, 'Towards a solar model', *Mem. Soc. Astron. Ital.* **55**, 13–35.
Gough, D. O.: 1993, 'Course 7. Linear adiabatic stellar pulsation', in Zahn, J.-P. and Zinn-Justin, J. (eds), *Astrophysical fluid dynamics, Les Houches Session XLVII*, Elsevier, Amsterdam, pp. 399–560.
Gough, D. O. and Kosovichev, A. G.: 1990, 'Using helioseismic data to probe the hydrogen abundance in the solar core', in Berthomieu G. and Cribier M. (eds), *Proc. IAU Colloquium No 121, Inside the Sun*, Kluwer, Dordrecht, pp. 327–340.
Gough, D. O. and Toomre, J.: 1991, 'Seismic observations of the solar interior', *Ann. Rev. Astron. Astrophys.* **29**, 627–685.
Gough, D. O. and Weiss, N. O.: 1976, 'The calibration of stellar convection theories', *Mon. Not. R. astr. Soc.* **176**, 589–607.
Gough, D. O., Kosovichev, A. G., Toomre, J., *et al.*: 1996, 'The seismic structure of the Sun', *Science* **272**, 1296–1300.
Grevesse, N. and Noels, A.: 1993, 'Cosmic abundances of the elements', in Prantzos, N., Vangioni-Flam, E. and Cassé, M. (eds), *Origin and evolution of the Elements*, Cambridge University Press, Cambridge, pp. 15–25.
Grevesse, N., and Sauval, A. J.: 1998, 'Standard Solar Composition', *Space Sci. Rev.*, this volume.
Iglesias, C. A. and Rogers, F. J.: 1996, 'Updated OPAL opacities', *Astrophys. J.* **464**, 943–953.
Iglesias, C. A., Rogers, F. J. and Wilson, B. G.: 1992, 'Spin-orbit interaction effects on the Rosseland mean opacity', *Astrophys. J.* **397**, 717–728.
Kippenhahn, R. and Weigert, A.: 1990, *Stellar structure and evolution*, Springer-Verlag, Berlin.

Kosovichev, A. G.: 1997, 'Inferences of element abundances from helioseismic data', in Habbal, S. R. (ed.), *Robotic Exploration Close to the Sun: Scientific Basis*, AIP Conf. Proc. 385, Amer.Inst. Phys., Woodbury, NY, pp. 159–166.

Kosovichev, A. G., Christensen-Dalsgaard, J., Däppen, W., Dziembowski, W. A., Gough, D. O. and Thompson, M. J.: 1992, 'Sources of uncertainty in direct seismological measurements of the solar helium abundance', *Mon. Not. R. astr. Soc.* **259**, 536–558.

Michaud, G. and Proffitt, C. R.: 1993, 'Particle transport processes', in Baglin, A. and Weiss, W. W. (eds), *Proc. IAU Colloq. 137: Inside the stars*, PASPC **40**, 246–259.

Mihalas, D., Däppen, W. and Hummer, D. G.: 1988, 'The equation of state for stellar envelopes. II. Algorithm and selected results', *Astrophys. J.* **331**, 815–825.

Noerdlinger, P. D.: 1977, 'Diffusion of helium in the Sun', *Astron. Astrophys.* **57**, 407–415.

Pérez Hernández, F. and Christensen-Dalsgaard, J.: 1994, 'The phase function for stellar acoustic oscillations – III. The solar case', *Mon. Not. R. astr. Soc.* **269**, 475–492.

Pijpers, F. P. and Thompson, M. J.: 1992, 'Faster formulations of the optimally localized averages method for helioseismic inversion', *Astron. Astrophys.* **262**, L33–L36.

Richard, O., Vauclair, S., Charbonnel, C. and Dziembowski, W. A.: 1996, 'New solar models including helioseismological constraints and light-element depletion', *Astron. Astrophys.* **312**, 1000–1011.

Rogers, F. J., Swenson, F. J. and Iglesias, C. A.: 1996, 'OPAL Equation-of-State Tables for Astrophysical Applications] *Astrophys. J.* **456**, 902–908.

Rosenthal, C. S.: 1997, 'Convective effects on mode frequencies', in Pijpers, F. P., Christensen-Dalsgaard, J. and Rosenthal, C. S. (eds), *SCORe'96: Solar Convection and Oscillations and their Relationship*, Kluwer, Dordrecht, p. 145–160.

Schou, J., Antia, H. M., Basu, S., *et al.*: 1998, 'Helioseismic studies of differential rotation in the solar envelope by the Solar Oscillations Investigation using the Michelson Doppler Imager', *Astrophys. J.* **505**, in the press.

Shibahashi, H. and Takata, M.: 1996, 'A seismic model deduced from the sound-speed distribution and an estimate of the neutrino flux', *Publ. Astron. Soc. Japan* **48**, 377–387.

Takata, M. and Shibahashi, H.: 1998, 'Solar models based on helioseismology and the solar neutrino problem', *Astrophys. J.* **504**, in the press.

Tomczyk, S., Streander, K., Card, G., Elmore, D., Hull, H. and Caccani, A.: 1995, 'An instrument to observe low-degree solar oscillations', *Solar Phys.* **159**, 1–21.

Turck-Chièze, S.: 1998, 'Composition and opacity in the solar interior', *Space Sci. Rev.*, this volume.

Turcotte, S., and Christensen-Dalsgaard, J.: 1998, 'Solar models with non–standard chemical composition', *Space Sci. Rev.*, this volume.

Vauclair, S.: 1998, 'Element settling in the solar interior', *Space Sci. Rev.*, this volume.

Vorontsov, S. V., Baturin, V. A. and Pamyatnykh, A. A.: 1991, 'Seismological measurement of solar helium abundance', *Nature* **349**, 49–51.

Address for correspondence: Institut for Fysik og Astronomi, Aarhus Universitet, DK-8000 Aarhus C, Denmark

SHORTCOMINGS OF THE STANDARD SOLAR MODEL

W.A. DZIEMBOWSKI
Warsaw University Observatory, Al. Ujazdowskie 4, 00-478 Warsaw, and Copernicus Astronomical Center, ul. Bartycka 18, 00-716 Warsaw, Poland

Abstract. The SSM, invented in early nineteen sixties, remains a useful construction. There are now much larger number of its predictions that may be compared with observations than when it was first introduced. Seismic sounding based on oscillations frequencies provides the best test of the physical input for modelling stellar evolution. The results of the test must be viewed as a support for the standard theory of stellar evolution. However, significant differences in the sound-speed, photospheric He abundance, and other parameters between the Sun and the current models remain. Shortcomings in the EOS and in treatment of convection have been revealed. The differences in the sound-speed in the radiative interior may be explained by small opacity errors but other explanations are possible. Results of seismic sounding support the idea that the element mixing in the outer part of the radiative interior occurred during a significant fraction of the Sun's life. Such mixing is considered as a possible explanation of the deficit of lithium. The shortcomings of SSM cannot explain the deficits of measured neutrino fluxes.

Key words: Sun: internal structure, oscillation, composition

Abbreviations: SSM – standard solar model; EOS – equation of state

1. Concept and Applications of SSM

The procedure of solar model construction, described in this volume by Christensen-Dalsgaard (1998), has not changed since early sixties. We still use the same solar data (mass, radius, luminosity, age, and heavy element to hydrogen abundance ratio) as input. However, their values have somewhat changed. As always, we rely on the best current stellar physics. In this respect there have been much more significant changes. The first thing that needs to be emphasised is the continuing importance of the SSM which follows from the fact that its shortcomings and failures, revealed by comparing its prediction with observations have significance beyond studies of the Sun's internal structure.

We now have millions of new solar observables – the oscillation frequencies – many of them known with better accuracy than the observables used as input data. One may thus consider alternative concept of solar models, in which observational data are used in different ways. Such attempts have been made. For instance, Weiss and Schlattl (1998) used as the input parameters selected quantities determined by of helioseismology instead of the age. Perhaps equal treatment of different observables is a better choice.

When the concept of SSM was introduced some 35 years ago, the main application was the prediction of the neutrino flux for the forthcoming Chlorine experiment. The failure to predict the measurements is no longer blamed on the shortcomings of the SSM. The widely preferred solution of the solar neutrino problem is a revision of one of the basic assumptions in the elementary particle physics which is zero rest mass of neutrinos. The important role of the SSM in this revision should not be forgotten. The context of solar neutrinos remains important. I briefly discuss it in the next section, Suzuki (1998) does it much more comprehensively in his review paper in this volume.

The main application of the SSM nowadays is in the seismic testing of physics underlying stellar evolution theory. This is a rather well established theory and certainly it is not among main challenges of astrophysics of our times. However, without stellar model calculations we cannot determine ages of stellar systems. Results of such calculations are also important in certain methods of measuring distances. In these applications accuracy matters. At the quantitative level much remains to be improved in stellar model calculations. There are some fundamental problems concerning the role of convective overshooting and rotation-induced mixing in the chemical evolution of stars which remain unsolved. These are discussed in this volume by S. Vauclair (1998) and J.-P. Zahn (1998). We have been witnessing in recent decades important progress in calculation of microscopic physics data for stellar models, such as equation of state and opacity reviewed in this volume by F. Rogers and Iglesias (1998) and W. Däppen (1998). However, these data are not free of uncertainties. Helioseismology provides the best way of testing all these aspects of stellar physics. To make use of this tool we must first construct a solar model.

One application of the SSM, which is particularly relevant of the subject of this meeting, is the determination of the helium abundance. In the SSM construction procedure the initial helium abundance, Y_0, is a parameter determined by fitting solar model luminosity and radius to the solar values. Until recently, Y_0 has been identified with the present photospheric abundance, $Y_\odot$. One of the most important achievement of helioseismology was showing that actually $Y_\odot \approx Y_0 - 0.03$. The amount was in a crude agreement with the models taking into account effect of element settling. This effect is now included in the SSM's. The agreement is much better but the difference is somewhat too large. It may be reduced if some element mixing beneath the convective zone base is allowed. This is a likely solution because it may also explain the deficit of lithium in the solar photosphere.

2. The Failures

The relative differences between parameters in SSMs and in the Sun revealed by seismology are at the level of $10^{-3} - 10^{-2}$. The problems these small differences reveal are properly termed as shortcomings. In the case of differences in the neutrino flux and photospheric lithium abundance the appropriate term is failures.

2.1. THE DEFICITS IN MEASURED NEUTRINO FLUXES

It is now prevailing point of view that the large differences between predicted and measured neutrino fluxes are not due to the inadequacies of solar models.

The argument, based on comparison of deficits determined with Homestake and Kamiokande detectors. was first put forward by Bahcall and Bethe (1990). The deficits according to recent data from different type of detectors (Turck-Chièze and Brun, 1997) are given in Table I. The ranges reflect primarily the differences in the SSM predictions.

Table I

Experiment	Neutrino energy threshold	Flux ratio measured/predicted
GALLEX & SAGE	0.233 MeV	0.51 – 0.56
Homestake	0.814 MeV	0.27 – 0.35
Kamiokande & SuperK.	$\sim$ 6 MeV	0.38 – 0.51

The argument against solution of the solar neutrino problem by means of modification of the solar model follows from fact that the largest deficit occurs for the intermediate energy neutrinos. The essence of so-called astrophysical solutions is lowering the temperature in the core of the solar model and this primarily lowers the production rate of the highest energy neutrinos. On the other hand, the deficits are naturally explained if neutrino conversion is allowed (Suzuki, 1998, in this volume).

Even if we accept that the astrophysical solution is no longer viable, there is still an application of the SSM in the context of neutrino experiments. The models are needed for accurate calculation of the neutrino production and conversion rates and which are used in estimates of neutrino parameters from experimental data.

2.2. THE DEFICITS IN PHOTOSPHERIC Li ABUNDANCE

This is in a sense the most severe failure of the SSM. The deficit in the case of the new generation models which include effects of heavy element settling is by a factor $\sim$ 50. The problem is not specific for the Sun, however only in this case we can measure the depth of convective zone, which is critical for quantifying the deficit.

Allowing a rotation-induced macroscopic mixing removes the discrepancy and otherwise has only a small influence on other properties of the solar model (Richard et al., 1996). The small changes are at the level which is detectable by means of helioseismology and in fact they tend to improve the agreement with the seismic model. I will discuss this point again later in this paper.

3. Shortcomings Revealed by Helioseismology

In Figure 1 I show the relative differences between the solar frequencies as determined from measurements with the MDI instrument on board of SOHO spacecraft

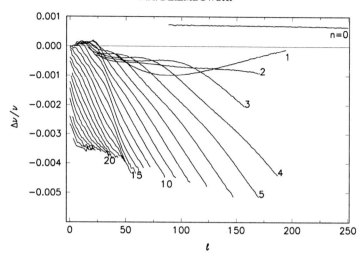

Figure 1. Differences between the solar(SOHO/MDI) and SSM(JCD) frequencies

(Rhodes *et al.*, 1998) and frequencies of oscillations calculated for a standard solar model (Model S0 of Christensen-Dalsgaard *et al.*, 1995, denoted here as JCD). The differences are small which is a strong indication that this is a good model of the Sun. Yet the difference are by 2 to 3 orders of magnitude greater than the measurement errors and they should be explained. Ideally, we would like to use the differences to identify specific shortcomings of the model, i.e., to disentangle inaccuracies in solar data from inadequacies in the input physics and in the standard assumptions of stellar evolution.

The nearly constant value of $\Delta \nu / \nu$ for f-modes ($n = 0$) was interpreted as the evidence that the seismic radius of the Sun is a little smaller than the photometric radius, which was adopted in the model. (Schou *et al.*, 1997). Similar inference from the GONG data was made by Antia (1998). Indeed, it is difficult to imagine another effect that could result in nearly constant relative frequency differences. The corresponding relative difference of the radius, which is -4.5×10^{-4}, may seem insignificant. In fact, it turns out quite important interpretation of the seismic differences in the sound-speed and in density inferred from p-mode frequencies (Basu, 1998).

The frequency differences for p-modes show a much more complicated behaviour. There are many potential contributors to the frequency differences. Not all of them may be blamed to the shortcomings of SSMs.

4. Contributions to $\Delta \nu_{\ell,n}$ for p-modes

It is fairly straightforward to derive an integral expression connecting frequency differences to difference in structural functions with the uncertainty in radius is taken into account. Starting from the equations for adiabatic oscillations written in dimensionless form, with R being the unit of length, and choosing

$$u = \frac{P}{\rho}\frac{R}{GM}$$

as the structural variable, we get, in a similar way as Christensen-Dalsgaard (1998) in this volume obtains his Eq.(17) (note, however, the difference in definition of u),

$$\left(\frac{\delta\nu}{\nu}\right)_i = \int_0^1 \mathcal{K}_{u,i}\frac{\delta_x u}{u}dx + \frac{F_{\text{surf}}(\nu)}{I_i} + \mathcal{J}_i\delta Y_\odot - 1.5\frac{\delta R}{R}, \qquad (1)$$

where $i \equiv (\ell, n)$ is a mode identifier, $x = r/R$, and

$$\mathcal{J}_i = \int_0^1 \left(\mathcal{K}_{\gamma,i}\frac{\partial \gamma_1}{\partial Y}\right) dx.$$

One additional simplification we made was ignoring $\delta_x\gamma_1$ in the radiative interior.

The $F_{\text{surf}}(\nu)/I_i$ term, where I_i denotes mode inertia, introduced to account for various effects localised near the surface. This term contains contributions from the real differences between the Sun and SSM but also a very significant contribution which results from the use of the adiabatic approximation in the calculation of model frequencies.

In almost all frequency inversions done so far the term in $\delta R/R$ was neglected and I also ignored this term in the inversions I did for this section. For historical reason I chose not the SOHO/MDI frequencies but rather an earlier data set which is combination of low-ℓ mode frequencies from BISON network (Elsworth et al., 1994) and frequencies for modes with ℓ from 3 to 150 from BBSO (Libbrecht et al., 1990). The inversions were done with use of two reference models. First model denoted as RS0 (model 0 of Dziembowski et al., 1994) was calculated ignoring element settling. These effects were taken into account in the second model (JCD). There are also differences in the opacity data and in the treatment of convection. In Figures 2 and 3 separate contributions to the frequency differences between the Sun and these two models are shown.

In the case of model RS0 the largest contribution is that of $F_{\text{surf}}(\nu)/I_i$. The two contributions which reveal the model shortcomings are comparable. The relatively large contribution from δY in the case of the p_1 and p_2 modes is the consequence of ignoring the gravitational settling. Here comes the historical part. When the first inversion of the BBSO set showed the exceptional departure for these modes, the first suspicion was that there was a systematic error in the procedure of deriving frequencies for modes in these two ridges. The authors of the data were even asked to reconsider their analysis. It took us some time to realise that the systematic error was on the side of our models which did not take into account element settling. The p_1 and p_2 mode frequencies are particularly sensitive to $Y_\odot$. We see in Fig. 3

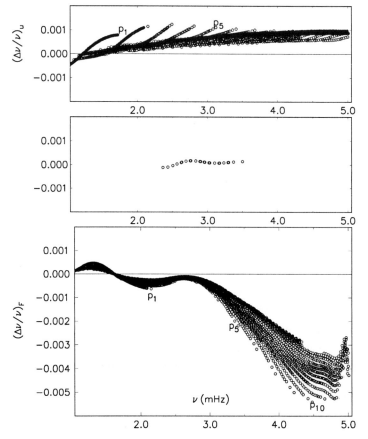

Figure 2. Contributions to the p-mode frequency differences between the Sun (BBSO&BISON data set) and RS0 model. $(\Delta\nu/\nu)_u$, $(\Delta\nu/\nu)_Y$, and $(\Delta\nu/\nu)_F$ denote contributions from the differences in u, differences in $Y_\odot$, and the $F(\nu)/I$ term (see Eq.1)

that the whole contribution nearly disappears once this effect is included in the reference model.

Somewhat surprisingly the contribution due to structural difference is larger for the more modern JCD model. The cause is partially revealed in the plots of $\delta_x u/u$ shown in Fig. 4. The two models differ in the treatment of convection. The JCD model was calculated with the use of standard mixing length theory while the RS0 model uses the Canuto (1990) formalism which implies higher efficiency of the convective energy transport and yields better agreement in the sound-speed. It should be stressed, however, that a comparison of the sound-speed does not provide a comprehensive test of convection theories.

The JCD model, on the other hand, describes better radiative interior of the Sun. The obsolete RS0 model looks also not too bad in this region. This, however, is to some extent accidental because two major shortcomings of RS0 model, which are ignoring gravitational settling and using obsolete opacity data, have effects in u.

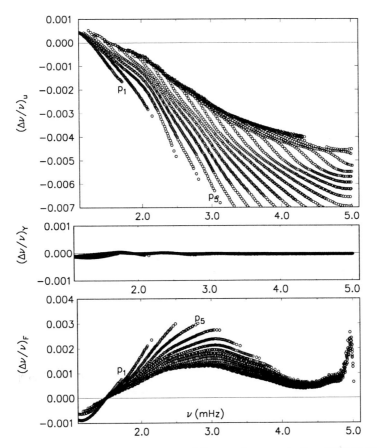

Figure 3. Contributions to the p-mode frequency differences between the Sun (BBSO&BISON data set) and SSM (JCD) (see Fig.2 caption)

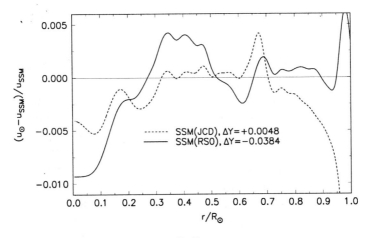

Figure 4. Relative differences in $u = \frac{P}{\rho}\frac{R}{GM}$ between the Sun and its two models

The former shortcoming is very clearly exposed in the large negative value of $\delta Y_\odot$. This emphasises the diagnostic value of this seismic observable.

There is a trivial way of removing the whole $\delta_x u$ in the radiative interior of JCD model. Tripathy *et al.* (1997) showed that this can be accomplished by *ad hoc* modification of the opacity coefficient. The relative modification does not exceed anywhere 5 percent, which is within the uncertainty of opacity calculations. It would be very important to reduce the uncertainty because at its current level it limits significantly the possibility of identification of specific shortcomings in modelling the Sun and other stars. With the present accuracy of frequency measurements we could obtain interesting limitation on solar age and the p+p reaction cross-section (e.g. Dziembowski *et al.*, 1994) if we knew the opacity coefficient more accurately.

The most conspicuous feature of $\delta_x u/u$ for JCD model – the spike of at $r \approx 0.68 R_\odot$ – may be eliminated by small (less than 1 percent) change of the opacity coefficient. Yet its form and localisation suggest a different explanation: Gough *et al.* (1996) pointed out that this feature may be eliminated by introducing helium mixing beneath the base of convective envelope. This could be the rotation-induced mixing postulated by Richard *et al.* (1996) to explain the lithium deficiency. The small positive value of $\delta Y_\odot$ supports this interpretation because such mixing works against helium settling.

5. Consequences of Inaccuracy of R

Basu (1998) showed that the small difference in solar radius found from the f-mode data has rather large consequences for the seismic corrections to the sound-speed and density inferred from p-mode frequencies. It is therefore important to know the precise value of R and to asses its uncertainty. The first question to ask is whether the concept of solar radius is well-defined at the required level of accuracy. The inversion with an adjustable radius was introduced as an attempt to answer this question. More specifically, we (Richard *et al.*, 1998) asked whether from p-modes we may deduce a significant and robust value of δR and how this value compares with that deduced from f-modes. The result is negative.

In Fig. 5 I show results of inversions of the two sets of two p-mode frequencies with and without δR term. It should be stressed that the f-mode data were excluded from the SOHO/MDI set because Eq.(1) is not applicable for f-modes. The value given in the figure for the SOHO/MDI p-modes is not very different from that inferred from f-modes (-0.00045). It is, however, very unstable. The highest ℓ value for p-modes in this data set is 194 (see Fig. 1). When modes with $\ell > 140$ were removed we obtained $\delta R/R = 0$. We deduced also a very different value from the BBSO& BISON set in which the highest degree is $\ell = 150$. One may speculate that it is just necessary to include p-modes of sufficiently high degree. However, there is a problem with application of Eq. (1) because it is valid only if the lower

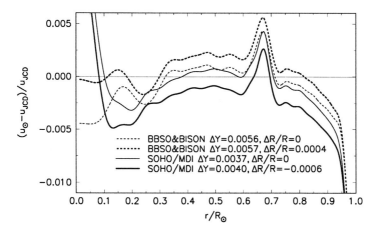

Figure 5. Comparison of the inversions for $\delta_x u/u$ of two data sets, with and without the δR term.

turning point is located sufficiently deep below the layer where the near surface uncertainties occur and this condition is not fulfilled for modes of high ℓ.

Fig.5 illustrates the consequences of the uncertainty of R. These are fairly large but fortunately the effect mostly reduces to a vertical shift of the $\delta_x u/u$ curves. Therefore the relative values are rather robust. Other seismic observables which are only weakly affected by the uncertainty of R are the depth of convective zone and the photospheric helium abundance. Of course, it is highly desirable to look for possible improvements in methods of helioseismic inversion.

6. How Accurate is the Seismic $Y_\odot$?

The value of $Y_\odot$ is one of most important seismic observables. We do not have better ways of measuring the helium abundance in the solar photosphere. Furthermore, this observable provides a constraint on the mixing in the radiative interior, which is the least understood effect in stellar evolution theory. According to the recent analysis of Basu (1998) we have $Y_\odot = 0.248 \pm 0.001$. The value has been confirmed by independent calculations of Richard *et al.* (1998), who quote, however, twice as big uncertainty. This bigger uncertainty reflects primarily the spread resulting from the use of different reference models calculated with different treatment of convection. Even with the 0.002 uncertainty the value of $Y_\odot$ is significantly less than that in JCD model.

It should be stressed that these two determinations of $Y_\odot$ were made with the use of the OPAL EOS and that neither Basu (1998) nor Richard *et al.* (1998) could estimate effects of uncertainties in the thermodynamical parameters. Actually, this could be a dominant source of the uncertainty of the seismic $Y_\odot$. With the use of the MHD EOS the value obtained by Basu was about 0.252, which is very different from the value 0.242 obtained by Richard *et al.* with use of the same EOS. I do

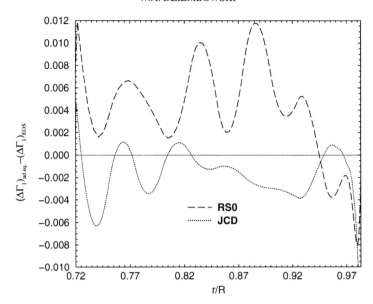

Figure 6. The difference $(\delta_x \gamma_1)_{\text{ad.eq.}} - (\delta_x \gamma_1)_{\text{EOS}}$ evaluated on the basis of the inversion of the SOHO/MDI frequency data with use of the two reference models employing different thermodynamical data. The MHD EOS is used in RS0 model and the OPAL EOS in JCD model

not know whether this discrepancy, which is contrasted with the good agreement obtained with the OPAL EOS, can be used as the argument in favour of credibility of the OPAL.

Fortunately there is a clean seismic test of the EOS in the adiabatic part of the solar convection zone (Dziembowski *et al.*, 1992; Christensen-Dalsgaard and Däppen 1992; Basu and Christensen-Dalsgaard, 1997). Having determined $\delta_x u/u$, we may evaluate $\delta_x \rho$ and $\delta_x P$ with the use of the linearized mechanical equilibrium condition. For the adiabatic stratification the same condition allows evaluation of a seismic correction to the adiabatic exponent. On the other hand, from EOS data we have $\gamma_1(\rho, P, Y)$; hence we may evaluate $(\delta_x \gamma_1)_{\text{EOS}}$ using seismic determinations of $\delta_x \rho$, $\delta_x P$ and, $\delta Y_\odot$. The difference between the two values of $\delta_x \gamma_1$ is a test of the EOS accuracy. In Fig. 6 (from Richard *et al.*, 1998) the result for the MHD (Däppen *et al.*, 1998) and the OPAL EOS (Rogers *et al.*, 1966) are shown. The latter equation of state passes this test better but not perfectly.

7. Conclusions

The Standard Solar Model is one of the great inventions of astrophysics. Without it the best evidence for neutrino mass would not be possible. Continuous refinement of SSM is important for accurate determination of the neutrino mass differences and the mixing angles.

The main astrophysical application of SSM today is in seismic testing various improvements of stellar physics by means of helioseismology. As long as observations reveal shortcomings of the SSM it is an indication that improvements in stellar physics are needed. Another important astrophysical application of the SSM is seismic measurement of $Y_\odot$.

There is no observational evidence that anything beyond refinements in standard modelling is needed. The most important is allowing some element mixing in the outer part of the radiative interior. The primary observational argument is the overabundance of lithium in current SSMs by factor ~ 50. This modification is also supported by seismic determination of the photospheric helium abundance and the sound-speed in the radiative interior near the base of the convective zone.

Results of seismic testing show that models calculated with the standard mixing length theory have higher sound-speed in the convective zone than the Sun. The essence of the required modification is an increase of the efficiency of the convective transport.

Seismology provides a clean test of the equation of state. The result points to a need of some refinement. The refinement is important for improving both stellar model calculation and the accuracy and reliability of seismic measurement of He abundance.

In the radiative interior we cannot disentangle in the seismic corrections to the sound-speed the contributions due to inadequacies in the opacity calculation from effects of element mixing, and effects of errors in the parameters like the solar age and nuclear reaction cross-section. Improving the accuracy of the opacity calculation would expand the usefulness of seismic sounding.

One solar parameter used in SSM construction that needs to be reconsidered in view of the comparison of predicted and solar frequencies is radius. The f-mode data point to a need of some downward revision. However, with inversions of the p-mode frequencies we could not confirm the revision. There is room for improvement in the methods of helioseismic inversions.

Acknowledgements

I thank Alosha Pamyatnykh for reading a preliminary version of this paper and suggesting corrections and the referee (J. Christensen-Dalsgaard) for his comments. The work was supported by KBN-2P304-013-07 grant.

References

Antia, H.M.: 1998, 'Estimate of solar radius from f-mode frequencies', *Astron. Astrophys.* **330**, 336–340.
Bahcall, J.N. and Bethe, H.A.: 1990, 'A solution of the Solar-Neutrino Problem', *Phys. Rev. Lett.* **65**, 2233–2238.

Basu, S.: 1998, 'Effects of Errors in the Solar Radius on Helioseismic Inferences', *Month. Not. Roy. astron. Soc.*, in the press.
Basu, S., Christensen-Dalsgaard, J.: 1997, 'Equation of state and helioseismic inversions', *Astron. Astrophys.* **337**, L5–L8.
Canuto, V.: 1990, 'The mixing length parameter α', *Astron. Astrophys.* **227**, 282–284.
Christensen-Dalsgaard, J. et al.: 1996, 'The current state of solar modelling', *Science* **272**, 1286–1292.
Christensen-Dalsgaard, J.: 1998, 'The "standard" Sun', *Space Sci. Rev.*, this volume.
Christensen-Dalsgaard, J., Däppen, W.: 1992, 'Solar oscillations and equation of state', *Astron. Astrophys. Rev.* **4**, 267–361.
Däppen, W.: 1998, 'Microphysics: equation of state', *Space Sci. Rev.*, this volume.
Däppen, W., Mihalas, D., Hummer, D.G. and Mihalas, B.: 1988, 'The equation of state for stellar envelopes. III. Thermodynamical quantities' *Astrophys. J.* **332**, 261–270.
Dziembowski, W.A., Goode, P.R., Pamyatnykh, A.A. and Sienkiewicz, R.: 1994, 'Seismic Model of the Sun's Interior'. *Astrophys. J.* **432**, 417–426.
Dziembowski, W.A., Pamyatnykh, A.A. and Sienkiewicz, R.: 1992, 'Seismological Tests of Standard Solar Models Calculated with New Opacities', *Acta Astronomica* **42**, 5–15.
Elsworth, Y., Howe, R., Isaak, G.R., McLeod, C.P., Miller, B.A., New, R., Speake, C.C. and Wheeler, S.J.: 1994, 'Solar p-mode frequencies and their dependence on solar activity: Recent results from the BISON network', *Astrophys. J.* **434**, 801–806.
Gough, D.O., et al.: 1996, 'The Seismic Structure of the Sun', *Science* **272**, 1296–1299.
Libbrecht, K.G., Woodard, M.F. and Kaufman, J.M.: 1990, 'Frequencies of solar oscillations', *Astrophys. J. Suppl.* **74**, 1129–1149.
Rhodes, E.J., Kosovichev, A.G., Schou, J., Scherrer, P.H. and Reiter, J.: 1997, Measurements of Frequencies of Solar Oscillations from the MDI Medium-ℓ' Program, *Solar Phys.* **175**, 287–310.
Richard, O., Vauclair, S., Charbonnel, C. and Dziembowski, W.A.: 1996, 'New Solar Models Including Helioseismological Constraints and Light Element Depletion', *Astron. Astrophys.* **312**, 1000–1013.
Richard, O., Dziembowski, W.A., Sienkiewicz, R. and Goode, P.R.:1998, 'On the Accuracy of Helioseismic Determination of Solar Helium Abundance', *Astronomy and Astrophysics*, submitted.
Rogers, F.J., Swenson, F.J. and Iglesias, C.A.: 1996, 'OPAL equation-of-state tables for astrophysical applications', *Astrophys. J.* **456**, 902–908.
Rogers, F.J. and Iglesias, C.A.: 1998, 'Opacity of stellar matter', *Space Sci. Rev.*, this volume.
Schou, J., Kosovichev, A.G., Goode, P.R., and Dziembowski, W.A.: 1997, 'Determination of the Sun's Seismic Radius from SOHO/MDI', *Astrophys. J. Lett.* **489**, L197–201.
Suzuki, Y.: 1998, 'Solar neutrinos', *Space Sci. Rev.*, this volume.
Tripathy, S.C., Basu, S. and Christensen-Dalsgaard, J.: 1997, 'Helioseismic determination of opacity correction', in J. Provost and F.-X. Schmider (eds.), *Sounding Solar and Stellar Interiors*, (Nice), poster.
Turck-Chièze, S. and Brun, A.S.: 1998, 'Spatial Seismic Constraints on Solar Neutrino Predictions', in W.Hampel (ed.), *Fourth International Neutrino Conference*, Heidelberg, in the press.
Vauclair, S.: 1998, 'Element settling in the solar interior', *Space Sci. Rev.*, this volume.
Weiss, A. and Schlattl, H.: 1998, 'The age of the most nearby star', *Astronomy and Astrophysics* **332**, 215–223.
Zahn, J.-P.: 1998, 'Macroscopic transport', *Space Sci. Rev.*, this volume.

Address for correspondence: Copernicus Astronomical Center, ul. Bartycka 18, 00-716 Warsaw, Poland

MICROPHYSICS: EQUATION OF STATE

To Veronika and her Hometown

WERNER DÄPPEN
Department of Physics and Astronomy
University of Southern California

Abstract. The equation of state is one of the three fundamental ingredients used to construct stellar models. The plasma of the interiors of stars such as the Sun is only slightly non-ideal. However, the extraordinary accuracy of the helioseismological data requires refined equations of state. It turned out to be necessary to include a Coulomb correction, commonly evaluated in the Debye-Hückel approximation. Higher-order non-ideal effects have implications as well, both for plasma physics and for solar physics. As a typical example, the recently studied thermodynamic consequence of excited states in compound particles is discussed. This effect is of considerable relevance in the helioseismic determination of the helium abundance in the solar convection zone.

Key words: helioseismology, equation of state, helium abundance, excited states

1. Introduction

Progress in the solar equation of state serves two purposes. For solar physicists on the one hand, a better equation of state will lead to a smaller uncertainty in solar models, which thus turn into a more reliable astrophysical tool, for instance, to tackle the solar neutrino problem. Plasma physicists on the other hand will recognize that a better equation of state can be found by an astrophysical experiment in a domain where there is not much laboratory competition.

The equation of state is one of the three fundamental ingredients used to construct stellar models. The other two are opacity and nuclear reaction rates. One star – the Sun – is very special in two respects. First, the methods of helioseismology allow us to obtain very accurate experimental data of the solar interior (in particular, sound speed and density). Second, in the solar convection zone, helioseismology presents an opportunity to isolate the the equation of state from opacity and nuclear reaction rates, since the stratification is essentially adiabatic and thus determined by thermodynamics.

The plasma of the interiors of "normal" stars, such as the Sun, is only slightly non-ideal. One would therefore think that at least for such stars finding a good equation of state is not too difficult. Indeed, simple models of the equation of state have been quite successful in many aspects of stellar structure and evolution. However, the extraordinary accuracy of the helioseismological data led to further refinement of the idea what a "good" equation of state for the Sun should be. This was recognized in the early 1980s and models with improved equations of state have since been used in helioseismic studies.

Several discrepancies between the experimental data and models with simple equations of state have been successfully identified as the signatures of various non-ideal phenomena. The most obvious of these phenomena deal with the Coulomb interaction between charged particles, pressure ionization, and the effects of the internal partition functions of bound systems on the thermodynamical properties of the solar plasma.

2. Helium Abundance and the Equation of State

Despite the fact that helium was first discovered in the solar spectrum, the helium abundance can not directly be obtained from spectroscopy. The existing, fairly strong, helium lines in the spectrum actually belong to He^+-ions. As for atomic helium lines, they are rather faint and their structure is so complex, that observation gives no precise information about the helium abundance. Any attempt to deduce the helium abundance from He^+ obviously depends on our uncertainties about the solar atmosphere which, among other, are related to the deviations from local thermodynamical equilibrium and the influence of radiation hydrodynamics.

Before the era of helioseismology the accepted method was based on a calibration of solar models, in which the helium abundance entered as an open parameter to be adjusted, essentially, to match the observed luminosity. The idea to determine the helium abundance directly through helioseismological observations, in contrast to calibration of models, was first proposed by Gough (1984). He realized that one can exploit the fact that the principal deviation of pressure-density relation from the ideal gas law is caused by partial ionization. Ionization lowers the adiabatic exponent

$$\gamma_1 = (\partial \ln p / \partial \ln \rho)_s \qquad (1)$$

from its ideal-gas value of 5/3. This lowering can be qualitatively understood in the following way. First, notice that γ_1 is approximately equal to the ratio of specific heat at constant pressure c_p over specific heat at constant volume c_v. Then, consider that in regions of partial ionization, both specific heats become much larger than their ideal gas values (which for c_v and c_p correspond to $\frac{3}{2}k_B$ and $\frac{5}{2}k_B$ per particle, respectively, k_B being Boltzmann's constant), because heat increments are used to ionize, with temperature changing little (the analogy is the latent heat at phase transitions of ordinary matter). However, the difference $c_p - c_v$ is, according to the first law of thermodynamics, a quantity that stays close to its ideal-gas value ($1\ k_B$), even when c_p and c_v each become large.

Thus, the ratio c_p/c_v and also γ_1 become smaller in zones of partial ionization. More specifically, the adiabatic exponent goes down to about 1.2 in the H-ionization zone and to about 1.55 in the second ionization zone of He. Therefore γ_1 demonstrates "dips" in its profile, corresponding to the ionization zones of hydrogen, helium and other elements. Quite clearly, the abundance of the respective species dictates how pronounced these dips are. Unfortunately, although the dip of the hydrogen ionization zone would be the most prominent of all, it cannot be used in practice, because it occurs in a region where our knowledge of the internal structure is very uncertain due to our ignorance on solar convection.

In the same way, the first helium-ionization zone is not useful either, because it coincides with the hydrogen ionization zone. However, the second helium ionization zone shows up in a clean and isolated way, and it occurs at a location where the stratification is very close to adiabatic. Most of our uncertainty associated with the details of convection is thus removed. Of course, any uncertainty in the equation of state will propagate in the quality of such a determination of the helium abundance. Fortunately, as Gough (1984) already recognized, helioseismology has the potential to probe the equation of state *and* the helium abundance at the same time. Several studies have made this point quantitatively (Gough, 1984; Däppen, 1987; Christensen-Dalsgaard and Däppen, 1992; Kosovichev *et al.*, 1992; Vorontsov *et al.*, 1994; Antia and Basu, 1994; Basu and Antia, 1995). These references contain the evidence why the equation of state is a sufficiently sensitive ingredient in an important astrophysical application. They also make the point that the present uncertainty in the equation of state is unacceptable. But none the less, the helioseismic helium abundance determination has led to an astrophysical success. Somewhat surprisingly it turned out that the helium abundance Y in the solar convection zone lies between 0.24 and 0.25 (Christensen-Dalsgaard *et al.*, 1996), substantially lower than the calibrated values for the age-zero Sun, which typically have a Y around 0.27. This discrepancy helped to draw attention to the importance of the effect of gravitational settling of helium and heavier elements. The net result were solar models that are in better agreement with helioseismological data. Settling in now contained in standard solar models (Christensen-Dalsgaard *et al.*, 1993; Christensen-Dalsgaard, 1998; Vauclair, 1998).

3. Equation of State Issues

3.1. Ideal and Non-ideal Plasmas

The simplest model is a mixture of nuclei and electrons, assumed fully ionized and obeying the classical perfect gas law. However, an *"ideal-gas"* equation of state can be more general. It may include deviations from the perfect gas law, namely ionization, radiation and degenerate of electrons, as long as the underlying microphysics of these additional effects is still ideal, that is, as long as it does not contain interactions. The "particles", however, can be classical or quantum, material or photonic. In such an ideal framework, bound systems (molecules, atoms, ions) are allowed to have internal degrees of freedom (excited states, spin). All such ideal effects can be calculated as exactly as desired.

One measure of non-ideality in plasmas is the so-called coupling parameter Γ. In a plasma where particles have average distance $\langle r \rangle$ from each other, we can define Γ as the ratio of average potential binding energy over mean kinetic energy $k_B T$ (in the simplest case of hydrogen; generalizations to other elements are straightforward)

$$\Gamma = \frac{\left\langle \frac{e^2}{\langle r \rangle} \right\rangle}{k_B T}. \qquad (2)$$

Plasmas with $\Gamma \gg 1$ are *strongly* coupled, those with $\Gamma \ll 1$ *weakly* coupled. A famous example of a strongly coupled plasmas is the interior of white dwarfs, where the coupling can become so strong to force crystallization. Weakly coupled are the interiors of stars with masses ranging from the slightly sub-solar ones to the largest.

As one can suspect, Γ is the dimensionless coupling parameter according to which one can classify theories. Weakly-coupled plasmas lend to systematic perturbative ideas (*e.g.* in powers of Γ), strongly coupled plasma need more creative treatments. Improvements in the equation of state beyond the model of a mixture of ideal gases are difficult. This has both conceptual and technical reasons. As a fundamental conceptual reason I mention the fact that in a plasma environment already the idea of isolated atoms (and compound ions) has to be abandoned. A technical reason is the difficulty encountered when specific non-ideal effect are modeled. The three principal non-ideal effects are related to: (i) the internal partition functions of bound systems, (ii) pressure ionization, and (iii) collective interactions of the charged particles. The internal partition functions contain the difficult problem of excited states, where and how they are to be cut off. They are an important element in determining the ionization balances. Pressure ionization has to be provided by non-ideal interaction terms, because ideal gases would unphysically recombine in the central regions of stars.

3.2. CHEMICAL AND PHYSICAL PICTURE

3.2.1. *Chemical Picture*

Most realistic equations of state that have appeared in the last 30 years belong to the chemical picture and are based on the free-energy minimization method. This method uses approximate statistical mechanical models (for example the nonrelativistic electron gas, Debye-Hückel theory for ionic species, hard-core atoms to simulate pressure ionization via configurational terms, quantum mechanical models of atoms in perturbed fields, *etc.*). From these models a macroscopic free energy is constructed as a function of temperature T, volume V, and the particle numbers $N_1, \ldots, N_m$ of the m components of the plasma. This free energy is minimized subject to the stoichiometric constraint. The solution of this minimum problem then gives both the equilibrium concentrations and, if inserted in the free energy and its derivatives, the equation of state and the thermodynamic quantities.

Obviously, this procedure automatically guarantees thermodynamic consistency. As an example, when the Coulomb pressure correction (to the ideal-gas contribution) is taken into account in the free energy (and not merely in the pressure), it affects both the pressure and the equilibrium concentration, *i.e.*, the degrees of ionization. In contrast, the mere inclusion of the pressure correction would be inconsistent with other thermodynamic quantities. In the chemical picture, perturbed atoms must be introduced on a more-or-less *ad-hoc* basis to avoid the familiar divergence of internal partition functions (see *e.g.* Ebeling *et al.*, 1976). In other words, the approximation of unperturbed atoms precludes the application of standard statistical mechanics, *i.e.* the attribution of a Boltzmann-factor to each atomic state. The conventional remedy of the chemical picture against this is a modification of the atomic states, *e.g.* by cutting off the highly excited states in function of density and temperature of the plasma. A currently popular equation of state realized in the chemical picture is based on an occupation probability formalism (Hummer and Mihalas, 1988; Mihalas *et al.*, 1988; Däppen *et al.*, 1988; Däppen *et al.*, 1987; hereinafter MHD).

Specifically, the internal partition functions Z_s^{int} of species s adopted by MHD are weighted sums

$$Z_s^{\text{int}} = \sum_i w_{is} g_{is} \exp\left(-\frac{E_{is}}{k_B T}\right). \tag{3}$$

Here, is label states i of species s. E_{is} are their energies, and the coefficients w_{is} are the occupation probabilities that take into account charged and neutral surrounding particles. In physical terms, w_{is} gives the fraction of all particles of species s that can exist in state i with an electron bound to the atom or ion, and $1 - w_{is}$ gives the fraction of those that are so heavily perturbed by nearby neighbors that their states are effectively destroyed. Perturbations by neutral particles are based on an excluded-volume treatment and perturbations by charges are calculated from a fit to a quantum-mechanical Stark-ionization theory (for details see Hummer and Mihalas, 1988).

3.2.2. *Physical Picture*

It is clear from the preceding subsection that the advantage of the chemical picture lies in the possibility to model complicated plasmas, and to obtain numerically smooth and consistent thermodynamical quantities. Nevertheless, the heuristic method of the separation of the atomic-physics problem from that of statistical mechanics is not satisfactory, and attempts have been made to avoid the concept of a perturbed atom in a plasma altogether. This has suggested an alternative description, the physical picture. In such an approach one expects that no assumptions about energy-level shifts or the convergence of internal partition functions have to be made. On the contrary, properties of energy levels and the partition functions should come out from the formalism.

There is an impressive body of literature on the physical picture. Important sources of information with many references are the books by Ebeling *et al.* (1976),

Kraeft et al. (1986), and Ebeling et al. (1991). However, the majority of work on the physical picture was not dedicated to the problem of obtaining a high-precision equation of state for stellar interiors. Such an attempt was made for the first time by a group at Livermore as part of an opacity project (Rogers, 1986; Iglesias and Rogers, 1995; Rogers et al., 1996 and references therein).

To explain the advantages of this approach for partially ionized plasmas, it is instructive to discuss the activity expansion for gaseous hydrogen. The interactions in this case are all short ranged and pressure is determined from a self-consistent solution of the equations (Rogers, 1981)

$$\frac{p}{k_B T} = z + z^2 b_2 + z^3 b_3 + \ldots \tag{4}$$

$$\rho = \frac{z}{k_B T}\left(\frac{\partial p}{\partial z}\right), \tag{5}$$

where $z = \lambda^{-3} \exp(\mu/k_B T)$ is the activity, $\lambda \equiv h/\sqrt{2\pi m_e k_B T}$ is the thermal (de Broglie) wavelength of electrons, μ is the chemical potential and T is the temperature. The b_n are cluster coefficients such that b_2 includes all two particle states, b_3 includes all three particle states, etc..

In contrast to the chemical picture, which is plagued by divergent partition functions, the physical picture has the power to avoid them altogether. An important example of such a fictitious divergence is that associated with the atomic partition function. This divergence is fictitious in the sense that the bound-state part of b_2 is divergent but the scattering state part, which is omitted in the Saha approach, has a compensating divergence. Consequently the total b_2 does not contain a divergence of this type (Ebeling et al., 1976; Rogers, 1977). A major advantage of the physical picture is that it incorporates this compensation at the outset. A further advantage is that no assumptions about energy-level shifts have to be made (see the previous subsection); it follows from the formalism that there are none.

As a result, the Boltzmann sum appearing in the atomic (ionic) free energy is replaced with the so-called Planck-Larkin partition function (PLPF), given by (it e.g. Ebeling et al., 1976; Kraeft et al., 1986; Rogers, 1986)

$$\text{PLPF} = \sum_{nl} (2l+1) \left[\exp(-\frac{E_{nl}}{kT}) - 1 + \frac{E_{nl}}{kT}\right]. \tag{6}$$

The PLPF is convergent without additional cut-off criteria as are required in the chemical picture. I stress, however, that despite its name the PLPF is not a partition function, but merely an auxiliary term in a virial coefficient (see, for example, Däppen et al., 1987).

3.3. COULOMB CORRECTION AND THE DEBYE-HÜCKEL APPROXIMATION

Debye-Hückel (DH) theory is based on the replacement of the long-range Coulomb potential by a screened potential (see e.g. Ebeling et al., 1976). This interaction leads to a negative pressure correction to the ideal-gas value. Under solar conditions, the relative correction culminates at about 8% in the outer part of the convection zone and it has another local maximum of about 1% in the core. Originally, the DH formalism included an additional ingredient of a fixed size for positive ions, inside of which the electrostatic potential is assumed to be constant. In astrophysical applications, such as MHD, this so-called τ-correction has been adopted in the form given by Harris *et al.* (1960) (see Gabriel, 1994; Baturin *et al.*, 1996). The systematic OPAL equation of state, however, comes close to the unmodified DH results under solar condition. This makes any τ-correction rather implausible, especially since in comparisons with observational data, OPAL seems to fare better than MHD (Christensen-Dalsgaard *et al.*, 1996), at least below the helium ionization zones where other effects might dominate (section 4).

I would like to emphasize that close attention to the DH theory is warranted, because it describes the main truly non-ideal effect under solar conditions. It was suggested in a number of early papers (*e.g.* Berthomieu *et al.*, 1980; Ulrich, 1982; Ulrich and Rhodes, 1983; Shibahashi *et al.*, 1983; Shibahashi *et al.*, 1984; Noels *et al.*, 1984) that improvements in the equation of state can reduce discrepancies between theory and observations. Later, Christensen-Dalsgaard *et al.* (1988) showed that the MHD equation of state reduced these discrepancies significantly for a large range of oscillation modes. It turned out that the Coulomb term is the dominant non-ideal correction in the hydrogen and helium ionization zones. This discovery led to an upgrade of the simple, but astrophysically useful Eggleton, Faulkner and Flannery (1973) (EFF) equation of state through the inclusion of the Coulomb interaction term (CEFF) (Christensen-Dalsgaard, 1991; Christensen-Dalsgaard and Däppen, 1992).

4. Beyond the Debye-Hückel Correction

Due to the relatively high temperature inside the Sun, the potential energy of the Coulomb interaction is small compared to the kinetic energy of particles, which allows us to believe that the DH theory makes quantitative sense at least asymptotically. But to estimate the possible error we need a physically based expression for next terms in the corresponding expansion. The OPAL equation of state contains such higher terms. An alternative is the path-integral based Feynman-Kac (FK) formalism of Alastuey and Perez (1992, 1996), Alastuey *et al.* (1994, 1995). Its application to solar model is in progress (Perez and Däppen, 1998).

New insight beyond DH might come through the equivalent description of the DH correction as a self energy of electrons and nuclei. A first study has revealed that the thermodynamic consequence of screened bound-state energies and a shifted continuum is hypersensitive on the details of the screening and the method with which the thermodynamic quantities are evaluated. No conclusion has been reached thus far but a preliminary study has shown that static screening alone comes close

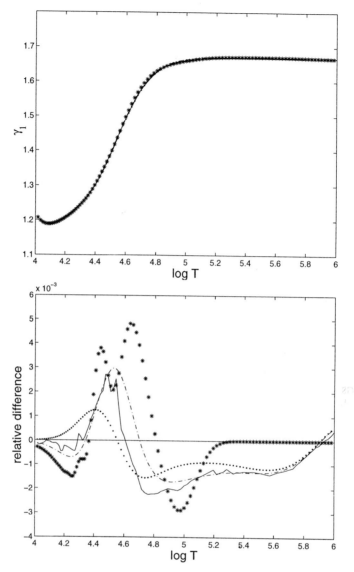

Figure 1. Top panel: absolute values of γ_1 for solar temperatures and densities of a hydrogen-only plasma. Linestyles: MHD – asterisks, MHD$_{GS}$ – dashed lines, MHD$_{PL}$ – dotted-dashed lines, MHD$_{PL,GS}$ – dotted lines, and OPAL – solid lines. *Bottom panel:* relative differences with respect to MHD$_{GS}$, in the sense $(\gamma_1 - \gamma_1[\text{MHD}_{GS}])/\gamma_1[\text{MHD}_{GS}])$, using the same line styles as in (a). The horizontal solid zero line, representing MHD$_{GS}$, is also shown. See text for the definitions of the different MHD versions.

to the observational data, leaving little room for any thermodynamic influence of dynamic screening (Arndt *et al.*, 1998). Screening has also been considered in connection with nuclear reaction rates, more specifically those responsible for the solar neutrino flux (*e.g.* Brüggen and Gough, 1997, and references therein).

Very recently, (Nayfonov and Däppen, 1998) examined the signature of the internal partition function in the equation of state. That study has revealed interesting features about excited states and their treatment in the equation of state. The MHD equation of state with its specific, density-dependent occupation probabilities (see section 3.2) is causing a characteristic "wiggle" in the thermodynamic quantities, most prominently in $\chi_\rho = (\partial \ln p / \partial \ln \rho)_T$, but equally present in the other thermodynamical quantities.

Figure 1 shows the result for γ_1 of a hydrogen-only plasma. The temperatures and densities are taken from a solar model. Density is implied but not shown in the figure (for more details and different thermodynamic variables and chemical compositions, see Nayfonov and Däppen, 1998). Five cases were considered: (i) MHD [standard MHD occupation probabilities (Hummer and Mihalas, 1988)], (ii) MHD$_{GS}$ [standard MHD internal partition function of hydrogen but truncated to the ground state (GS) term], (iii) OPAL: OPAL tables [version of November 1996 (Rogers et al., 1996)], (iv) MHD$_{PL}$ [MHD internal partition function of hydrogen, but replaced by the Planck-Larkin partition function, Eq. (6)], (v) MHD$_{PL,GS}$ [MHD$_{PL}$ truncated to the ground state term]. The effect of the inclusion of the excited states in the internal partition function is manifest in the differences between MHD and MHD$_{GS}$, and between MHD$_{PL}$ and MHD$_{PL,GS}$, respectively. The effect of the different occupation probability of the ground and excited states shows up in the difference between MHD and MHD$_{PL}$, and between MHD$_{GS}$ and MHD$_{PL,GS}$. It was found that the presence of excited states is crucial. Also, the wiggle, which is a genuine neutral-hydrogen effect, is present despite the fact that most of hydrogen is already ionized. The qualitative picture does not change when helium is added (Nayfonov and Däppen, 1998).

It seems that this effect of excited states has already been observed in the Sun. Figure 2 shows the result of a very recent inversion (Basu et al., 1998), based on the solar oscillation frequencies obtained from the SOI/MDI instrument on board the SOHO spacecraft during its first 144 days in operation (Rhodes et al., 1997). The difference between γ_1 in the Sun and that of various calibrated solar models is shown. The models alternate between equations of state [MHD, OPAL], different values for the solar radius, and formalism for convection [standard mixing length theory (MLT), Canuto and Mazitelli (1991) (CM) formalism]. The models are specified in Table I, where in addition, the calibrated values of the surface helium abundance Y_s and the depth of the convection zone r_{cz} are given. All models, except M9, assume the gravitational settling of helium and the heavy elements.

In contrast to earlier results (Basu and Christensen-Dalsgaard, 1997), which had uniform resolution throughout the Sun, this time the focus is on the 20% uppermost layers. It appears that in the top layers, the MHD models give a more accurate description of the Sun than the OPAL models. Since the difference in γ_1 between MHD and OPAL is the wiggle of Fig. 1, the observed preference of the MHD model in the upper region could indicate a validation of an MHD-like treatment of the exited states. Fig. 2 confirms that below the wiggle region, OPAL fares better than

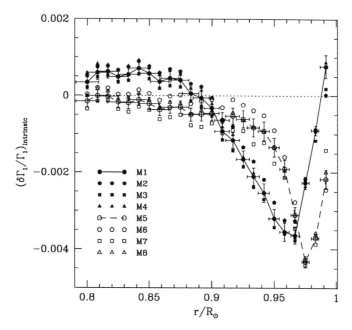

Figure 2. Relative difference between γ_1 obtained from an inversion of helioseismological data and γ_1 of the solar models listed in Table I, in the sense "Sun – model". (In the figure, γ_1 is denoted by the upper case Γ_1.) Only the so-called *"intrinsic"* difference in γ_1 is shown, that is, the part of the difference due to the equation of state (from Sarbani Basu, *private communication*).

Table I
Properties of the solar models of Fig. 2.

Model	EOS	Radius Mm	Convective Flux	Y_s	$r_{cz}/R_\odot$
M1	MHD	695.78	CM	0.2472	0.7145
M2	MHD	695.99	CM	0.2472	0.7146
M3	MHD	695.51	CM	0.2472	0.7145
M4	MHD	695.78	MLT	0.2472	0.7146
M5	OPAL	695.78	CM	0.2465	0.7134
M6	OPAL	695.99	CM	0.2465	0.7135
M7	OPAL	695.51	CM	0.2466	0.7133
M8	OPAL	695.78	MLT	0.2465	0.7135
M9	OPAL	695.78	CM	0.2646	0.7268

MHD (Christensen-Dalsgaard *et al.*, 1996). There, the internal partition functions of OPAL *à la* Planck-Larkin, and the absence of a τ-correction (section 3.3) might be the better choice.

The reversal of fortune in favor of MHD in the upper part of the Sun (above 0.98 solar radii) could be due to the different implementations of many-body interactions in the two formalisms. Since density decreases in the upper part, OPAL by its nature of a systematic expansion, inevitably becomes itself more accurate; but MHD might, by its heuristic approach (and by luck!), have incorporated even finer, higher-order effects.

Let me add a word of caution, though. It could appear tempting to produce a "combined" solar equation of state, with MHD for the top part and OPAL for the lower part. However, such a hybrid solution is fraught with danger. For instance, it is known that patching together equations of state can introduce spurious effects (Däppen et al., 1993). It seems that the right way is to improve MHD and OPAL in parallel and independently, guided by the progress of helioseismology.

As we have seen in section 2, the helium-abundance determination procedure is quite sensitive to the details of internal partition functions. Since the wiggle is located in the He II ionization zone although it is a pure hydrogen effect, it has a very important bearing on the astrophysically relevant helioseismic helium-abundance determination in the solar convection zone. Specifically, it would suddenly make the value based on the MHD equation of state ($Y=0.246$) more likely than the one based on OPAL ($Y=0.249$) (Basu and Antia, 1995).

Acknowledgements

I thank the entire team of ISSI for organizing such a productive and truly interdisciplinary meeting. Some work reported here was supported in part by the grant AST-9618549 of the National Science Foundation and in part by the SOHO Guest Investigator grant NAG5-6216 of NASA.

References

Alastuey, A., and Perez, A.: 1992, *Europhys. Lett.* **20**, 19–24.
Alastuey, A., Cornu, F., and Perez, A.: 1994, *Phys. Rev. E* **49**, 1077.
Alastuey, A., Cornu, F., and Perez, A.: 1995, *Phys. Rev. E* **51**, 1725.
Alastuey, A., and Perez, A.: 1996, *Phys. Rev. E* **53**, 5714.
Antia, H.M., and Basu, S.: 1994, *Astrophysical Journal* **426**, 801.
Arndt, A., Däppen, W., and Nayfonov, A.: 1998, *Astrophysical Journal* **498**, in press.
Basu, S. and Antia, H.M.: 1995, *Monthly Notices of the RAS* **276**, 1402.
Basu, S., and Christensen-Dalsgaard, J.: 1997, *Astron. Astrophys* **322**, L5–L8.
Basu, S., Däppen, W. and Nayfonov, A.: 1998, in *Proc. SOHO6-GONG98 Workshop*, S. Korzennik (ed.), ESA-SP, in press.
Baturin, V.A., Däppen, W., Wang, X., Yang, F.: 1996, in *Proc. 32nd Liège International Astrophysical Colloquium "Stellar Evolution: What should be done"*, M. Gabriel and A. Noels (eds.), 33.
Brüggen, M. and Gough, D.O.: 1997, *Astrophys. J.* **488**, 867.
Berthomieu, G., Cooper, A.J., Gough, D.O., Osaki, Y., Provost, J. and Rocca, A.: 1980, in *Lecture Notes in Physics*, Vol. 125: Nonradial and Nonlinear Stellar Pulsation, H.A. Hill and W. Dziembowski (eds.), Springer, Berlin, 307–312.
Canuto, V. M., Mazzitelli, I.: 1991, *Astrophys. J.* **370**, 295.

Christensen-Dalsgaard, J.: 1991, In *Lecture Notes in Physics*, Vol. 388: Challenges to Theories of the Structure of Moderate-mass Stars, D.O. Gough and J. Toomre (eds.), Springer, Heidelberg, 11–36.
Christensen-Dalsgaard, J.: 1998, *Space Sci. Rev.*, this volume.
Christensen-Dalsgaard, J., and Däppen, W.: 1992, *Astron. Astrophys. Review* **4**, 267.
Christensen-Dalsgaard, J., Däppen, W., and Lebreton, L.: 1988, *Nature* **336**, 634.
Christensen-Dalsgaard, J., Proffitt, C.R. and Thompson, M.J.: 1993, *Astrophys. J.* **403**, L75.
Christensen-Dalsgaard, J., Däppen, W., and the GONG team: 1996, *Science* **272**, 1286–1292.
Däppen, W.: 1987, in *Strongly coupled plasma physics*, F. Rogers and H. DeWitt (eds.), NATO-ASI Ser., Plenum, New York, 179–182.
Däppen, W.: 1996, *Bull. Astr. Soc. India* **24**, 151.
Däppen, W., Anderson, L.S. and Mihalas, D.: 1987, *Astrophys. J.* **319**, 195.
Däppen, W., Mihalas, D., Hummer, D.G., and Mihalas, B.W.: 1988, *Astrophys. J.* **332**, 261.
Däppen, W., Gough, D.O., Kosovichev, A.G. and Rhodes, E.J., Jr.: 1993, in *Proc. IAU Symposium No 137: Inside the Stars*, W. Weiss and A. Baglin (eds.), PASP Conference Series Vol. 40, 304–306.
Ebeling, W., Kraeft, W.D., and Kremp, D.: 1976, *Theory of Bound States and Ionization Equilibrium in Plasmas and Solids*, Akademie Verlag, Berlin, DDR.
Ebeling, W., Förster, A., Fortov, V.E., Gryaznov, V.K. and Polishchuk, A.Ya.: 1991, *Thermodynamic Properties of Hot Dense Plasmas*, Teubner, Stuttgart, Germany.
Eggleton, P. P., Faulkner, J. and Flannery, B. P.: 1973, *Astron. Astrophys.* **23**, 325.
Gabriel, M.: 1994, *Astron. Astrophys.* **292**, 281.
Gough, D.O.: 1984, *Mem. Soc. Astron. Ital.* **55**, 13–35.
Harris, G.M., Roberts, J.E. and Trulio, J.G.: 1960, *Phys. Rev.* **119**, 1832.
Hummer, D.G., Mihalas, D.: 1988, *Astrophys. J.* **331**, 794.
Iglesias, C.A., Rogers, F.J.: 1995, *Astrophys. J.* **443**, 460.
Kosovichev, A.G., Christensen-Dalsgaard, J., Däppen, W., Dziembowski, W.A., Gough, D.O., and Thompson, M.J.: 1992, *Mon. Not. R. astr. Soc.* **259**, 536.
Kraeft W.D., D. Kremp, W. Ebeling and Röpke G.: 1986, *Quantum Statistics of Charged Particle Systems*, Plenum, New York.
Mihalas, D., Däppen, W., and Hummer, D.G.: 1988, *Astrophys. J.* **332**, 815.
Nayfonov, A. and Däppen, W.: 1998, *Astrophys. J.* **499**, in press.
Noels, A., Scuflaire, R., and Gabriel, M.: 1984, *Astron. Astrophys.* **130**, 389–396.
Perez, A. and Däppen, W.: 1998, *Astrophysical Journal*, submitted.
Rhodes E. J., Kosovichev A. G., Schou J., Scherrer P. H., and Reiter, J.: 1997, *Solar Phys.* **175**, 287.
Rogers, F.J.: 1977, *Phys. Lett.* **61A**, 358.
Rogers, F.J.: 1981, *Phys. Rev.* **A24**, 1531.
Rogers, F.J.: 1986, *Astrophysical Journal* **310**, 723.
Rogers, F.J., Swenson, F.J., and Iglesias, C.A.: 1996, *Astrophysical Journal* **456**, 902.
Shibahashi, H., Noels, A., and Gabriel, M.: 1983, *Astron. Astrophys.* **123**, 283–288.
Shibahashi, H., Noels, A., and Gabriel, M.: 1984, *Mem. Soc. Astron. Ital.* **55**, 163–168.
Ulrich, R.K.: 1982, *Astrophys. J.* **258**, 404–413.
Ulrich, R.K., and Rhodes, E.J.: 1983, *Astrophys. J.* **265**, 551–563.
Vauclair, S.: 1998, *Space Sci. Rev.*, this volume.
Vorontsov, S.V., Baturin, V.A., Gough, D.O., and Däppen, W.: 1994, in *Proc. IAU Colloquium 147: the equation of state in astrophysics*, G. Chabrier and E. Schatzmann (eds.), Cambridge University Press, 545–549.

Address for correspondence: Werner Däppen, Department of Physics and Astronomy, University of Southern California, Los Angeles, CA 90089-1342, U.S.A.

OPACITY OF STELLAR MATTER

FORREST J. ROGERS AND CARLOS A. IGLESIAS
Lawrence Livermore National Laboratory

Abstract. New efforts to calculate opacity have produced significant improvements in the quality of stellar models. The most dramatic effect has been large opacity enhancements for stars subject to large amplitude pulsations. Significant improvement in helioseismic modeling has also been obtained. A description and comparisons of the new opacity efforts are given.

1. Introduction

Opacity has long been an issue in understanding stars. As long ago as 1926 Eddington identified opacity as one of two clouds obscuring stellar model calculations. At that time it was thought that bound-bound absorption was not a significant source of opacity. It was another 40 years before Cox and Stewart (1962; 1965; 1970) included bound-bound transitions and obtained increases in the Rosseland mean opacity exceeding a factor of three in some cases. The Cox-Stewart opacities greatly improved the quality of stellar models and remained the standard for more than a quarter century. This work continued to be modified and improved by Cox and others at Los Alamos (Cox and Tabor, 1976; Huebner et al., 1977). A detailed description of this first generation Los Alamos opacity (LAO1) is given by Huebner (1986).

Even though the LAO1 opacities helped elucidate many features of stars, a number of observations continued to resist explanation. For example, period ratios in classical Cepheid models were too low, the mechanism for pulsation in β-Cephei stars could not be identified, the calculated Li abundance in dwarf stars of the Hyades cluster was much less than observed, and simulations underestimated wind-driven mass loss in classical Novae. A number of studies found that these problems are sensitive to changes in opacity (Fricke, Stobie, and Strittmatter, 1971; Petersen, 1974; Stellingswerf, 1978). However, the opacity increases needed seemed unrealistically large; 300% in the case of the classical Cepheids and the β-Cephei stars. Simon (1982) determined that increasing the opacity for temperatures above 1×10^5 K would be sufficient to resolve the Cepheid and β-Cephei problems. He speculated that problems with heavy element opacities could be responsible and issued a plea for their reinvestigation. A group at Los Alamos (Magee et al., 1984) were the first to respond. They concluded that such large increases in opacity were inconsistent with atomic physics. Nevertheless, two completely new efforts to calculate the opacity were undertaken.

One of these is the Opacity Project (OP), led by M. Seaton at Univ. College London, the other is our effort at Lawrence Livermore National Laboratory, known as

OPAL. These efforts have obtained large increases in the opacity (Iglesias, Rogers and Wilson, 1987; 1992; Iglesias and Rogers, 1991; 1996; Rogers and Iglesias, 1992; Seaton *et al.*, 1994) which helped resolve a number of long-standing puzzles (Rogers and Iglesias, 1994). The differences between OP and OPAL opacities are generally small compared to the differences with the older LAO1. An important exception is with solar interior opacities where OPAL obtained modest increases over LAO1, while OP is 40% lower. The decrease seems incompatible with helioseismology (Bahcall and Glasner, 1994; Tripathy, Basu and Christensen-Dalsgaard, 1997) and has been attributed to approximations in the OP calculations (Iglesias and Rogers, 1995). A second generation of Los Alamos opacities (LAO2) have recently been released (Magee *et al.*, 1998). In addition to the new stellar interior opacity calculations, there have been several new efforts to calculate surface opacities (Sharp, 1993; Alexander and Ferguson, 1994).

2. Brief Description of OPAL and OP

The calculation of opacity involves four distinct disciplines: equation of state (EOS), atomic physics, spectral line broadening, and plasma collective effects. The LAO1 opacities were calculated with an ad hoc model of the EOS and mostly hydrogenic approximations to the atomic physics. The new OP and OPAL opacity efforts are based on improved theoretical methods in all four of the disciplines mentioned above. In the following the improved physics and its impact on opacity are briefly described. More detailed accounts can be found in Rogers and Iglesias (1992), Iglesias and Rogers (1996), and the book by the OP team (Seaton *et al.*, 1995)

Calculation of the EOS is logically the first step in the calculation of opacity, since this gives the state occupation numbers. Typically this part of the calculation has been based on simple ad hoc methods. New EOS methods used by OPAL (Rogers, 1986; Rogers, Swenson, and Iglesias, 1996) and OP (Däppen *et al.*, 1987; Hummer and Mihalas, 1988) have been instrumental to theoretical interpretations of the helioseismic data (Däppen, 1998; Christensen-Dalsgaard, 1998; Dziembowski, 1998).

Although differences in EOS models have in general not significantly affected astrophysical opacities, differences in bound state occupation numbers are a primary reason for OPAL opacity enhancements near the base of the solar convection zone (Iglesias and Rogers, 1991). It is also one of the reasons OP gets a smaller opacity than LAO1 and OPAL in this region (Iglesias and Rogers, 1995; see also Sec. 3).

By far the most significant effect on opacity has come from improved calculations of bound-bound absorption that include much more detailed atomic data. The OPAL and OP groups chose different approaches for this part of the calculation. The goal of OPAL was solely to calculate opacity, whereas OP had the additional aim to produce a general purpose atomic database. A continuation of that effort known as the Iron Project is still in progress (Bautista and Pradhan, 1997). For the required atomic data the OPAL group developed a parametric potential method that is fast enough to allow on-line calculations, while achieving accuracy comparable to single configuration Dirac-Fock self-consistent field calculations (Rogers, Iglesias, and Wilson, 1988). This on-line capability provides flexibility to study easily the effects of atomic physics approximations; e.g. angular momentum coupling or data averaging methods. By contrast, the OP group uses first principle (non-relativistic) methods to construct detailed atomic databases (Seaton, 1987; Seaton et al., 1994). The large increase in the iron opacity obtained with the LS coupling scheme compared to calculations that neglect term splitting suggested that fine structure is also important (Rogers and Iglesias, 1992). OPAL opacities calculated since 1992 include spin-orbit effects in full intermediate coupling (Iglesias, Rogers and Wilson, 1992), while OP (Seaton et al., 1994) uses an approximate method that does not include spin changing transitions (see Fig. 1 of Rogers and Iglesias, 1994). On the other hand, the OPAL calculation assumes single configurations, while OP includes configuration-interaction effects in both the bound-bound and bound-free calculations. Configuration-interaction is most important for atoms and near neutral ions.

The OPAL calculations include degeneracy and plasma collective effects in the free-free absorption using a screened form of the parametric potentials, whereas these effects are neglected in OP. Both OPAL and OP include collective effects in Thomson scattering (Boercker, 1987). The OPAL spectral line broadening for one, two, and three electron ions is computed with a suite of codes provided by Lee (1988) that include linear Stark theory. For all other transitions the OPAL calculations use Voigt profiles where the Gaussian width is due to Doppler broadening and the Lorentz width is due to natural plus electron impact collision broadening (Dimitrievic and Konjevic, 1980). The OP approach is similar (Seaton, 1987) except that for spectral lines not subject to linear Stark effect OP uses widths from quantum-mechanical close coupling calculations (Seaton, 1988), which are similar to those used by OPAL.

The improved line broadening has in general had a small effect on opacity. One important exception is Stark broadening of hydrogen. LAO1 used the theory of Griem (1960) which gives lines that are much too broad compared to experiment (Wiese, 1972). The OP and OPAL hydrogenic lines agree well with the data and result in an opacity reduction for Population II compositions around $\log T = 4.8$. Cox (1990) showed that this reduction in opacity in conjunction with a modest increase in opacity for $\log T \approx 5.3$ removes several long-standing puzzles in models of RR-Lyrae stars.

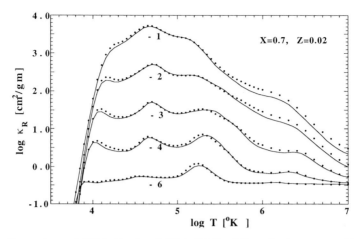

Figure 1. Comparison of OPAL (dots) and OP (solid lines) Rosseland mean opacities at constant values of log R for the element distribution used by Seaton *et al.* (1994) where X is the hydrogen mass fraction and Z is the metallicity.

3. New Opacity Data

The latest OPAL and OP calculations of the Rosseland mean opacity, κ_R, are compared in Fig. 1 for various values of log R, where R = density$/T_6^3$ and T_6 is temperature in Megakelvin. Although OPAL includes 19 elements while OP includes 15, the extra elements do not substantially affect the comparison. Both codes predict similar, but slightly shifted, bumps near $\log T = 5.3$ due to millions of transitions originating in M-shell iron. This feature is completely absent from the older Los Alamos opacities (see Fig. 2 of Rogers and Iglesias, 1994). This is the reason LAO1 failed to explain a wide range of stellar phenomena. Recently, the iron opacity bump was instrumental in predicting pulsational instability (Charpinet *et al.*, 1996; 1997) in a newly discovered class of sdB subsequently confirmed by observations. The large increases in opacity predicted in the region of the iron bump have been corroborated by laboratory experiments. The first measurements were in the correct temperature range but at higher densities then occur in Cepheids (DaSilva *et al.*, 1992; Springer *et al.*, 1992; Winhart *et al.*, 1995). Recently, Springer *et al.* (1997) measured the iron opacity at conditions comparable to those in stellar envelopes. The OPAL frequency dependent opacities are in good agreement with all these experiments. Similar comparisons using OP data have not been reported.

Figure 2 compares the ratio of OPAL to OP Rosseland mean opacities as a function of temperature for $\log R = -1.5, -3.5$, and -5.5 for mixture having $X = 0.7$, and $Z = 0.02$. The largest differences occur near solar conditions, i.e., $\log R = -1.5$. In this case, the ratio is near unity for $\log T < 5.6$, but for higher temperatures the ratio increases rapidly to around 1.35. The major part of the discrepancy can be traced to incomplete photoionization data in OP (Iglesias and Rogers, 1995). Figure 3 shows the He-like carbon photoionization cross-sections

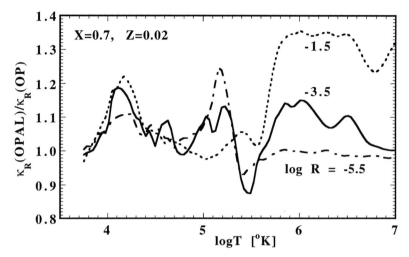

Figure 2. Ratio of OPAL to OP κ_R along a track that is close to solar.

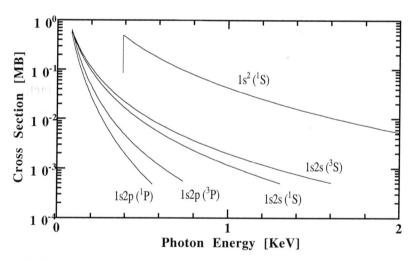

Figure 3. OP Photoionization cross-sections vs. photon energy for various configurations in helium-like carbon.

from K and L shell states (Canuto et al., 1993). It is obvious that the 1s2s and 1s2p photoionization cross-sections do not include photoionization of the 1s electron, which should produce an edge in the vicinity of the 1s2 threshold. Figure 4 shows the OPAL monochromatic opacity for a representative solar mixture as a function of $h\nu/kT$ for $\log T = 6$ and density of 0.01 g/cm^3, with and without the missing inner shell data in OP (Iglesias and Rogers, 1997). The impact on the Rosseland mean from the missing inner shell photoionization data is similar in magnitude to the discrepancy shown in Fig. 2

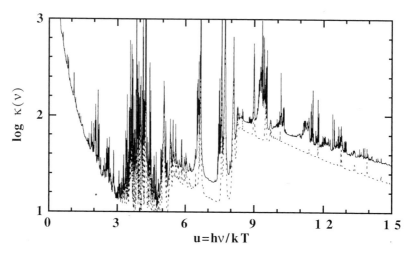

Figure 4. $\log \kappa(\nu)$ vs. $h\nu/kT$ with (solid line) and without (dashed line) the inner shell photoionization data missing from OP.

Another source of discrepancy between OP and OPAL has been traced to an approximation in OP that affects the occupation numbers. Hummer and Mihalas (1988) assumed that the Holtsmark electric microfield, valid for randomly distributed ions, determines the probability a state is localized. Furthermore, in order to reduce computational expense they adopted an approximate form of the Holtsmark function. In a real plasma however, the Coulomb interaction modifies the ion distribution and causes the microfield distribution to peak at lower values of the field strength relative to Holtsmark Consequently, the probability that a state is dissolved by the electric microfield fluctuations is reduced. Iglesias and Rogers (1995) show: 1) the OP approximation to Holtsmark is poor; 2) using the more realistic APEX microfield (Iglesias *et al.*, 1985) significantly increases the OP occupation numbers (as defined in Rogers, 1986) for high lying states, bringing them closer to OPAL. Since the Hummer and Mihalas procedure and OPAL are based on different physical assumptions, the two calculations can not be expected to agree exactly. Although these differences have only a small effect on the EOS, they may affect the opacity.

4. Sources of Missing Opacity

Tsytovich *et al.* (1996) suggested that a number of effects listed, in Table I, have been neglected in existing calculations of solar interior opacities. They estimated that these effects could decrease the solar center opacity by as much as 14%. More recent work indicates that the potential decrease is somewhat smaller. Iglesias (1997a) showed that, contrary to the claims of Tsytovich *et al.* (1996), the effect of Raman resonance broadening (labeled *a*) is already included in the fre-

Table I
Effects and size of opacity correction obtained by Tsytovich et al. (1996).

Mechanism	$\delta\kappa_R/\kappa_R$ (in %)
Broadening of Raman resonance (a)	−2.0
Relativistic collective scattering	−0.1
Stimulated scattering and frequency diffusion	−3.0
Collective Bremsstrahlung (b)	−0.1
Relativistic Bremsstrahlung (c)	−4.7
Density inhomogeneity	−0.1
Refractive index (d)	−0.1
Quantum recoil scattering	−0.7
Ion correlations (d)	−1.5
Electron degeneracy (d)	−2.0
Total	−14.3

quency integration over the dynamic structure factor in existing calculations. The new expressions for electron-ion Bremsstrahlung (labeled *b*) reported by Tsytovich *et al.* result from a misinterpretation of earlier work and actually reproduce existing results (Iglesias, 1997b) while their calculations of inverse Bremsstrahlung (labeled *c*) predict incorrectly both the sign and magnitude of the relativistic correction (Iglesias, 1996). Finally, Iglesias and Rose (1997) showed that the major corrections to Bremsstrahlung and Thomson scattering (labeled *d*) given by Tsytovich *et al.* are already included in the OPAL calculations and that remaining corrections not currently included are small. The unlabeled mechanisms in Table I have not been independently evaluated, so there remains a potential 3.9% reduction in the solar interior from these sources.

In addition to the effects proposed by Tsytovich *et al.*, there are a number of other possible sources of opacity not included in OP and OPAL; e.g., electron-electron Bremsstrahlung (Maxon, 1967) and two-photon absorption (More and Rose, 1991). Furthermore, the current calculations are carried out in the single atom approximation. At high density there could be effects due to particle clustering beyond those already included in line-broadening.

In addition to new physics, some aspects of the current calculations may need improvement. For example, measurements by Glenser and Kunze (1996) of 2s-2p transitions in Li-like B indicate that, in this specific case, OPAL and OP may underestimate the line widths by a factor of two. Griem, Ralchenka, and Bray (1998) have challenged the experimental results, so that this issue is unresolved. Fortuitously, solar opacities are not very sensitive to line widths (Iglesias and Rogers, 1991), but a factor of two change in the line widths could affect Cepheid Variable opacities by 10% (Rogers and Iglesias, 1992).

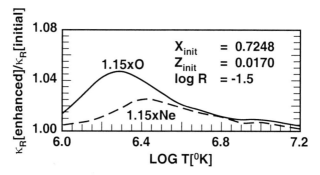

Figure 5. Effect on κ_R of enhancing O and Ne abundances by 15%.

5. Composition

Until recently solar models have considered convection to be the only mechanism for material motion. However, gravitational settling and thermal diffusion have been found to improve agreement with observations and are becoming part of the standard model (Bahcall and Pinsonneault, 1992; 1994; Guenther and Demarque, 1997; Guzik and Swenson, 1998). Diffusion is found to produce abundance changes of order 10% in the solar interior. Radiatively driven diffusion can also be an important source of material motion in some stars, e.g., hot white dwarfs. A number of calculations of radiation acceleration have recently appeared (Richer *et al.*, 1998; Seaton, 1997). Using these new results Turcotte *et al.* (1998) have verified that radiation driven diffusion is small in the sun. Even so, it has been suggested that it should also be included in the best solar models (Christensen-Dalsgaard, 1998).

In addition to element diffusion, there are uncertainties in the observed abundances (Grevesse; this volume). Historically, improvements in the photospheric abundances have reduced the differences with meteoritic determinations. Figure 5 illustrates the affect on the opacity of increasing the O and Ne abundances individually by 15% for a track that is close to solar. These elements are seen to make their largest contribution at temperature around 2×10^6 K, near the base of the solar convection zone. The effects are not large, but would show up in comparisons with helioseismic data. The most recent measurement of the solar neon abundance (Feldman and Widing, 1990; Feldman, 1998) give a value that is 9% higher than Grevesse and Noels (1993) with an uncertainty of 15% due to atomic physics limitations.

6. Conclusion

The new opacity data has favorably impacted modeling of a broad range of stellar properties. This success provides a strong motivation to extend the calculations to cover a broader range of applications. For example, the temperature density range of white dwarfs and other dense stellar objects are partly beyond the range of the current tables, the elemental composition is not adequate to model s-process stars that have significant amounts of elements heavier than Fe, there are many applications requiring frequency dependent opacity data such as radiative levitation (Seaton, 1997; Richer *et al.*, 1997).

In the specific case of the sun, current opacity tables only allow for changes in the total Z. To facilitate the process of adding diffusion to the best standard solar models (SSM) (see Christensen-Dalsgaard, 1998) it will be necessary to provide opacity tables that allow for variation of individual element abundances. Due to the stringent requirements set by helioseismology, and as abundance determinations improve, even small sources of opacity not included in the current calculations will need to be considered. This will make the sun a formidable opacity experiment.

Acknowledgments

This work was performed under the auspices of the U. S. Department of Energy by the Lawrence Livermore National Laboratory under Contract W–7405–Eng–48.

References

Alexander, D. R., and Ferguson, J. W.: 1994, *ApJ* **437**, 879.
Bahcall, J. N. and Glasner, A.: 1994, *ApJ* **437**, 484.
Bahcall, J. N. and Pinsonneault, M. H.: 1992, *Rev. Mod. Phys.* **64**, 885.
Bahcall, J. N. and Pinsonneault, M. H.: 1994, *Rev. Mod. Phys.* **67**, 781.
Bautista, M. A., and Pradhan, A. K.: 1997, *A&AS* **126**, 365.
Boercker, D. B.: 1987, *ApJ* **316**, L95.
Canuto *et al.*: 1993, *A&A* **275**, L5.
Charpinet , S. Fontaine, G., Brassard, P., and Dorman, B.: 1996, *ApJ* **471**, L103.
Charpinet , S. Fontaine, G., Brassard, P., Chayer, P., Rogers, F. J., Iglesias, C. A., and Dorman, B.: 1997, *ApJL* **483**, L123.
Christensen-Dalsgaard, J.: 1998, *Space Sci. Rev.*, this volume.
Cox, A. N. and Stewart, J. N.: 1962, *ApJ* **67**, 113.
Cox, A. N. and Stewart, J. N.: 1965, *ApJS* **11**, 22.
Cox, A. N. and Stewart, J. N.: 1970, *ApJS* **19**, 261.
Cox, A. N. and Tabor, J. E.: 1976, *ApJS* **31**, 271.
Cox, A. N.: 1991 *ApJ* **381**, L7138.
Däppen, W., Anderson, L. and Mihalas, D.: 1987, *ApJ* **319**, 195.
Däppen, W.: 1998, *Space Sci. Rev.*, this volume.
DaSilva, L. B., *et al.*: 1992, *Phys. Rev. Lett* **69**, 438.
Dimitrievic, M. S. and Konjevic, N.: 1980, *J. Quant. Spectrosc. Rad. Transf.* **24**, 451.
Dziembowski, W.: 1998, *Space Sci. Rev.*, this volume.
Eddington, A. S.: 1926, The Internal Composition of Stars, Cambridge: Cambridge Univ. Press.
Feldman, U. and Widing, K. G.: 1990. *ApJ* **363**, 292.
Feldman, U.: 1998, *Space Sci. Rev.*, this volume.
Fricke, K., Stobie, R. S., and Strittmatter, P. A.: 1971, *MNRAS* **154**, 23.
Glenzer, S. and Kunze, H.-J.: 1996, *Phys. Rev. A* **53**, 2225.

Gould, R. J.: 1990, *ApJ* **362**, 284.
Grevesse, N. and Noels, A.: 1993, in *Origin and Evolution of the Elements*, eds. N. Prantzos, E. Vangioni-Flam, and M. Cassé, Cambridge: Cambridge Univ. Press.
Griem, H. R.: 1960, *ApJ* **132**, 883.
Griem, H. R., Ralchenko, Y. V., and Igor, B.: 1997, *Phys. Rev. E* **56**, 7186.
Guenther, D. B., and Demarque, P.: 1997, *ApJ* **484**, 937.
Guzik, J. A. and Swenson, F. J.: 1997, *ApJ* **491**, 967.
Huebner, W. F., Merts, A. L., Magee, N. H., and Argo, M. F.: 1977, Los Alamos Scientific Report La-647
Hummer, D. G., and Mihalas, D.: 1988, *ApJ.* **331**, 794.
Iglesias, C. A.: 1996, *Phys. Lett.* **466**, L115.
Iglesias, C. A.: 1997a, *J. Quant. Spectrosc. Rad. Transf.* **58**, 141.
Iglesias, C. A.: 1997b, *J. Plasma Phys.* **58**, 381.
Iglesias, C. A., DeWitt, H. E., Lebowitz, J. L., MacGowan, D., and Hubbard, W. B.: 1985, *Phys. Rev. A.* **31**, 1698.
Iglesias, C. A., Rogers, F. J., and Wilson, B. G.: 1987, *ApJL* **322**, L45.
Iglesias, C. A. and Rogers, F. J.: 1991, *ApJ* **371**, 40.
Iglesias, C. A. and Rogers, F. J.: 1996, *ApJ* **464**, 943.
Iglesias, C. A., Rogers, F. J., and Wilson, B. G.: 1992, *ApJ* **397**, 717.
Iglesias, C. A. and Rose, S.: 1997, *ApJ* **466**, L115.
Lee, R. W.: 1988, *J. Quant. Spectrosc. Rad. Transf.* **40**, 561.
Magee, N. H., Merts, A. L. and Huebner, W. F.: 1984, *ApJ* **283**, 264.
Magee, N. H., and Clark, R. E. H.: 1998, in *Proc. of the International Conference on Atomic and Molecular Data and their Applications*, Gaithersburg, Maryland, in press.
Maxon, M. S., and Gorman, G. E: 1967, **163**, 156.
Mihalas, D., Däppen, W., and Hummer, D. G.: 1988, *ApJ* **331**, 815.
More, R. and Rose, S.: 1991, *Radiative Properties of Hot Dense Matter*, eds.
Nakagawa, M., Kohyama, Y., and Itoh, N.: 1987, *ApJL* **63**, 661.
Petersen, J. O.: 1974, *A&A* **34**, 309.
Richer, J., Michaud, M. G., Rogers, F., Iglesias, C., Turcotte, S., and LeBlanc, F.: 1997, *ApJ* **492**, 833.
Rogers, F. J.: 1986, *ApJ.* **310**, 723.
Rogers, F. J. and Iglesias, C. A.: 1992, *ApJS* **79**, 507.
Rogers, F. J. and Iglesias, C. A.: 1994, *Science* **263**, 50.
Rogers, F. J., Swenson, F. J., and Iglesias, C. A.: 1996, *ApJ* **456**, 902.
Rogers, F. J., Wilson, B. G., and Iglesias, C. A.: 1988, *Phys. Rev.* **A38**, 5007.
Seaton, M. J.: 1987, *J. Phys. B.* **20**, 6363.
Seaton, M. J.: 1988, *J. Phys. B.* **21**, 3033.
Seaton, M. J.: 1990, *J. Phys. B.* **23**, 3255.
Seaton, M. J., Yan, Y. D., Mihalas, D., and Pradhan, A. K.: 1994, *MNRAS* **266**, 805.
Seaton *et al.*: 1995, *The Opacity Project*, Bristol: Institute of Physics Publishing.
Seaton, M. J.: 1997, *MNRAS* **289**, 700.
Sharp, C. M.: 1993, in *Inside the Stars*, eds. Weiss, W. W. and Baglin, A., ASP Conf. Ser. **49**, 263.
Simon, N. R.: 1982, *ApJ* **260**, L87.
Springer, P. T. *et al.*: 1992, Phys. Rev. Lett., 69, 3735.
Springer, P. T. *et al.*: 1997, J. Quant. Spectrosc. Rad. Transf., 58, 927.
Stellingswerf, W. F.: 1978, *ApJ* **83**, 1184.
Tripathy, S. C. Basu, S. and Christensen-Dalsgaard, J.: 1997, in *Proc. of IAU Symp.* **181**, eds. Schmider, F. X. and Provost, J.
Tsytovich , V. N., Bingham, R., De Angelis, U., and Forlani, A.: 1996, *Physica Scripta* **54**, 313.
Turcotte, S., Richer, J., Michaud G., Iglesias, C. A., and Rogers F. J.: 1998, in press.
Wiese, W. L., Kelleher, D. E., and Paquette, D. R.: 1972, *Phys. Rev.* **A6**, 1132.
Winhart, G. *et al.*: 1995, *J. Quant. Spectrosc. Rad. Trans.* **54**, 43.

ELEMENT SETTLING IN THE SOLAR INTERIOR

SYLVIE VAUCLAIR

Laboratoire d'Astrophysique Observatoire Midi-Pyrénées Toulouse, France

Abstract. Element settling inside the Sun now becomes detectable from the comparison of the observed oscillation modes with the results of the theoretical models. This settling is due, not only to gravitation, but also to thermal diffusion and radiative acceleration (although this last effect is small compared to the two others). It leads to abundance variations of helium and heavy elements of $\cong 10\%$ below the convective zone. Although not observable from spectroscopy, such variations lead to non-negligible modifications of the solar internal structure and evolution. Helioseismology is a powerful tool to detect such effects, and its positive results represent a great success for the theory of stellar evolution. Meanwhile, evidences are obtained that the element settling is slightly smoothed down, probably due to mild macroscopic motions below the convective zone. Additional observations of the abundances of both ^{7}Li and ^{3}He lead to specific constraints on these particular motions.

Key words: Sun: abundances; helioseismology; diffusion processes; element settling

1. Introduction

The importance of element settling inside the stars during their evolution is now widely recognized as a "standard process" (see, for example, Vauclair, 1998). As soon as condensed from interstellar clouds, the self-gravitating spheres built density, pressure and temperature gradients which force the various chemical species present in the stellar gas to move with respect to one another. This process, first introduced by Michaud (1970) to account for the chemically peculiar stars, is believed to be the reason for the large abundance variations observed in main-sequence type stars, horizontal branch stars and white dwarfs (Vauclair and Vauclair, 1982).

Inside the convective regions, the rapid macroscopic motions mix the gas components and force their homogenisation. The chemical composition observed in the external regions of cool stars is thus affected by the settling which occurs below the outer convective zones. As the settling time scales vary in first approximation like the inverse of the density, the expected variations are smaller for cooler stars, which have deeper convection zones. While some elements can see their abundances vary by several orders of magnitude in the hottest Ap stars, the maximum expected variations in the Sun are not larger than $\cong 10\%$.

Such variations cannot be observed in the solar atmosphere by spectroscopy. In the present days however, due to helioseismology, we know the internal structure of the Sun with a high degree of precision. Evidences for the occurrence of element settling are found. Abundance variations of the order of a few percent now become indirectly detectable, by comparisons of the theoretical computations with the results of the inversion of pulsating modes. We have entered a new area in this respect.

2. Theory of Element Settling

2.1. THE DIFFUSION EQUATION

What we use to call "microscopic" diffusion of the chemical elements in stars represents a competition between two kinds of processes. First the individual atoms want to move under the influence of the local gravity (or pressure gradient), thermal gradient, radiative acceleration and concentration gradient. Second their motion is slowed down due to collisions with the other ions as they share the acquired momentum in a random way. This competition leads to selective element settling inside the stars.

The computations of this settling process are based on the Boltzmann equation for dilute collision-dominated plasmas. At equilibrium the solution of the equation is the Maxwellian distribution function. We consider here situations where the distribution is not Maxwellian, but where the deviations from the Maxwellian distribution are very small.

Two different methods have been used to solve the Boltzmann equation in the framework of this approximation. The first method relies on the Chapman-Enskog procedure (described in Chapman and Cowling, 1970), using convergent series of the distribution function. This procedure is applied to binary mixtures, leading to expressions with successive approximations for the binary diffusion coefficients. For the diffusion of charged particles in a plasma, a ternary mixture approximation is introduced, including the electrons. This method was widely used in the first computations of diffusion processes in stars (see Vauclair and Vauclair, 1982). More recently, similar methods have still been used by many authors, for example Bahcall and Loeb (1990), Proffitt and Michaud (1991), Michaud and Vauclair (1991), Bahcall and Pinsonneault (1992), Charbonnel, Vauclair, Zahn (1992), Richard et al. (1996, hereafter RVCD). The second method is that of Burgers (1969), in which separate flow and heat equations for each component of a multi-component mixture are solved simultaneously. Descriptions of this method may be found for example in Cox, Guzik and Kidman (1989), Proffitt and VandenBerg (1991), Richer and Michaud (1993), Thoul, Bahcall and Loeb (1994).

In the formalism of RVCD, which has been used for the results given below, the local abundances of the elements are given in terms of their concentrations, solutions of equations of the following type:

$$\frac{\partial c_i}{\partial t} = D'_{1i}\frac{\partial^2 c_i}{\partial m_r^2} + \left(\frac{\partial D'_{1i}}{\partial m_r} - V'_{1i}\right)\frac{\partial c_i}{\partial m_r} - \left(\frac{\partial V'_{1i}}{\partial m_r} + \lambda_i\right)c_i$$

where c_i, the concentration of element i, is given in terms of the stellar mass fraction m_r, λ_i is the nuclear reaction rate, and D'_{1i} is given by:

$$D'_{1i} = \left(4\pi\rho r^2\right)^2 (D_T + D_{1i})$$

in which D_T represents the effective macroscopic diffusion coefficient and D_{1i} the microscopic diffusion coefficient of element i relative to element 1 (here hydrogen). V'_{1i} is given by:

$$V'_{1i} = \left(4\pi\rho r^2\right) V_{1i}$$

with:

$$V_{1i} = -D_{1i}\left[\left(A_i - \frac{Z_i}{2} - \frac{1}{2}\right)\left(\frac{m_H}{kT}\frac{GM}{R^2}\right) - \alpha_{1i}\nabla\ln T\right]$$

The thermal diffusion coefficient α_{1i} is computed using the formalism of Paquette et al. (1986); A_i and Z_i represent the atomic mass number and charge of element i and m_H is the atomic hydrogen mass. M and R stand for the stellar mass and radius.

A normalisation condition on the mass fractions of all the elements has to be added to correct for the center-of-mass displacement.

The diffusion equation has to be solved simultaneously for all the considered elements. The order of magnitude of the time scales generally implies the computation of many iterations of the diffusion process for a single evolutionary time step. For each computation of a new model along the evolutionary track, the tables of abundances inside the star have to be transferred for every element, as a function of the internal mass. For the model consistency, these abundance profiles must be taken into account in the interpolation of the opacity tables.

For the Sun, the whole process of complete time evolution has to be iterated several times from the beginning, with small adjustments in the original helium mass fraction and mixing length parameter, to obtain the right Sun and the right age (luminosity and radius with a relative precision of at least 10^{-4}).

2.2. THE TREATMENT OF COLLISIONS

The diffusion time scales are direct functions of the collision probabilities for the considered species. A good treatment of collisions is thus necessary to obtain the abundance variations with a high degree of precision.

For the diffusion of neutral atoms in a neutral gas, the "hard sphere approximation" is used. For ions moving in a neutral medium, or neutrals moving in a plasma, the polarisation of the neutrals have to be taken into account. For collisions between charged ions, problems similar to those encountered for the equations of state have to be solved. The basic question concerns the divergence of the coulomb interaction cross sections. In the first computations of diffusion, the "Chapman and Cowling approximation" was used, assuming a cut-off of the cross section equal to the Debye shielding length. Average values of the shielding factor were used for analytical fits of the resulting diffusion coefficients.

Paquette et al. (1986) proposed a more precise treatment of this problem. They pointed out that the Debye shielding length has no physical meaning as soon as it is smaller than the inter-ionic distance. They proposed to introduce a screened coulomb potential with a characteristic length equal to the largest of the Debye length and inter-ionic distance, and they gave tables of collision integrals which can be used in the computations of diffusion processes in the stellar gases. The Paquette *et al.* approximation should be generally used in the computations of stellar structure. It may however in most cases be replaced by an analytical expression given by Michaud and Proffitt (1992).

2.3. THE RADIATIVE ACCELERATION

Many authors have computed the gravitational and thermal diffusion of helium and heavier elements in the Sun with various approximations: see, for example, Michaud and Vauclair (1991) and references therein; Cox, Guzik and Kidman (1989); Bahcall and Pinsonneault (1992); Proffitt (1994); Thoul, Bahcall and Loeb (1994); Christensen-Dalsgaard, Proffitt and Thompson (1993); RVCD. In all cases the radiative accelerations were neglected.

For the first time, Turcotte *et al.* (1998) have consistently computed the radiative accelerations on the elements included in the OPAL opacities. They have found that, contrary to current belief, the effect of radiation can, in some cases, be as large as $\cong 40\%$ that of gravity below the solar convective zone. This is important only for metals however, and not for helium. When the radiative accelerations are neglected, the abundances of most metals change by $\cong 7.5\%$ if complete ionisation is assumed below the convection zone, and by $\cong 8.5\%$ if detailed ionisation rates are computed. When the radiative accelerations are introduced, with detailed ionisation, the results lie in-between. The resulting effect on the solar models is small and can be neglected (while it becomes important for hotter stars).

3. Evidence of Element Settling Inside the Sun from Helioseismology

Solar models computed in the old "standard" way, in which the element settling is totally neglected, do not agree with the inversion of the seismic modes. This result has been obtained by many authors, in different ways (see Gough *et al.*, 1996, and references therein). There is a characteristic discrepancy of order one percent, just below the convective zone, between the sound velocity computed in the models and that of the seismic Sun. Introducing the element settling reconciliates the two results. Fig. 1 shows an example of solar models obtained with and without element settling in the computation. These models have been obtained with the Toulouse code, as described in Charbonnel, Vauclair and Zahn (1992) and RVCD. Some improvements have been introduced in the treatment of the opacities and equation of state, as described in Richard, Vauclair and Charbonnel (1998).

These computations are compared to the results of helioseismology in collaboration with the Warsaw group (Dziembowski *et al.*, 1994). The values of the function $u = P/\rho$ as obtained from the inversion of the solar oscillation modes (seismic Sun) are displayed together with the results of our models.

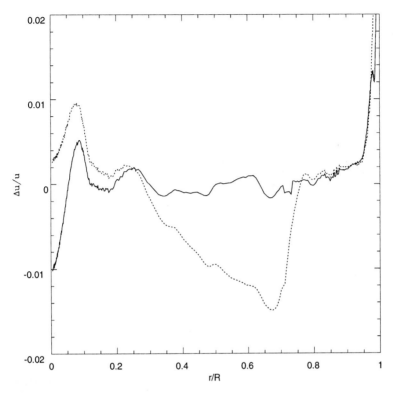

Figure 1. Comparison of the $u = P/\rho$ function in the "seismic Sun" and in our models. Dotted line: without element settling; solid line: with settling computed for helium and 14 other elements (from Richard, Vauclair and Charbonnel, 1998)

It could be possible to reduce the discrepancy between the sound velocity in the old solar standard models and in the seismic Sun by adding other effects than element settling. For example changes in the opacities could possibly lead to similar results. However, as already pointed out, element settling must not be considered as a new parameter added in the computations. It represents second order effects in the physics of auto-gravitational spheres, precisely and without any free parameter. Introducing element settling in the standard models means improving the physics. The fact that these new models lie closer to the seismic Sun than the old ones is quite encouraging and may be considered as a proof that the physical improvements are correct.

4. Discussion: Necessity of Mild Mixing, ^{7}Li and ^{3}He

Although the introduction of pure element settling in the solar models considerably improves the consistency with the "seismic Sun", some discrepancies do remain, particularly below the convective zone where a "spike" appears in the sound velocity (see Gough *et al.*, 1996). The helium profiles directly obtained from helioseismology (Basu 1997; Antia and Chitre, 1997) show indeed a helium gradient below the convection zone which is smoother than the gradient obtained with pure settling. Furthermore, standard solar models including element settling do not reproduce the observed abundances of lithium.

The abundance determinations in the solar photosphere show that lithium has been depleted by a factor of about 140 compared to the protosolar value while beryllium is generally believed to be depleted by a factor 2. These values have widely been used to constraint the solar models (e.g. RVCD). However, while the lithium depletion factor seems well established, the beryllium value is still subject to caution. Balachandran and Bell (1998) argue that the beryllium depletion is not real due to an underestimate of the opacity of the continuum in the abundance determinations. Their new treatment leads to a solar value identical to the meteoritic value.

In RVCD, a mild mixing below the convection zone, attributed to rotation-induced shears (Zahn, 1992), was introduced to account for the lithium and beryllium depletion. It was shown that such a mixing may also wipe out the spike in the sound velocity, leading to more consistent solar models than the standard ones, computed with pure element settling. In these models the abundance of ^{3}He increased in the convection zone, due to a small dredge up from the tail of the ^{3}He peak.

Observations of the ^{3}He/^{4}He ratio in the solar wind and in the lunar rocks (Geiss, 1993; Gloecker and Geiss, 1996; Geiss and Gloecker, 1998) show that this ratio may not have increased by more than $\cong 10\%$ since 3 Gyr in the Sun, which is in contradiction with the results of RVCD. While the occurrence of some mild mixing below the solar convective zone is needed to explain the lithium depletion and helps for the conciliation of the models with helioseismological constraints, the ^{3}He/^{4}He observations put a strict constraint on its efficiency.

Vauclair and Richard (1998) have tried several parameterizations of mixing below the solar convection zone, which could reproduce both the ^{7}Li and the ^{3}He constraints. The only way to obtain such a result is to postulate a mild mixing, which would be efficient down to the lithium nuclear burning region but not too far below, to preserve the original ^{3}He abundance. A mixing effect decreasing with time, as obtained with the rotation-induced shear hypothesis, helps to obtain a ^{7}Li destruction without increasing ^{3}He too much, as the ^{3}He peak itself builds up during the solar life. A "cut-off" od the mixing process must however be postulated at some depth just below the ^{7}Li destruction layer.

The various kinds of mixing processes which may take place below the solar convection zone are summarized in J.P. Zahn's review (this conference). Here we have tested an effective diffusion coefficient varied as a parameter that we have adjusted to reproduce the observed abundances. We found that the observations of ^{7}Li and ^{3}He are well accounted for with a very mild mixing described by a diffusion coefficient not larger than $10^3 \text{cm}^2\text{s}^{-1}$, vanishing at about two scale heights below

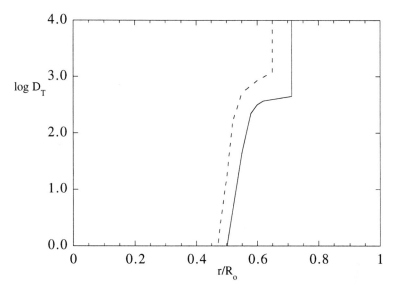

Figure 2. Example of profiles of the macroscopic diffusion coefficient below the solar convection zone, which can account for both the ^{7}Li and ^{3}He constraints (after Vauclair and Richard, 1998). In this case D_T decreases with time. The profiles are shown at 1.22 Gyr (dashed line) and 4.6 Gyr (solid line). ^{7}Li is then depleted by a factor 140 while ^{3}He does not increase by more than 5% during the whole solar life. In these models, beryllium is not depleted by more than 20%

the bottom of the convection zone (figure 2). In this case ^{7}Li is destroyed by a factor 140 and ^{3}He is not increased by more than $\cong$ 5%. Meanwhile, beryllium is not destroyed by more than $\cong$ 20%.

In summary the best solar models must include the effect of element settling, which represents an improvement on the physics, without any free parameter added. These models can be considered as the new "standard" models. They cannot however reproduce the ^{7}Li depletion and they lead to a spike in the sound velocity, compared to the seismic Sun, just below the convective zone. These two observations suggest the presence of some mild mixing in this region of the internal Sun. Adding the very strict constraint on the abundance of ^{3}He as given by Geiss and Gloecker (1998) leads to a precise description of the allowed profile of the macroscopic diffusion coefficient below the convective zone. This result can be taken as a challenge for the hydrodynamicists.

References

Antia, H.M., and Chitre, S.M.: 1997, *A&A* **000**, 000.
Bahcall, J.N., and Loeb, A.: 1990, *ApJ* **360**, 267.
Bahcall, J.N., and Pinsonneault, M.H.: 1992, *Reviews of Modern Physics* **64**, 885.
Balachandran, S., and Bell, R.A.: 1998, preprint.
Basu, S.: 1997, *Mon. Not. R. Astron. Soc.* **288**, 572.
Burgers, J.M..: 1969, *Flow Equations for Composite Gases*, New York: Academic Press.

Charbonnel, C., Vauclair, S., and Zahn, J.P.: 1992, *A&A* **255**, 191.
Chapman, S., and Cowling,T.G.: 1970, *The Mathematical Theory of Non-Uniform Gases*, Cambridge University Press, 3rd ed.
Christensen-Dalsgaard, J., Proffitt, C.R., and Thompson, M.J.: 1993, *ApJ* **408**, L75.
Cox, A.N., Guzik, J.A., and Kidman, R.B.: 1989, *ApJ* **342**, 1187.
Dziembowski, W.A., Goode, P.R., Pamyatnikh, A.A., and Sienkiewicz, R.: 1994, *ApJ* **432**, 417.
Geiss, J.: 1993, *Origin and Evolution of the Elements*, Prantzos, Vangioni-Flam, and Cassé (eds.), Cambridge Univ. Press, 90.
Gloecker, G., and Geiss, J.: 1996, *Nature* **381**, 210.
Geiss, J., and Gloecker, G.: 1998, *Space Sci. Rev.*, this volume.
Gough, D.O., Kosovichev, A.G., Toomre, J., Anderson, E., Antia, H.M., Basu, S., Chaboyer, B., Chitre, S.M., Christensen-Dalsgaard, J., Dziembowski, W.A., Eff-Darwich, A., Elliott, J.R., Giles, P.M., Goode, P.R., Guzik, J.A., Harvey, J.W., Hill, F., Leibacher, J.W., Monteiro, M.J.P.F.G., Richard, O., Sekii, T., Shibahashi; H., Takata, M., Thompson, M.J., Vauclair, S., and Vorontosov, S.V.: 1996, *Science* **272**, 1296.
Michaud, G.: 1970, *ApJ* **160**, 641.
Michaud, G., and Proffitt, C.R.: 1992, in *Inside the Stars*, W.W. Weiss and A. Baglin (eds.), vol. 371, IAU, San Francisco: PASPC, pp. 246–259.
Michaud, G., and Vauclair, S.: 1991, in *Solar Interior and Atmosphere*, A.N. Cox, W.C. Livingston, and M.S. Matthews (eds.), The University of Arizona Press, p. 304.
Paquette, C., Pelletier, C., Fontaine, G., and Michaud, G.: 1986, *ApJS* **61**, 177.
Proffitt, C.R.: 1994, *ApJ* **425**, 849.
Proffitt, C.R., and Michaud, G.: 1991, *ApJ* **380**, 238.
Proffitt, C.R., and VandenBerg, D.A.: 1991, *ApJS* **77**, 473.
Richard, O., Vauclair, S., and Charbonnel, C.: 1998, preprint.
Richard, O., Vauclair, S., Charbonnel, C., and Dziembowski, W.A.: 1996, *A&A* **312**, 1000.
Richer, J., and Michaud, G.: 1993, *ApJ* **416**, 312.
Thoul, A.A., Bahcall, J.N., and Loeb, A.: 1994, *ApJ* **421**, 828.
Turcotte, S., Richer, J., Michaud, G., Iglesias, C.A., and Rogers, F.J.: 1998, *ApJ* **000**, 000.
Vauclair, S.: 1998, *Space Sci. Rev.*, in press.
Vauclair, S., and Vauclair, G.: 1982, *ARA&A* **20**, 37.
Vauclair, S., and Richard, O.: 1998, preprint.
Zahn, J.P.: 1992, *A&A* **265**, 115.

Address for correspondence: Sylvie.Vauclair@obs-mip.fr

MACROSCOPIC TRANSPORT

Large-scale advection, turbulent diffusion, wave transport

JEAN-PAUL ZAHN
Observatoire de Paris, 92195 Meudon, France

Abstract. While the solar convection zone is very well mixed by its turbulent motions, chemical composition gradients build up in the radiative interior due to microscopic diffusion and settling, and to nuclear burning. Standard models, which ignore any type of macroscopic transport, cannot explain the depletion of lithium in solar-type stars, as they evolve; neither do they account for the observed profile of molecular weight at the base of the solar convection zone.

Such macroscopic transport can be achieved through thermally driven meridian currents, through turbulent diffusion generated by differential rotation and possibly through gravity waves. These processes transport also angular momentum, and therefore the internal rotation profile of the Sun provides a crucial test for their relative importance. So does also the behavior of tidally locked binaries, which appear to destroy less lithium than single stars of the same mass.

Key words: Stellar interiors - stellar rotation - turbulent transport

1. Introduction

In standard stellar models, only the convective regions are assumed to be well mixed. In the stable radiative regions, the chemical composition evolves in time due both to microscopic diffusion and gravitational settling (cf. the preceding review by Vauclair, 1998), and to nuclear reactions. But these processes alone cannot explain the depletion of lithium observed in solar-type stars, and they produce a molecular weight gradient below the solar convection zone which disagrees with that measured through helioseismology. Therefore one is lead to conclude that other transport mechanisms operate in the Sun, at least in the vicinity of the convection zone, and that they are presumably of macroscopic nature.

A few of such mechanisms have been examined in some detail, and we shall review them here. The first which comes in mind is the mixing induced by the convection zone itself, in the form of overshooting and penetration. But two other causes of mixing have been identified in radiation zones: one is the turbulence generated by the differential rotation, and the other the large scale circulation caused by the thermal unbalance in a rotating star. Finally, internal gravity waves emitted by the turbulent convection may play an important role in the transport of angular momentum, and perhaps also in the transport of chemical elements.

2. Convective Penetration, and the Extent of the Convection Zone

In stellar models, one uses commonly the Schwarzschild criterion to fix the boundary of a convective region. This criterion indicates where the acceleration of the convective motions vanishes, but nothing prevents these from penetrating beyond into the adjacent stable region. The first attempt to estimate the amount of overshooting was made by Saslaw and Schwarzschild (1965), but it was Shaviv and Salpeter (1973) who correctly took into account the fact that the stratification in the stable region is rendered almost adiabatic by the convective penetration. The next step was a series of one-dimensional simulations by Schmitt *et al.* (1984), who assumed that the penetration proceeds in the form of downwards directed plumes carrying all the heat flux; by exploring the relevant parameter space, they found that the extent of penetration scales as

$$d \propto f^{1/2} w_i^{3/2} \tag{1}$$

where w_i is the initial vertical velocity of the plumes (at the top of the stable zone) and f their spatial filling factor. What fluid dynamicist called plume (panache in French!) is a structured vertical flow which lasts much longer than the time it takes for the fluid to travel through it; examples of such plumes abound in the Earth atmosphere (above chimneys, volcanos, etc.) and they can be easily generated in the laboratory.

The problem of convective penetration in stellar interiors stars was further clarified by Zahn (1991), who emphasized that such plumes enforce an almost adiabatic stratification as long as their Péclet number Péc $= w\ell/K$ remains larger than unity (K is the thermal diffusivity, w the local velocity and ℓ the depth of penetration). Below the solar convection zone, the plumes start with a Péclet number of order 10^6; once they have decelerated to the point where Péc ≈ 1, the temperature gradient quickly adjusts to the radiative temperature gradient. It is then easy to show that the extent of adiabatic penetration follows the law discovered empirically by Schmitt *et al.*:

$$\frac{d_{\text{pen}}}{H_P} = f^{1/2} w_i^{3/2} \left[\frac{3}{2} g \, K \, \chi_P \, \nabla_{\text{ad}} \right]^{-1/2}, \tag{2}$$

with the classical notations g, H_P and ∇_{ad} for the gravity, the pressure scale-height and the adiabatic gradient, and $(\chi_P = \partial \ln \chi / \partial \ln P)_{\text{ad}}$ being the logarithmic derivative of the radiative conductivity with respect to pressure, at constant entropy.

In the meanwhile such downwards directed plumes had been found indeed in 2 and 3-dimensional simulations of strongly stratified convection (Hurlburt *et al.*, 1986; Stein and Nordlund, 1989; Malagoli *et al.*, 1990). Recently these plumes have been detected also in the Sun, thanks to time-distance tomography performed with the SOHO/MDI data (Kosovichev and Duvall, 1997). In the numerical simulations, they are clearly responsible for the penetration below a convective region, and the amount of penetration is in good agreement with the theoretical arguments leading to (2), as was shown by Hurlburt et al. (1994). Whether these plumes are able to cross the whole solar convection zone is still an open question, although this appears quite plausible, based on prescriptions validated in atmospheric sciences, as was advocated by Rieutord and Zahn (1995).

Unfortunately, relation (2) is not of great help when it comes to predict the extent of penetration in the Sun, since it requires the knowledge of the initial velocity w_i. One is then left with mixing-length arguments, which suggest that this extent should to be of the order of the scale-height of the radiative conductivity

$$d_{\text{pen}} \approx \left| \frac{d \ln \chi}{dr} \right|^{-1}, \qquad (3)$$

a distance which at the base of the convection zone amounts to about half of the pressure scale-height. Numerical simulations of penetrative convection confirm that the penetration is of that order, but they must be interpreted with care. The reason is that even with present-days computers it is difficult to reach a Péclet number much larger than unity.

We have used on purpose the term 'penetration' to designate this spread of the turbulent convection zone into the stable layer below, where it establishes an almost adiabatic – though slightly sub-adiabatic – stratification $\nabla \lesssim \nabla_{\text{ad}}$. That extra layer is mixed as thoroughly as the rest of the convection zone. It matches the radiation zone below with a thermal boundary layer, in which the temperature gradient adjusts to the radiative gradient ∇_{rad}, and whose thickness is approximately (Zahn, 1991)

$$d_{\text{th}} \approx K^{1/2} \left(\frac{H_P}{g} \right)^{1/4}. \qquad (4)$$

In the Sun, this layer is extremely thin: 1 km! We thus conclude that the convection zone ends abruptly, with a very steep stable temperature gradient, which acts like a barrier to prevent its crossing by isolated turbulent eddies.

On the contrary, sub-photospheric convection in a A-type star has a Péclet number which is less than unity (which makes it much easier to simulate than solar convection). Although convection is rather vigorous, it is unable to enforce a nearly adiabatic stratification, but the motions 'overshoot' far into the radiative region below because they exchange heat so fast with the surroundings that they are barely decelerated by the buoyancy force (Toomre et al., 1981; Freytag et al., 1996). We thus make the distinction between *convective penetration* and *overshooting*. Convective penetration occurs a high Péclet number; it renders the temperature gradient adiabatic over some depth in the stable region but causes no mixing beyond. Overshooting operates a low Péclet number; it has little effect on the temperature stratification, but induces mixing in the adjacent radiative layer. Obviously, the results of simulating convection near the surface of a A-type star cannot be applied to the base of the solar convection zone.

Let us come back to the Sun. A penetration as strong as predicted by (3) would cause a near discontinuity in the temperature gradient which would be easily detected through helioseismology. Since this is not the case, it was concluded that d_{pen} is less than 1/4 of the pressure scale-height (Roxburgh and Vorontsov, 1994); according to Basu (1997), who uses improved data, $d_{\text{pen}} \approx 0.05 \, H_P$.

Two explanations may be invoked for this severe disagreement. First, since there is no reason why all plumes should have the same initial velocity w_i, they do not terminate at the same depth; hence this thermal interface probably looks like the surface of a wavy ocean. Therefore the vertical temperature gradient which is sensed by helioseismology represents a horizontal average; it varies more smoothly than hinted by the near discontinuity in the thermal adjustment layer, and hence it could go below the detection limit. Such smoothened temperature profiles have been considered by Christensen-Dalsgaard et al. (1995). Second, in most numerical simulations so far the Coriolis force has been ignored, although it probably plays a non negligible role at the base of the convection zone, where the convective turnover time is of same order as the rotation period ($\approx$ 1 month). Three-dimensional calculations of non-stationary convection performed by Julien et al. (1997) show indeed that the vertical progression of a convective layer is slowed down by rotation. We shall learn soon whether this means that the Coriolis force inhibits penetration below a stationary convection zone; to that purpose calculations are being carried out with a code simulating the rotating solar convection zone in spherical shell geometry (Elliott et al., 1998).

3. The Solar Tachocline

One of the surprising results obtained through helioseismology is the detection of a layer of strong shear between the differentially rotating convection zone of the Sun and its uniformly rotating interior (Brown et al., 1990). This *tachocline* plays presumably a crucial role in the solar dynamo, since it can easily build up a strong toroidal field from a rather weak poloidal field. The thinness of the layer – it measures only $0.05 \, R_\odot$ according both to Charbonneau et al. (1997) and to Corbard et al. (1998) – poses a challenging problem, for the reason we are going to explain.

The differential rotation $\widehat{\Omega}(\theta)$ imposed by the convection zone on the top of the radiative interior causes there a horizontal temperature gradient $\partial T / \partial \theta$ through the thermal wind balance, and this temperature gradient tends to spread inwards on a thermal time-scale, due to radiative diffusion. But the stable temperature stratification hinders the spread, because angular momentum can only be redistributed through a meridian flow, in the absence of any turbulence (the microscopic viscosity plays a negligible role), and the vertical component of this flow works against buoyancy. The progression of the tachocline behaves then as a "hyper-diffusion", such as encountered earlier in the solar spin-down problem (Howard et al., 1967).

Treating the tachocline as a thin boundary layer (Spiegel and Zahn, 1992), the evolution of the differential rotation $\widehat{\Omega}$ is governed by

$$\frac{\partial \widehat{\Omega}}{\partial t} = -K \left(\frac{2\Omega}{N}\right)^2 r^2 \frac{\partial^4 \widehat{\Omega}}{\partial r^4}, \tag{5}$$

K being the thermal diffusivity and N the buoyancy frequency:

$$N^2 = \frac{g}{H_P} \left[\left(\frac{\partial \ln T}{\partial \ln P}\right)_{\rm ad} - \frac{d \ln T}{d \ln P} \right]. \tag{6}$$

In other words the thickness of the layer increases as

$$\frac{h}{r} \simeq \left(\frac{t}{t_{\rm ES}}\right)^{1/4}, \tag{7}$$

with a time-scale equal to the local Eddington-Sweet time (see below in §4)

$$t_{\rm ES} = \frac{r^2}{K} \left(\frac{\Omega}{N}\right)^2. \tag{8}$$

According to this prediction, the tachocline would reach the depth $r \simeq 0.3 R_\odot$ in the present Sun, which is in contradiction with the helioseismic data.

What prevents the tachocline from spreading that far? Isotropic turbulence would contribute also to the deepening of the layer; it would operate on a viscous timescale r^2/ν, with ν being the turbulent diffusivity. But the radiative spread can be stopped through *anisotropic* turbulence, provided that the contrast between horizontal and vertical transport be greater than the square of the aspect ratio: $\nu_h/\nu_v \gg (r/h)^2$. Then a stationary regime is achieved, in which the advection of angular momentum is balanced by horizontal diffusion, such that (Spiegel and Zahn, 1992)

$$2\Omega \cos\theta\, r \sin\theta\, v = \frac{\nu_h}{\sin\theta} \frac{\partial}{\partial \theta}\left(\sin^3\theta \frac{\partial \widehat{\Omega}}{\partial \theta}\right), \tag{9}$$

where v is the horizontal (latitudinal) component of the flow velocity. The thickness of the tachocline depends on the ratio between the two diffusivities, that of heat and that of momentum:

$$\frac{h}{r} \simeq \left(\frac{\Omega}{N}\right)^{1/2} \left(\frac{K}{\nu_h}\right)^{1/4} \quad \text{or} \quad h \simeq 20,000 \left(\frac{\kappa}{\nu_h}\right)^{1/4} \text{ km}. \tag{10}$$

We see that a horizontal viscosity of about 10^5 cm^2 s^{-1}, implying a Reynolds number $Re \simeq 10^3$, is sufficient to comply with the observed upper limit of $h \leq 50,000$ km.

Unfortunately, it has not been possible so far to predict from first principles the value of the turbulent viscosity ν_h. In Spiegel and Zahn's treatment it was assumed for simplicity that ν_h is constant through the whole layer; if one wishes a more realistic prescription, a reasonable guess would be that it scales with the differential rotation which produces it, as hinted by some laboratory experiments.

The tachocline is the seat of a meridian circulation consisting of two counter-rotating cells in each hemisphere, with an upflow at mid-latitude. The net vertical flux of angular momentum vanishes, since the advection through the circulation is exactly balanced by the turbulent diffusion in the horizontal direction, as stated in (9). Numerical simulations have been carried out recently by Elliott (1997) with a two-dimensional code; they confirm the main properties derived from boundary layer theory, and yield roughly the same estimate for the thickness of the tachocline, in terms of the horizontal viscosity ν_h.

What may be the cause of this anisotropic turbulence? The latitudinal rotation profile $\Omega(\theta)$ is likely to be linearly stable, according to Rayleigh's criterion, which in this spherical geometry takes the form established by Watson (1981). However any shear flow is liable to finite amplitude instabilities, provided the Reynolds number is high enough, as demonstrated by many laboratory experiments (plane or cylindrical Couette and Poiseuille flows). The turbulence produced by such shearing flows is genuinely three-dimensional (cf. Lesieur, 1997), but in the presence of restoring forces it can display a pronounced anisotropy. We expect such anisotropy to occur in the tachocline, because the stable stratification limits the size and the velocity of the turbulent eddies in the vertical direction. We refer to Michaud and Zahn (1998) for an estimate of that anisotropy, which they find of order 100 or more.

Beside the smoothing of differential rotation in latitude, which allows for the stationary regime, this anisotropic turbulence has another important effect on the transport of chemicals. In the absence of such turbulence, the advection through the 2-cell circulation would proceed on a rather short time-scale

$$t_{\rm adv} \approx 10^7 \left(\frac{h}{20{\rm Mm}}\right)^4 \left(\frac{\Delta\Omega}{\Omega}\right)^{-1} {\rm yrs}, \qquad (11)$$

where $\Delta\Omega/\Omega \approx 1/10$ is the differential rotation imposed on the tachocline. But this anisotropic turbulence erodes the advective transport, in a manner described by Chaboyer and Zahn (1992), and it turns it into a vertical diffusion whose time-scale is about 100 times larger.

That some mixing actually occurs below the solar convection zone is suggested by the profile of sound velocity which is determined through helioseismology. Helium and heavy elements are known to settle there, as described by Vauclair (1998). But when all microscopic processes are accounted for, in what is now called the *standard solar model*, a difference persists with the solar profile of sound velocity, which in that region peaks at a few 10^{-3}. This is 10 times larger than the precision of solar inversion, and is therefore interpreted as a genuine discrepancy. One way to remove it is to postulate some macroscopic mixing just in the tachocline, as was shown by Richard *et al.* (1996); more precise calculations are under way to model

Busse, F.H.: 1981, 'Do Eddington-Sweet circulations exist?', *Geophys. Astrophys. Fluid Dynamics* **17**, 215.
Brown, T.M, Christensen-Dalsgaard, J., Dziembowski, W.A., Goode, P., Gough, D.O. and Morrow, C.A.: 1989, 'Inferring the Sun's internal angular velocity from observed p-mode frequency splittings', *Astrophys. J.* **343**, 526.
Chaboyer, B., and Zahn, J.-P.: 1992, 'Effect of horizontal turbulent diffusion on transport by meridional circulation', *Astron. Astrophys.* **253**, 143.
Chaboyer, B., Demarque, P., Guenther, D.B. and Pinsonneault, M.H.: 1995, 'Rotation, diffusion and overshoot in the Sun', *Astrophys. J.* **446**, 435.
Charbonneau, P., and MacGregor, K.B.: 1993, 'Angular momentum transport in magnetized stellar radiative zones. II. The solar spin-down', *Astrophys. J.* **417**, 762.
Charbonneau, P., Christensen-Dalsgaard, J., Henning, A., Schou, J., and Thompson, M.J.: 1997, 'Angular momentum transport in the Sun through meridian circulation', *IAU Symposium 181/* (ed. J. Provost and F.X. Schmider). in press.
Christensen-Dalsgaard, J., Monteiro, M., and Thompson, M.J.: 1995, 'Helioseismic estimation of convective overshoot in the Sun, *Monthly Not. Roy. Astron. Soc.* **276**, 283.
Corbard, T., Berthomieu, G., Provost. J., and Morel, P.: 1992 'Inferring the equatorial solar tachocline from frequency splittings', *Astron. Astrophys.* **330**, 1149.
Eddington, A.S.: 1925, 'Circulating currents in rotating stars', *Observatory* **48**, 73.
Elliott, J.R.: 1997, 'Aspects of the solar tachocline', *Astron. Astrophys.* **327**, 1222.
Elliott, J.R., Miesch, M.S., and Toomre, J.: 1998, 'Rotating spherical shell convection', preprint.
Freytag, B., Ludwig, H.-G., and Steffen, M.: 1996, 'Hydrodynamical models of stellar convection', *Astron. Astrophys.* **313**, 497.
Gough, D.: 1998, 'Solar interior', *Space Sci. Rev.*, this volume.
Howard, L.N., Moore, D.W., and Spiegel, E.A.: 1967, 'Solar spin-down problem', *Nature* **214**, 1297.
Hurlburt, N.E., Toomre, J., and Massaguer, J.: 1986, 'Nonlinear compressible convection penetrating into stable layers and producing internal gravity waves', *Astrophys. J.* **311**, 563.
Hurlburt, N.E., Toomre, J., Massaguer, J.M., and Zahn, J.-P.: 1994, 'Penetration below a convection zone', *Astrophys. J.* **421**, 245.
Julien, K., Werne, L., Legg, S., and McWilliams, J.: 1997, 'The effect of rotation on convective overshoot', *Solar Convection and Oscillations and their Relationship* (F.P. Pijpers, J. Christensen-Dalsgaard and C.S. Rosenthal, eds.; Kluwer, Astrophysics and Space Science Library vol. 225), 231.
Kosovichev, A.G., and Duvall, T.L. Jr.: 1997, 'Acoustic tomography of solar convective flows and structures', *Solar Convection and Oscillations and their Relationship* (F.P. Pijpers, J. Christensen-Dalsgaard and C.S. Rosenthal, eds.; Kluwer, Astrophysics and Space Science Library vol. 225), 241.
Kumar, P., and Quataert, E.J.: 1997, 'Angular momentum transport by gravity waves and its effect on the rotation of the solar interior', *Astrophys. J.* **475**, L143.
Lesieur, M.: 1997 *Turbulence in Fluids* (3d edition; Kluwer Acad. Publ.)
Malagoli, A., Cattaneo, F., and Brummel, N.H.: 1990, 'Turbulent supersonic convection in three dimensions', *Astrophys. J.* **361**, L33.
Matias, J., and Zahn, J.-P.: 1997, 'Angular momentum transport in the Sun through meridian circulation', *IAU Symposium 181* (ed. J. Provost and F.X. Schmider), in press.
Mestel, L.: 1953, 'Rotation and stellar evolution', *Monthly Not. Roy. Astron. Soc.* **113**, 716.
Michaud, G., and Zahn, J.-P.: 1998, 'Turbulent transport in stellar interiors', *Theor. Comput. Fluid Dynamics*, in press.
Pinsonneault, M.H., Kawaler, S.D., Sofia, S. and Demarque, P.: 1989, 'Evolutionary models of the rotating Sun', *Astrophys. J.* **338**, 424.
Press, W.H.: 1981, 'Radiative and other effects from internal waves in stellar interiors', *Astrophys. J.* **245**, 286.
Richard, O., Vauclair, S., Charbonnel, C. and Dziembowski, W.A.: 1996, 'New solar models including helioseismological constraints and light-element depletion', *Astron. Astrophys.* **312**, 1000.
Rieutord, M., and Zahn, J.-P.: 1995, 'Turbulent plumes in stellar convective envelopes', *Astron. Astrophys.* **296**, 127.

Roxburgh, I.W., and Vorontsov, S.V.: 1994, 'Seismology of the solar envelope: the base of the convection zone as seen in the phase shift of acoustic waves', *Mon. Not. Roy. Astron. Soc.* **268**, 880.

Saslaw, W.C., and Schwarzschild, M.: 1965, 'The overshoot region at the bottom of the solar convection zone', *Astrophys. J.* **142**, 1468.

Schatzman, E.: 1962, 'A theory for the role of magnetic activity during star formation', *Ann. Astrophys.* **25**, 18.

Schatzman, E.: 1993,. 'Transport of angular momentum and diffusion by the action of internal waves', *Astron. Astrophys.* **279**, 431.

Schatzman, E.: 1996, 'Diffusion process produced by random internal waves', *J. Fluid Mech.* **322**, 355.

Schmitt, J.H.M.M., Rosner, R., and Bohn, H.U.: 1984, 'The overshoot region at the bottom of the solar convection zone', *Astrophys. J.* **282**, 316.

Shaviv, G., and Salpeter, E.: 1973, 'Convective overshooting in stellar interior models', *Astrophys. J.* **184**, 191.

Skumanich, A.: 1972, 'Time scales for CaII emission decay, rotational braking and lithium depletion', *Astrophys. J.* **171**, 563.

Spiegel, E.A., and Zahn, J.-P.: 1992, 'The solar tachocline', *Astron. Astrophys.* **265**, 106.

Stein, R.F., and Nordlund, A.: 1989, 'Topology of convection beneath the solar surface', *Astrophys. J.* **342**, L95.

Sweet, P.A.: 1950, 'The importance of rotation in stellar evolution', *Monthly Not. Roy. Astron. Soc.* **110**, 548.

Talon, S. and Zahn, J.-P.: 1996, 'Anisotropic diffusion and shear instabilities', *Astron. Astrophys.* **317**, 749.

Thompson, M.J., Toomre, J., Anderson, E.R., Antia, H.M., Berthomieu, G., Burtonclay, D., Chitre, S.M., Christensen-Dalsgaard, J., Corbard, T., DeRosa, M., Genovese, C.R., Gough, D.O., Haber, D.A., Harvey, J.W., Hill, F., Howe, R., Korzennik, S.G., Kosovichev, A.G., Leibacher, J.W., Pijpers, F.P., Provost, J., Rhodes Jr., E.J., Schou, J., Sekii, T., Stark, P.B., and Wilson, P.R.: 1996, 'Differential rotation and dynamics of the solar interior', *Science* **272**, 1300.

Thorburn, J.A., Hobbs, L.M., Deliyannis, C.P. and Pinsonneault, M.H.: 1993, 'Lithium in the Hyades. I - New observations', *Astrophys. J.* **415**, 150.

Toomre, J., Zahn, J.-P., Latour, J. and Spiegel, E.A.: 1976, 'Stellar convection theory. II. Single mode study of the second convection zone in an A type star.' *Astrophys. J.* **207**, 545.

Vauclair, S.: 1998, 'Element settling in the solar interior', *Space Sci. Rev.*, this volume.

Vogt, H.: 1925, 'Zum Strahlungsgleichgewicht der Sterne', *Astron. Nachr.* **223**, 229.

Watson, M.: 1981, 'Shear instability of differential rotation in stars', *Geophys. Astroph. Fluid Dyn.* **16**, 285.

Zahn J.-P.: 1974, 'Rotational instabilities and stellar evolution.' *Stellar Instability and Evolution* (ed. P. Ledoux, A. Noels and R.W. Rogers; Reidel), 185.

Zahn J.-P.: 1991, 'Convective penetration in stellar interiors', *Astron. Astrophys.* **252**, 179.

Zahn J.-P.: 1992, 'Circulation and turbulence in rotating stars', *Astron. Astrophys.* **265**, 115.

Zahn, J.-P.: 1994, 'Rotation and lithium depletion in late-type binaries', *Astron. Astrophys.* **288**, 829.

Zahn, J.-P., Talon, S. and Matias, J.: 1997, 'Angular momentum transport by internal waves in the solar interior', *Astron. Astrophys.* **322**, 320.

SOLAR NEUTRINOS

Y. SUZUKI

Kamioka Observatory, Institute for Cosmic Ray Research,
The University of Tokyo, Higashi-Mozumi, Kamioka 506-12, Japan.

Abstract. The current status of solar neutrino experiments is reviewed. All the experimental measurements show deficits of solar neutrinos. Non monotonic suppression indicates that the problem may naturally be explained by neutrino oscillations, but not by modifying solar models. A new experiment shows very promising results. We hope that a definite answer to the question of whether solar neutrinos are oscillating will be obtained in the very near future.

Key words: sun, neutrino, neutrino mass, neutrino oscillation, MSW effect

1. Introduction

Solar neutrinos are produced by nuclear reaction chains in the central core of the sun. The net reaction is

$$4p \rightarrow {}^4He + 2e^+ + 2\nu_e + 26.1\,\text{MeV}.$$

The solar neutrino flux depends upon the temperature, chemical composition, cross section of the nuclear reactions, and so on. Therefore those neutrinos were first thought to be a good probe of the solar interior.

Several tens of billions of those neutrinos traverse a square cm of the earth each second. The first solar neutrino experiment done by R. Davis *et al.* (Cleveland *et al.*, 1998) observed only less than 30% of the flux predicted by a solar model. This is called a solar neutrino problem suggesting either problems with solar models or problems with our understanding of neutrinos—a coupled problem of neutrinos and the sun.

Since then, solar neutrinos have been detected by four other experiments (Hirata *et al.*, 1990; Hirata *et al.*, 1991a; Hirata *et al.*, 1992; Hirata *et al.*, 1991b; Fukuda *et al.*, 1996; Abdurashitov *et al.*, 1994; Hampel *et al.*, 1996; Anselmann *et al.*, 1994; Fukuda *et al.*, 1998). All of the experiments have observed significantly less flux than that predicted by the standard solar models (SSMs) (Bahcall and Pinsonneault, 1995; Turck-Chieze and Lopes, 1993). One characteristic of the problem is that the suppression rate measured at each experiment is non-monotonic with neutrino energy. Detailed studies on this problem strongly suggest that the deficits are not easily explained by changing solar models, but are naturally explained by neutrino oscillations (Hata and Langacker, 1997; Mikheyev and Smirnov, 1985; Wolfenstein, 1978; Glashow *et al.*, 1987), namely by finite neutrino masses.

It is, however, strongly desirable to obtain direct and solar model independent evidence of the neutrino oscillations. A distortion of the energy spectrum, or short-term time variations like a possible daytime and nighttime flux difference, are such kinds of observations. Solar neutrino fluxes are supposed to be stable over several million years, and if any time variations are to be found, that would also be an indication of neutrino mass and mixing, or neutrino magnetic moments (Okun *et al.*, 1986; Lim and Marciano, 1988; Akhmedov, 1988). An anomaly in the ratio of charged current interactions to neutral current interactions would also be a model independent evidence of neutrino oscillations.

Such studies can be done by the on-going Super-Kamiokande, SNO (Chen, 1986; Ewan *et al.*, 1987) and Borexino (Arpesella *et al.*, 1992) and others.

2. Neutrino Oscillation

If neutrinos have finite masses and non-zero mixings among mass eigen-states, then neutrinos produced in flavor eigenstates may oscillate among different states with the following probability (for a two neutrino oscillation case),

$$P(\nu_\alpha \to \nu_\beta) = \sin^2 2\theta \sin^2 \left(1.27 \cdot \Delta m^2 \cdot L/E\right),$$

where E (MeV) is the energy of neutrino and L (m) is the distance between the source and the detector, and Δm^2(eV^2) is the squared mass difference of neutrino mass eigenstates: L and E are determined by the experimental arrangement. If $\Delta m^2 \gg E/L$, then $P(\nu_\alpha \to \nu_\beta) \sim \frac{1}{2} \sin^2 2\theta$, and if $\Delta m^2 \ll E/L$, then experiments hardly see the oscillation effect and are able to set the upper limit by $\sin 2\theta \cdot \Delta m^2 \ll 0.8[E/L]\sqrt{P(\nu_\alpha \to \nu_\beta)}$. If $\Delta m^2 \sim E/L$, experiments are maximally sensitive. For example, for a configuration $E \sim 1$ GeV and $L \sim 10$ to $10,000$ km (in the case of the atmospheric neutrinos), $\Delta m^2 \sim 10^{-1} \sim 10^{-4}$ eV^2 and for $E \sim 10$ MeV and $L \sim 150,000,000$ km (for 'Just So' solar neutrino oscillation (Glashow *et al.*, 1987)), $\Delta m^2 \sim 10^{-11 \sim -10}$ eV^2.

When neutrinos pass through a medium, the electron neutrinos obtain an additional potential, $\sqrt{2}G_F n_e$, through the charged current forward scattering amplitude. (Here G_F is the Fermi coupling constant and n_e is the number density of electrons.) The equation of neutrino propagation for the case of $\nu_e \to \nu_\mu$ is,

$$i\frac{d}{dt}\begin{pmatrix}\nu_e \\ \nu_\mu\end{pmatrix} = \begin{pmatrix} -\frac{\Delta m^2 \cos 2\theta}{4p} + \sqrt{2}G_F n_e & \frac{\Delta m^2 \sin 2\theta}{4p} \\ \frac{\Delta m^2 \sin 2\theta}{4p} & \frac{\Delta m^2 \sin 2\theta}{4p} \end{pmatrix}\begin{pmatrix}\nu_e \\ \nu_\mu\end{pmatrix}.$$

The mixing angle in the matter becomes,

$$\tan 2\theta_m = \frac{\tan 2\theta_V}{1 - (2p\sqrt{2}G_F n_e)/\Delta m^2 \cos 2\theta_V}.$$

If $(2p\sqrt{2}G_F n_e)/(\Delta m^2 \cos 2\theta_V) = 1$ (resonance condition), then the mixing angle in the matter becomes maximal, even though the vacuum mixing angle is small. If the neutrino propagates adiabatically through the resonance region, an electron neutrino converts into the other flavor, ν_μ in this example. 10 MeV neutrinos produced in the sun's core satisfy the resonance condition when then pass through the sun ($0 < n_e/N_A < 100$) if

$$\Delta m^2 \leq 1.6 \times 10^{-4} \mathrm{eV}^2.$$

The adiabatic condition for 10 MeV neutrinos is,

$$\Delta m^2 \frac{\sin^2 2\theta}{\cos 2\theta} \geq 6.3 \times 10^{-8} \mathrm{eV}^2.$$

These conditions determine the parameter region explaining the 'observed' conversion rate.

Searches for neutrino masses have been done not only by the neutrino oscillation searches, but also by other methods like direct mass searches and double beta decay experiments although the indication of the finite neutrino masses only come from oscillation experiments so far. The current upper limit on the neutrino mass from the experiments looking directly for the evidence of finite neutrino masses are, $m_{\nu_\tau} \leq 18.2 \mathrm{~MeV}/c^2$ (Buskulic et al., 1995; Albrech et al., 1988; Albrech et al., 1992), $m_{\nu_\mu} \leq 170 \mathrm{~keV}/c^2$ (Assamagan et al., 1996; Assamagan et al., 1994) and $m_{\nu_e} \leq 3.5 \sim 5.6 \mathrm{~eV}/c^2$ (Weinheimer et al., 1993; Belesev et al., 1995).

The best limits for Majorana neutrino masses of $0.48 \sim 1.5 \mathrm{~eV}^2$ (see, for example, Klapdor-Kleingrothaus, 1996) were obtained by the double beta decay experiments (we have included the uncertainty of the nuclear matrix element).

As you see above, neutrino oscillations are the only way to approach small mass (difference) well below 1 eV, and are sensitive to the very wide Δm^2 range down to 10^{-12} eV^2.

3. Solar Neutrino Experiments

The recent results of all current solar neutrino experiments are shown in table I*.

The five solar neutrino experiments—the Homestake experiment (Cleveland et al., 1998), Kamiokande (Hirata et al., 1990; Hirata et al., 1991a; Hirata et al., 1992; Hirata et al., 1991b; Fukuda et al., 1996), SAGE (Abdurashitov et al., 1994), GALLEX (Hampel et al., 1996; Anselmann et al., 1994), and Super-Kamiokande (Fukuda et al., 1998)—use three different techniques and therefore have different energy thresholds: 233 keV for the gallium experiments (SAGE and GALLEX), 814 keV for the Chlorine experiment (Homestake) and 7(6.5) MeV for the water Cherenkov experiments (Kamiokande and Super-Kamiokande). It is an advantage

---

* Those latest results are obtained from the presentation given at Neutrino98 at Takayama, June 4th to 9th, 1998, by K. Lande (Homestake), G.N. Gavrin (SAGE), T. Kirsten (GALLEX) and Y. Suzuki (Super-Kamiokande).

Table I

Recent results of the solar neutrino experiments. The Super-Kamiokande result shown in this table is based on 504 days of data presented at Neutrino98 while those results discussed in this report is based on the 374 days of data presented at this workshop. Those flux results agree with each other. Note also that the most recent standard solar models (Bahcall et al., 1998; Turck-Chièze, 1998) make the discrepancy between the prediction and the measurement slightly smaller than BP95, especially of ^{8}B-neutrino flux. See text for more details.

Experiment	Reactions (for detection)	Threshold (MeV)	Results (Ratio to the SSM(BP95))
Homestake	$\nu_e + {}^{37}\text{Cl} \to e + {}^{37}\text{Ar}$	0.814	$0.275 \pm 0.017 \pm 0.017$
Kamiokande	$\nu_e + e \to \nu_e + e$	7.0	$0.424 \pm 0.029 \pm 0.050$
SAGE	$\nu_e + {}^{71}\text{Ga} \to e + {}^{71}\text{Ge}$	0.233	$0.486^{+0.050 \, +0.028}_{-0.052 \, -0.029}$
GALLEX	$\nu_e + {}^{71}\text{Ga} \to e + {}^{71}\text{Ge}$	0.233	0.57 ± 0.06
Super-K	$\nu_e + e \to \nu_e + e$	6.5(5.0)	$0.368^{+0.008 \, +0.013}_{-0.007 \, -0.011}$

that the experiments have different energy thresholds and therefore sensitivity for the different regions of the solar neutrino spectrum.

The pioneering experiment—the Homestake experiment—started to take data in 1970, and lasting more than 25 years, first indicated the so called 'solar neutrino problem': the observed neutrino flux is smaller than that calculated from the standard solar models (Bahcall and Pinsonneault, 1995; Turck-Chièze and Lopes, 1993). The result of the flux measurement of the Homestake experiment has been persistent to be around 25 to 30% of the SSM: the most recent value reported is 0.275±0.017±0.017 of that predicted by the SSM of BP95.

The Kamiokande experiment, which showed a first evidence that the neutrinos are really coming from the sun's direction, stopped its solar neutrino observation on the 6th of February in 1995. The final result (Fukuda et al., 1996) for the ^{8}B-solar neutrino flux based on the data taken from 2079 days of effective live time is $2.80 \pm 0.19(\text{stat}) \pm 0.33(\text{syst}) \times 10^6$ cm^{-2}s^{-1}, which is $0.424 \pm 0.029 \pm 0.050$ times SSM$_{BP95}$.

Kamiokande showed that there is no evidence of time variations in accordance with the solar activity within its experimental sensitivity. They also see that there are no significant seasonal variations and daytime/nighttime flux difference (Fukuda et al., 1996). Note, however, that the results on the time variations are limited by the statistics and systematic uncertainties. A time variation smaller than the experimental sensitivities is not excluded. For example, we expect a semi-annual variation of about 6% due to the earth's eccentricity around the sun, which has not been positively identified by Kamiokande, but is expected to be observed by new experiments like Super-Kamiokande.

The SAGE (started in 1990) and GALLEX (started in 1991) experiments which both use Ga as a target are able to measure much of the lowest energy solar neutrinos. They observed similar results: the flux ratio to the SSM of BP95 is

$0.486^{+0.050}_{-0.52}{}^{+0.028}_{-0.029}$ and 0.57 ± 0.06 for SAGE (Abdurashitov et al., 1994) and GALL-EX (Hampel et al., 1996; Anselmann et al., 1994), respectively. Those experiments performed artificial neutrino source experiments using ^{51}Cr ($e^- + {}^{51}\text{Cr} \rightarrow {}^{51}\text{V} + \nu$) which provides mono-energetic neutrinos of 746 keV (81%), 751 keV (9%), 426 keV (9%) and 431 keV (1%). The GALLEX collaboration performed two such experiments between September and December, 1994 and between October and February, 1996. The first experiment using 1.69 ± 0.03 MCi of the ^{51}Cr sources gave the ratio of measurement to predicted to be 1.01 ± 0.10; the second using 1.86 ± 0.02 MCi showed the ratio to be 0.84 ± 0.11 (Hampel et al., 1998). The combined result is 0.93 ± 0.08. SAGE shows the ratio to be 0.95 ± 0.12. These experiments demonstrate that the overall efficiency estimates of the gallium experiments are correct within ~10%.

After this workshop two standard solar models are published (Bahcall et al., 1998; Turck-Chièze, 1998; Brun et al., 1998). Both of them include the results of recent extensive studies on the nuclear fusion cross sections (Adelberger et al., 1998). New models give a slightly lower flux value for the ^{8}B-neutrinos than BP95 due to the different choice of the p+^{7}Be cross section. The interpretation of the flux measurement by the new models will be discussed at the end of section 6. Although the efforts to improve the SSMs (especially to reduce the uncertainties on the nuclear cross sections) are in progress, the best way to resolve the solar neutrino problem is to derive a conclusion from a measurement independent of any solar model calculations, which will be described in section 7.

4. Implications of the Flux Measurements

Let us briefly summarize what we have learned so far from the flux measurements. Implications from the solar model independent studies will be discussed in later section.

First, these are the consequence of the solar neutrino flux measurement assuming that neutrino masses are zero, namely we assume that what we observe on the earth is what the sun has produced.

1) Neutrinos are really coming from the sun. This is proven by the directional measurement done by Kamiokande and Super-Kamiokande. We can see the image of the sun by means of neutrinos as shown in figure 3.

2) The results of the gallium experiments provided a direct evidence that the pp fusion process exists in the sun.

3) ^{8}B neutrinos exist. This was demonstrated by the Cl-experiment having a combined sensitivity for ^{7}Be and ^{8}B neutrinos, and was unambiguously shown by the Kamiokande measurement. This shows that the very end of the pp-chain is really taking place in the core of the sun, and the p+^{7}Be nuclear reaction process is active in the sun.

4) ^{7}Be-neutrinos are probably largely missing, which is implied by a simple comparison of the experimental results between Kamiokande and the Homestake experiment. It is also implied by the results of the gallium experiments with a minimum assumption of the luminosity constraint. (With help from the Kamiokande result the argument will become stronger.)

5) By a global comparison to SSM (this is a simple interpretation of the experimental results assuming no neutrino oscillation($m_\nu = 0$)), the following statements may be appropriate: (a) the pp-neutrino flux is almost ok, (b) ^{7}Be neutrinos are mostly missing, and (c) the ^{8}B neutrino flux is ∼40∼50%.

Possibilities to explain this implication by modifying the SSMs were discussed by many authors and as a general consequence, with a stronger support from the recent helioseismological observations, it is very difficult to explain the experimental results by changing the SSMs.

All the experimental results can be very naturally explained by assuming finite neutrino masses, namely the phenomena of neutrino oscillations.

The implications for neutrino oscillation parameters have been presented by many authors*. There are three parameter regions of neutrino oscillations which are able to consistently explain the flux measurements in the framework of neutrino oscillation: small mixing angle solutions and large mixing angle solutions for the MSW effect, and vacuum oscillations. For the case of the vacuum, $\Delta m^2 \gg E/L$ is not the case since the suppression for this would be constant and therefore $\Delta m^2 \simeq E/L$, called "Just So oscillations" (Glashow et al., 1987).

The flux suppression as a function of the neutrino energy of typical solutions in those parameter regions are shown in figure 1, together with the solar neutrino flux of the standard solar model (Bahcall and Pinsonneault, 1995). For the large angle solutions one expects a positive effect of the day/night flux difference, and for the small mixing angle one expects distortion for the ^{8}B neutrino spectrum as a function of energy. The ^{7}Be flux is very sensitive to the exact parameters of vacuum oscillations. In figures 6 and 7, we show the allowed parameter regions which obtained by the total flux measurements of each experiment taken from Hata and Langacker (1997).

5. New Generation Experiments

The three different 'allowed' parameter regions may be distinguished by future results from Super-Kamiokande, SNO, and Borexino in a model independent way. Once we have resolved the neutrino problem, we will be able to study the sun by neutrinos.

SNO will be able to measure neutral current, electron energy shape of the charged current interactions, day-night flux difference and time variations. A devi-

---

* See for example Hata and Langacker (1997), and also see hep-ph/9807216 for a most recent analysis including an updated result of Super-Kamiokande.

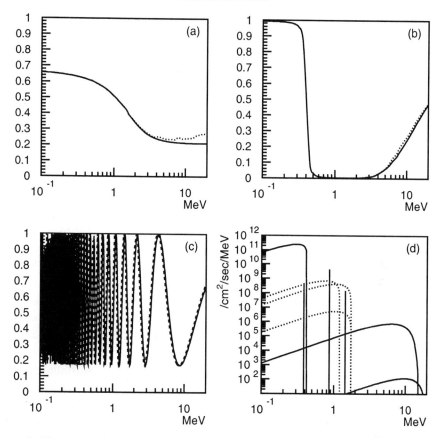

Figure 1. The suppression of the solar neutrino flux for a typical parameter of the three possible oscillation parameter regions: a) a large mixing angle solution, b) a small mixing angle solution and c) a Just So vacuum oscillation solution. In a) and b), the solid lines show the suppressions at daytime and the dotted lines show those at the nighttime. The regeneration effect through the earth is seen in a large mixing angle solution. The predicted solar neutrino spectrum is shown in d). The spectrum shown in the solid lines is for neutrinos from the pp-chain and that in the dotted line for neutrinos from the CNO cycle. See the text book (Bahcall, 1989) for more details.

ation from the expectation of the ratio of the charged current to the neutral current unambiguously show the existence of neutrino oscillations.

SNO is now in the final stage of the construction and hope to turn on in 1998. SNO uses 1,000 tons of D_2O as a target shielded by 7,000 tons of H_2O viewed by 9,600 8-inch PMTs. The solar neutrinos can be measured through; (1) Charged current: (inverse beta decay) $\nu_e + d \to p + p + e^-$ (11 events/day for SSM), (2) Neutral Current: (deuteron disintegration) $\nu_x + d \to p + n + \nu_x$ (3 ~ 9 events/day), and neutrons are detected either with $n+d \to t+\gamma$, $n+Cl \to Cl+\gamma$ or $n+^3He \to t+p$, and (3) Electron Scattering: $\nu_e + e \to \nu_e + e$ (1 event/day).

The liquid scintillator experiment, Borexino located at the Gran Sasso underground laboratory aims to achieve a 250 keV threshold to measure the flux of ^7Be

neutrinos through $\nu + e \to \nu + e$ scattering and could measure the day-night effect and also any other time variations. Borexino will have particular sensitivity to the vacuum oscillations since ^{7}Be neutrinos are monochromatic.

6. Super-Kamiokande

The neutrino oscillation studies in Super-Kamiokande (SK) will be performed not only with high statistics and better quality but with different approaches: model independent and flux calculation independent analyses are the key issue for the SK experiment.

Super-Kamiokande, a gigantic imaging water Cherenkov detectors located at 1000 m underground (2,700 m water equivalent), is the powerful tool to understand neutrino properties and study proton decay. The volume of SK is 50,000 tons (11,146 PMTs are arranged for the inner volume of 32,000 tons). SK is not only bigger than Kamiokande in its size, but also has better energy, position and directional resolutions: 14%, 87 cm, and 26° for 10 MeV electrons, respectively. 20 inch diameter PMTs are arranged on the inner surface of the detector with the density of 2 PMTs / 1 m^2 which covers the 40% of the inner surface. This arrangement compensates the light loss (due to attenuation) of the longer travel distance of the Cherenkov light than Kamiokande. The inner detector is surrounded by the outer detector of 2 m thick active water slab viewed by 1,185 20 cm PMTs. This provides a shield against external gamma rays and neutrons, and positively identifies incoming particles like through-going cosmic ray muons, particles exiting the detector and so on.

The relativistic charged particles traversing faster than the light velocity in the water emit Cherenkov lights with an angle of

$$\cos \theta = \frac{1}{\beta n},$$

where $\theta = 42°$ for particles with $\beta = 1$, and n is the refractive index of water. Photons in the ring of Cherenkov light are detected by the PMTs. The hit pattern, pulse height and timing information are used to reconstruct event vertex, direction, energy and particle species.

Super-Kamiokande (SK) started operation on the 1st of April in 1996 and has been continuously taking data with the cumulative live-time close to 90%. Most of the down time for taking data is spent for calibrations, especially using the electron LINAC (Nakahata *et al.*, 1998), which is used to determine the absolute energy scale of the detector.

The LINAC is placed near the SK tank. Its energy covers the range between 5 and 15 MeV, exactly matching the range of the energy spectrum of solar neutrinos.

The LINAC was also used to obtain the angular resolution and the vertex resolution. The beam energy of the LINAC was calibrated by a germanium detector

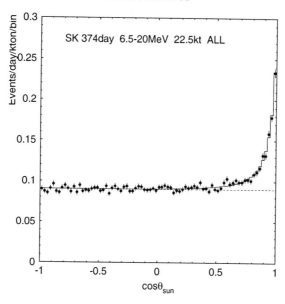

Figure 2. The $\cos\theta_{sun}$ distribution. The forward peak caused by the solar neutrino interaction is prominent.

which was calibrated by using the monochromatic electrons selected by a magnetic spectrometer. The energy scale error for the total flux measurement is estimated to be ±0.85%. The details of the calibration will be found in Nakahata et al. (1998).

The data presented here is based on 374.2 days of data taken between May 31, 1996 and Oct 20, 1997.

The expected solar neutrino event rate is 37 events per day for SSM_{BP95} while the number of event triggered per day is about one million. The most serious background is cosmic ray muons (∼3 Hz) and subsequent β and γ rays originating from the decay of their spallation products. Those muons can be removed easily by using the information of the visible energy in the inner volume and the outer detector. The spallation products were removed by using the distance and time correlation between the spallation products and the parent muons. There is also a significant amount of background entering from outside of the detector, mostly γ rays. Those backgrounds were removed by applying a fiducial volume cut (2 m from the surface of the inner PMTs). After selecting the data between 6.5 MeV and 20 MeV, we obtained 196 events/day as a final sample.

The $\cos\theta_{sun}$ distribution of the final sample is shown in figure 2 and the two dimensional plot of those events, in a coordinate system where the sun always sits at the center, is shown in figure 3: this figure may be called a Neutrino Heliograph. From figure 2 we obtained the number of solar neutrinos by using a maximum likelihood method. We assumed the shape of the ^8B neutrino spectrum of Bahcall (1996) and for the likelihood calculation the data in the $\cos\theta_{sun}$ distributions are

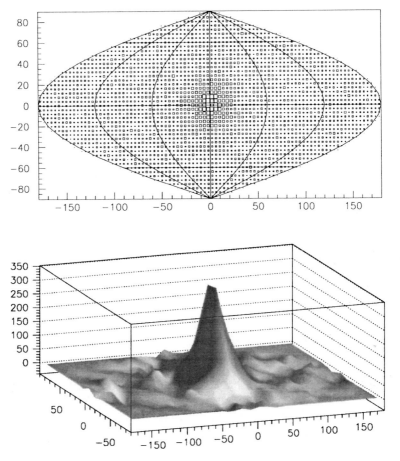

Figure 3. The image of the sun measured by the solar neutrino: Neutrino Heliograph. Upper and lower plots are different presentations of the same data.

further divided into energy bins. The number of events thus obtained is 4,951.8 events in 374.2 days between 6.5 and 20 MeV in 22.5 kton fiducial volume.

The solar ^{8}B neutrino flux is:

$$\phi_{^8B} = 2.37^{+0.06}_{-0.05}(\text{stat.})^{+0.09}_{-0.07}(\text{syst.}) \times 10^6 \text{ cm}^{-2}\text{s}^{-1},$$

Note that KM observed 597 events in 2079 days. The measured flux is $\phi_{^8B}$ = $2.80 \pm 0.19(\text{stat.}) \pm 0.33(\text{syst.}) \times 10^6 \text{ cm}^{-2}\text{s}^{-1}$. The result from SK agrees with that from KM within experimental errors.

The ratio to the prediction of the standard solar model of BP95 is,

$$\frac{\text{DATA}_{\text{SK}}}{\text{SSM}_{\text{BP95}}} = 0.358^{+0.009}_{-0.008}(\text{stat.})^{+0.014}_{-0.010}(\text{syst.}).$$

If we use the recent ^{8}B-flux values from BP98 ($5.15 \times 10^6 \text{ cm}^{-2}\text{s}^{-1}$ (Bahcall *et al.*, 1998)) and TC98 ($4.82 \times 10^6 \text{ cm}^{-2}\text{s}^{-1}$ (Turck-Chieze, 1998)), then the central

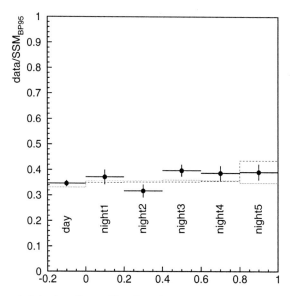

Figure 4. The day and night-time fluxes. The night data are divided into 5 bins with $\Delta \cos\theta = 0.2$. The expected day-night fluxes for $\Delta m^2 = 2.82 \times 10^{-5} \text{eV}^2$, $\sin^2 2\theta = 0.66$ (dashed line) and $\Delta m^2 = 6.31 \times 10^{-6} \text{eV}^2$, $\sin^2 2\theta = 9.12 \times 10^{-3}$ (dotted line) are shown. These are representative values for the large and small mixing angle solutions.

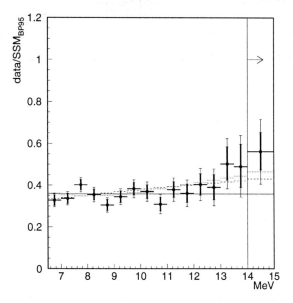

Figure 5. The spectrum of the recoil electrons. Note that this is the ratio of measured to expected. The thick parts of the error bars show the statistical errors and the extensions with the thin bars show the combined error of statistics and systematics. Also shown are a typical small angle solution ($\Delta m^2 = 6.31 \times 10^{-6} \text{eV}^2$ and $\sin^2 2\theta = 9.12 \times 10^{-3}$ (dotted line)) and a typical just so solution ($\Delta m^2 = 7.08 \times 10^{-11} \text{eV}^2$ and $\sin^2 2\theta = 0.83$ (dashed line)).

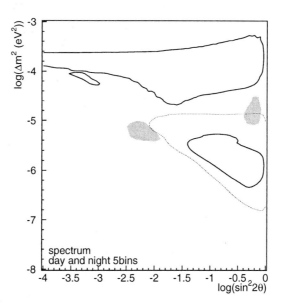

Figure 6. The confidence contour for the MSW effect by the SK data. The day/night flux difference and the electron energy distribution are used. No absolute flux information is used. The solid line shows the 95% excluded region by the energy spectrum and the dotted line shows the 95% excluded region by day/night data. inside the closed sections is excluded. The shaded regions are those obtained by Hata and Langacker (1997).

value of the ratio becomes, $DATA_{SK}/SSM_{BP98} = 0.46$ and $DATA_{SK}/SSM_{TC98} = 0.49$. There is still more than a factor of 2 discrepancy.

7. Implications from the Solar Model Independent Measurements

The daytime and nighttime flux difference has solar model independent information on the large angle MSW solutions. The day and night flux ratio obtained by SK is,

$$\frac{(\text{Day} - \text{Night})}{(\text{Day} + \text{Night})} = -0.031 \pm 0.024(\text{stat.}) \pm 0.014(\text{syst.}).$$

Combining the statistic and the systematic errors in quadrature, the deviation is 1.1σ. There are no significant day-night flux difference. The effect of the regeneration through the earth depends upon the path length and the density profile of the earth through which the solar neutrinos pass. We therefore divided the night-data ($\cos\theta_z < 0.0$, the nadir is the z-coordinate) into five bins ($\Delta\cos\theta_z = 0.2$). We show in figure 4 the fluxes of day time and nighttime divided into 5 bins. The typical day/night fluxes expected from a large angle solution ($\Delta m^2 = 2.82 \times 10^{-5} \text{eV}^2$,

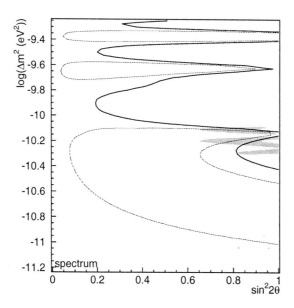

Figure 7. The confidence contour for the Just So oscillation parameter space, obtained by the electron energy distribution. The solid line shows the region excluded by 99% C.L. and the dotted line shows the allowed region at 95% C.L. The shaded regions are allowed regions obtained by Hata and Langacker (1997).

$\sin^2 2\theta = 0.66$) and a small angle solution ($\Delta m^2 = 6.31 \times 10^{-6} \text{eV}^2$, $\sin^2 2\theta = 9.12 \times 10^{-3}$) of MSW effect are also plotted in figure 4.

By using these data we can study the neutrino oscillation hypothesis. The confidence contour obtained is shown in figure 6 at 95% C.L. It is quite interesting that even with this 'small' data sample we can draw an excluded region where half of the large angle allowed regions obtained by Hata and Langacker (1997) are excluded.

The recoil electron energy spectrum is a key issue to untangle the solution of the solar neutrino problem, since the suppression of flux due to neutrino oscillation may have a strong energy dependence. It is sensitive especially to the small angle solutions and the Just So vacuum oscillations. In figure 5, the spectrum of the recoil electrons is shown. The current data indicate that the distribution is consistent with expectation, but fit better to a small angle or just so solutions (that is, a non-flat distribution), although statistically this is not significant. In figure 5, we also plotted the expected energy distortion for a typical small angle ($\Delta m^2 = 6.31 \times 10^{-6} \text{eV}^2$ and $\sin^2 2\theta = 9.12 \times 10^{-3}$) and a just so solution ($\Delta m^2 = 7.08 \times 10^{-11} \text{eV}^2$ and $\sin^2 2\theta = 0.83$).

From the energy spectrum, we define confidence contours for the neutrino oscillation parameters. For the MSW regions, we have obtained an excluded region with

95% C.L. which is shown as regions surrounded by the solid curves in figure 6. For the just so region we have obtained excluded region at 99% confidence level, but we get an allowed region at 95% C.L. as shown in figure 7. However, we should not take this too seriously at this time not only because the statistics is not sufficient and the indication is very marginal but also because the SK experiment will soon increase the statistics very rapidly at least for the next couple of years. We need to wait another year or so before the experiment can make a definite conclusion.

References

Abdurashitov, J.N. et al.: 1994, *Phys. Lett. B* **328**, 234.
Adelberger, E. et al.: 1998, astro-ph/9805121.
Akhmedov, E.Kh.: 1988, *Phys. Lett. B* **213**, 64.
Albrech, H. et al.: 1988, *Phys. Lett. B* **202**.
Albrech, H. et al.: 1992, *Phys. Lett. B* **292**.
Anselmann, P. et al.: 1994, *Phys. Lett. B* **327**, 377.
Arpesella, C. et al.: 1992, Borexino proposal, Vols. 1 and 2, eds. Bellini, G., Raghavan, R. et al., (Univ. of Milano, Milano).
Assamagan, K. et al.: 1994, *Phys. Lett. B* **335**, 231.
Assamagan, K. et al.: 1996, *Phys. Rev. D* **53**, 2065.
Bahcall, J.N.: 1989, *Neutrino Astrophysics*, Cambridge University Press.
Bahcall, J.N. and Pinsonneault, M.: 1995, *Rev. Mod. Phys.* **67**, 781.
Bahcall, J.N. et al.: 1996, *Phys. Rev. C* **54**, 411.
Bahcall, J.N., Basu, S., and Pinsonneault, M.H.: 1998, astro-ph/9805135.
Belesev, A.I. et al.: 1995, *Phys. Lett. B* **350**, 263.
Brun, A.S., Turck-Chièze, S., and Morel, P.: 1998, astro-ph/9806272, to appear in *ApJ* **506**, part 2.
Buskulic, D. et al.: 1995, *Phys. Lett. B* **359**, 585.
Chen, H.H.: 1986, *Phys. Rev. Lett.* **55**, 1534.
Cleveland, B.T. et al.: 1998, *ApJ* **496**, 505.
Ewan, G.T. et al.: 1987, *Sudbury Neutrino Observatory Proposal, – SNO* **87-12**.
Fukuda, Y. et al.: 1996, *Phys. Rev. Lett.* **77**, 1683.
Fukuda, Y. et al. (Super-Kamiokande Collaboration): 1998, *Phys. Rev. Lett.*, in press.
Glashow, L. et al.: 1987, *Phys. Lett. B* **190**, 199.
Hampel, W. et al.: 1996, *Phys. Lett. B* **388**, 364.
Hampel, W. et al.: 1998, *Phys. Lett. B* **420**, 114.
Hata, N. and Langacker, P.: 1997, IASSNS-ASR 97/29, UPR-751T, hep-ph/9705339, May.
Hirata, K.S. et al.: 1990, *Phys. Rev. Lett.* **65**, 1297.
Hirata, K.S. et al.: 1991a, *Phys. Rev. D* **44**, 2241.
Hirata, K.S. et al.: 1991b, *Phys. Rev. Lett.* **66**, 9.
Hirata, K.S. et al.: 1992, *Phys. Rev. D* **45**, 2170E.
Klapdor-Kleingrothaus, H.V.: 1996, in *Proceedings of the International Conference on Neutrino Physics and Astrophysics*, Helsinki, *Nucl. Physics. B (Proc. Suppl.)* **48**, 216.
Lim, C.S. and Marciano, W.J.: 1988, *Phys. Rev. D* **37**, 1368.
Mikheyev, S.P. and Smirnov, A.Y.: 1985, *Sov. Jour. Nucl. Phys.* **42**, 913.
Nakahata, M. et al.: 1998, *Nucl. Instr. Methods.*, in press.
Okun, L.B. et al.: 1986, *Sov. J. Nucl. Phys.* **44**, 440.
Turck-Chièze, S. and Lopes, I.: 1993, *Ap. J.* **408**, 347.
Turck-Chièze, S.: 1998, *Space Sci. Rev.*, this volume.
Weinheimer, Ch. et al.: 1993, *Phys. Lett. B* **300**, 210.
Wolfenstein, L.: 1978, *Phys. Rev. D* **17**, 2369.

LITHIUM DEPLETION IN THE SUN: A STUDY OF MIXING BASED ON HYDRODYNAMICAL SIMULATIONS

T. BLÖCKER
Max-Planck-Institut für Radioastronomie, Bonn, Germany

H. HOLWEGER and B. FREYTAG
Institut für Theoretische Physik und Astrophysik, Universität Kiel, Germany

F. HERWIG
Astrophysikalisches Institut Potsdam, Germany

H.-G. LUDWIG
Astronomical Observatory, Copenhagen, Denmark

M. STEFFEN
Institut für Theoretische Physik und Astrophysik, Universität Kiel, Germany
Astrophysikalisches Institut Potsdam, Germany

Abstract. Based on radiation hydrodynamics modeling of stellar convection zones, a diffusion scheme has been devised describing the downward penetration of convective motions beyond the Schwarzschild boundary (overshoot) into the radiative interior. This scheme of exponential diffusive overshoot has already been successfully applied to AGB stars. Here we present an application to the Sun in order to determine the time scale and depth extent of this additional mixing, i.e. diffusive overshoot at the base of the convective envelope. We calculated the associated destruction of lithium during the evolution towards and on the main-sequence. We found that the slow-mixing processes induced by the diffusive overshoot may lead to a substantial depletion of lithium during the Sun's main-sequence evolution.

1. Introduction

Since lithium is destroyed already at temperatures of $2.5 \cdot 10^6$ K by nuclear burning in stellar interiors, its surface abundances can be considerably affected by a sufficiently deep reaching surface convection zone. The solar Li problem is the long-standing conflict between the observed photospheric Li depletion of the Sun by 2.15 dex (Anders and Grevesse, 1989) and the predictions of stellar evolution models based on the standard mixing-length prescription. The latter show only moderate Li depletion during the pre main-sequence (PMS) phase (0.3–0.5 dex) whereas the depletion during the main-sequence evolution is negligible. In contrast, observations of open clusters indicate that effective Li depletion takes place on the main-sequence. Consequently, in order to account for the observations, at least one additional mixing mechanism must operate in the radiative regions below the bottom of the surface convection zone.

Suggested solutions include *mass loss* to expose depleted matter from the interior (e.g. Schramm *et al.*, 1990), *microscopic diffusion* leading to a leakage of Li out of the surface convection zone (e.g. Michaud, 1986), mixing due to *internal gravity waves* arising from pressure fluctuations in convective flows (e.g. Press, 1981), rotationally induced mixing by *meridional circulation*, and rotationally induced mixing due to *shear instabilities* associated with differential rotation (e.g. Zahn, 1992; or Pinsonneault *et al.*, 1992; Deliyannis and Pinsonneault, 1997 ("Yale" models)).

Among G and K stars in general, the current situation appears still controversial. For example, Stephens *et al.* (1997) conclude that the combined evidence of Li and Be in G and K stars rules out all but the "Yale" models. On the other hand, Martin and Claret (1996) criticize the angular momentum loss law adopted in these models, and find that rotation inhibits, rather than enforces, depletion.

For more details on the question of Li depletion see, e.g., Pinsonneault (1997) or Chaboyer (1998) and references therein.

2. Our Approach to the Solar Li Problem

Based on radiation hydrodynamics modeling of stellar convection zones, a diffusion scheme has been devised describing the mixing process due to the downward penetration of convective flows into the radiative interior (Freytag *et al.*, 1996).

Freytag *et al.* (1996) have investigated the interface between stellar surface convection zones and the radiative interior in favourable cases where modeling of the entire convection zone and adjacent overshoot region has been possible: white dwarfs and A-type main-sequence stars. Below the classical overshoot layers (characterized by a well defined (anti)correlation between velocity and temperature fluctuations), the numerical models show an extended region where the rms velocity fluctuations decrease exponentially with depth. The existence of this low-amplitude velocity field is in the end simply a consequence of the conservation of mass. Randomly modulated by deep-reaching plumes, the resulting flow gives rise to *diffusive mixing without significant temperature perturbations*. The nature of the underlying velocity field is fundamentally different from that of propagating gravity waves: while the latter represent oscillating motions with amplitudes increasing with depth, the former are the extension of closed convective flows decaying into the stable layers (for details see Freytag *et al.* 1996). Extended overshoot leads to slow mixing of a total mass that can exceed that of the convection zone proper by a large factor. It is much more efficient than microscopic diffusion, but otherwise similar. The corresponding depth-dependent diffusion coefficient can be derived from the hydrodynamical models and expressed in terms of an efficiency parameter f, the ratio of the scale heights of rms velocity and pressure.

Fig. 1 illustrates the corresponding hydrodynamical simulation of the shallow convection zones of an A-type star. Two convective plumes penetrate deeply into radiative regions: The Schwarzschild border of convective instability is located at an height of ≈ -6000 km whereas the plumes extend down to -8000 km, corresponding to about one pressure scale height. Convection carries up to 30 % of the total energy flux.

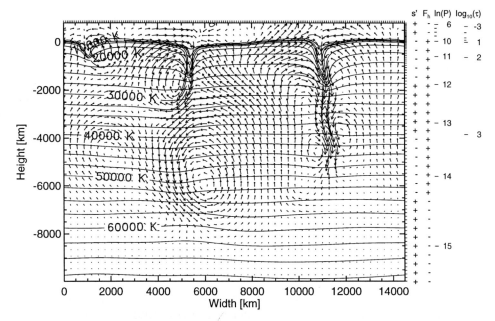

Figure 1. Snapshot of a hydrodynamical simulation for an A-type star with $T_{\text{eff}} = 7500$ K, $\log g = 4.4$ and solar metallicity, comprising the photosphere and two distinct convection zones: the upper one is due to combined HI/HeI ionization, the lower one is produced by HeII ionization. The velocity field is indicated by pseudo streamlines integrated over $t_{\text{int}} = 50$ s (maximum velocity $v_{\text{max}} = 18.9 \text{ km s}^{-1}$), contour lines indicate the temperature field in steps of 2500 K. Small tick marks at the top and the right show the positions of the computational grid points (182 × 95). Four columns at the right show some horizontally averaged quantities: the sign of the vertical entropy gradient, the sign of the vertical enthalpy flux, the natural logarithm of the mean pressure, and the logarithm of the optical depth. A negative entropy gradient defines a convectively unstable region. Note that the stable layer separating both convection zones (s':+) is completely mixed due to overshoot. Convection carries up to 30 % of the total energy flux.

In the case of the Sun we are not yet able to include the entire convective envelope in our simulation box and, thus, do not know f. However, we know its approximate value for surface convection of A-type stars, $f \approx 0.25$, and main-sequence core convection, $f = 0.02$, respectively (Freytag *et al.*, 1996; Herwig *et al.*, 1997). In the following, we will address the question of *whether this slow mixing, with a "reasonable" choice of f, can be considered a viable alternative to rotational mixing for explaining the depletion of Li during the main-sequence phase of the Sun.*

3. Application to Stellar Evolution Calculations

The scheme of exponential diffusive overshoot has already successfully been applied in stellar evolution calculations to core and deep envelope convection (Herwig *et al.*, 1997). Introducing one single efficiency parameter, f, it was possible to account for the observed width of the main-sequence as well as for important properties of AGB stars, namely efficient dredge-up processes to produce carbon stars at low luminosities as required by observations, and the formation of ^{13}C within the intershell region during the thermal pulses as the neutron source ($^{13}C(\alpha,n)^{16}O$) for the s-process. Since these calculations deal with the very deep stellar interior, f was found to be considerably smaller than for the shallow surface convection of A-type stars, namely 0.02, in accordance with the corresponding quasi-adiabatic conditions which allow only small growth rates for convective perturbations. Note, that *only* with the inclusion of additional mixing processes dredge up was obtained, and that sufficient amounts of ^{13}C are formed *only* due to *slow* mixing schemes. For more details, see Herwig *et al.* (1997) or Blöcker (1998).

Abundance changes due to nuclear burning (nuc) and mixing (mix) are calculated according to

$$\frac{dX_i}{dt} = \left(\frac{\partial X_i}{\partial t}\right)_{nuc} + \frac{\partial}{\partial m_r}\left[\left(4\pi r^2 \rho\right)^2 D \frac{\partial X_i}{\partial m_r}\right]_{mix} \quad (1)$$

with X_i being the mass fraction of the respective element, m_r the mass coordinate, and D the diffusion coefficient. Nuclear burning is treated with a detailed nucleosynthesis network. The choice of D depends on the mixing model. Within convectively unstable regions according to the Schwarzschild criterion we follow Langer *et al.* (1985) and adopt $D_{conv} = 1/3\, v_c l$ with l being the mixing length and v_c the average velocity of the convective elements according to the mixing length theory (Böhm-Vitense, 1958). The depth-dependent diffusion coefficient of the extended overshoot regions is given by Freytag *et al.* (1996):

$$D_{over} = D_0 \exp\frac{-2z}{H_v}, \quad H_v = f \cdot H_p, \quad (2)$$

where z denotes the distance from the edge of the convective zone ($z = |r_{edge} - r|$ with r: radius), and H_v is the velocity scale height of the overshooting convective elements at r_{edge}, given as a fraction of the pressure scale height H_p. Consequently, f expresses the efficiency of the mixing process. For D_0 we take the value of D_{conv} near the convective boundary r_{edge}. Note that D_0 is well defined because D_{conv} drops almost discontinuously at r_{edge}.

4. Model Calculations

Solar models were evolved from the birthline in the HRD through the pre-main-sequence (PMS) and main-sequence phase up to an age of 10 Gyr. The calculations are based on the code described by Blöcker (1995) and Herwig *et al.* (1997). We use the most recent opacities of Iglesias *et al.* (1992) and Iglesias and Rogers (1996) complemented with the low-temperature tables of Alexander and Ferguson (1994). With an initial composition of $(Y, Z) = (0.277, 0.02)$, we get a mixing

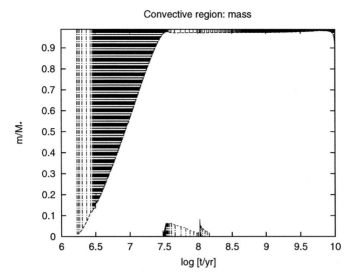

Figure 2. Extent (in mass units) of the solar convection zone as a function of time. Note the transition ($\log t \approx 7.5$) from deep-reaching PMS convection to shallow surface convection on the main-sequence.

length parameter of $\alpha = 1.66$ to fit solar radius and luminosity at $t = 4.6$ Gyr. At the solar age, the *depth* of the convection zone is 0.282 $R_\odot$, a value slightly lower than the currently adopted helioseismic value of 0.287 $R_\odot$.

Convection zones in PMS models are deep-reaching and massive. Fig. 2 illustrates that during the PMS evolution the depth of the convection zone changes rapidly in mass (compared to the time scales of the main-sequence evolution). We cannot assume that f is constant during this phase and that it has the same value as on the main-sequence. Thus, an initial ZAMS model was generated by evolving a PMS model with properly adjusted (mixing) parameters to fit the 'observed' Li depletion of 0.3 dex in accordance with results from young open clusters (Jones *et al.*, 1997).

The structural and nuclear evolution was calculated for a total of ten *main-sequence* $1M_\odot$ *models,* each with a fixed value of f ranging between $f = 0.02$ and 0.31. The dependence of structural properties of the models on f was found to be negligible. Apart from taking into account mixing, these are standard models and do not include any effects of rotation, microscopic diffusion, internal gravity waves, accretion, magnetic fields, or mass loss.

5. Results and Discussion

Fig. 2 shows the time evolution of the mass of the convection zone in a $1M_\odot$ model during the pre-main-sequence (PMS) and main-sequence evolution. Each vertical dash-dotted line corresponds to one stellar model (for most of the time the

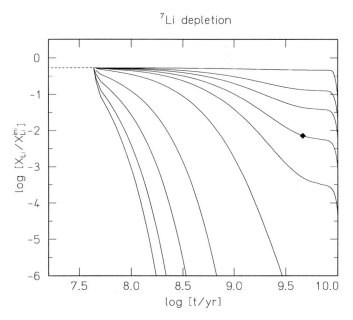

Figure 3. Lithium depletion vs. time. The dashed line refers to the (late) PMS phase, whereas the set of solid curves corresponds to different values of the mixing efficiency parameter, $f = 0.02$, 0.05, 0.06, 0.07, 0.08, 0.10, 0.15, 0.20, 0.26 and 0.31 (from top to bottom), applied during the main-sequence evolution. The diamond refers to the solar depletion of −2.15 dex.

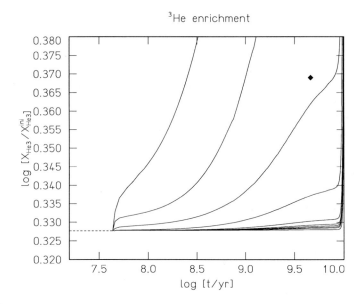

Figure 4. ^{3}He enrichment vs. time. The dashed line refers to the (late) PMS phase, whereas the set of solid curves corresponds to different values of the mixing efficiency parameter, $f = 0.02$, 0.05, 0.06, 0.07, 0.08, 0.10, 0.15, 0.20, 0.26 and 0.31 (from bottom to top), applied during the main-sequence evolution. The diamond indicates a 10% increase during the solar main-sequence evolution (see text).

models are so closely spaced that a continuous band appears). The deep-reaching convection of the PMS evolution corresponds to a track in the HRD starting at the Hayashi limit. There is a well-defined transition to the shallow, low-mass surface convection zone of the main-sequence star ($\log t \approx 7.5$).

Fig. 3 shows the predicted depletion of lithium during the main-sequence life of a $1M_\odot$ star (*solid lines*). The set of solid curves corresponds to different values of the mixing efficiency parameter, $f = 0.02, 0.05, 0.06, 0.07, 0.08, 0.10, 0.15, 0.20, 0.26$ and 0.31. (from top to bottom). The present solar value, -2.15, is represented by a diamond, and is close to the $f = 0.07$ curve. The PMS value has been adjusted to -0.3 dex according to observations of young open clusters (Sect. 4).

Fig. 4 illustrates the steady dredge-up of ^{3}He, an intermediate product of the p-p chain that has accumulated around $m \approx 0.6 M_\odot$ during the first few Gyr of main-sequence evolution. The different curves have the same meaning as in Fig. 3, but in this case f increases upwards. The ZAMS value of ^{3}He is the sum of primordial ^{3}He and D, the latter being converted into ^{3}He during D burning in the PMS phase.

Bochsler *et al.* (1990) have used solar-wind data to constrain the enrichment of photospheric ^{3}He from the ZAMS to the present. They derive an upper limit of 10 to 20 % and emphasize that this isotope is a sensitive tracer for mixing processes. The amount of mixing predicted by our $f = 0.07$ model (which reproduces the observed Li depletion) is less than 1 %, compatible with their upper limit. We notice that a 10 % increase of ^{3}He would imply a mixing efficiency of $f \approx 0.2$ (see Fig. 4) which, according to our model, leads to total destruction of Li.

6. Conclusions

We have shown that *slow mixing*, a diffusion process related to extended convective overshoot that operates in the almost radiative layers underneath a surface convection zone, may lead to substantial depletion of lithium during the main-sequence evolution of a $1M_\odot$ star. The mixing efficiency parameter required to reproduce the observed *depletion of lithium in the Sun*, $f \approx 0.07$, is intermediate between the parameter range of 0.25 ± 0.05 inferred by Freytag *et al.* (1996) from hydrodynamical models of the shallow surface convection of main-sequence A stars and the value of $f = 0.02$ derived empirically by Herwig *et al.* (1997) from stellar evolution calculations for core and deep envelope convection.

We believe that the existence of an exponentially decaying velocity field below a surface convection zone is a general feature of overshoot. Although the results of the simulations for A-type stars can certainly not be readily applied to the solar case, the basic situation seems not entirely different: as in A-type stars, a substantial fraction of the total flux is in fact carried by radiation in the lower part of the solar convection zone. In this study we have attributed the depletion of Li exclusively to extended overshoot. If other mixing processes should prove to contribute as well, the efficiency of overshoot will be smaller than derived here, i.e. $f < 0.07$.

However, in view of the success of the simple mixing model presented here we feel that the potential importance of slow mixing in the context of stellar convection deserves further study.

References

Alexander, D.R. and Ferguson, J.W.: 1994, 'Low-temperature Rosseland opacities', *ApJ* **437**, 879–891.
Anders, E. and Grevesse, N: 1989, 'Abundances of the elements: Meteoritic and solar', *Geochim. Cosmochim. Acta* **53**, 197–214.
Blöcker, T.: 1995, 'Stellar evolution of low and intermediate mass stars: I. Mass loss on the AGB and its consequences for stellar evolution', *A&A* **297**, 727–738.
Blöcker, T.: 1998, 'Theory of AGB evolution', *Space Science Reviews*, in press.
Bochsler, P, Geiss, J. and Maeder, A.: 1990, 'The abundance of He-3 in the solar wind – A constraint for models of solar evolution', *Solar Phys.* **128**, 203–215.
Böhm-Vitense, E.: 1958, 'Über die Wasserstoffkonvektionszone in Sternen verschiedener Effektivtemperaturen und Leuchtkräfte', *Z. f. Astrophys.* **46**, 108–143.
Chaboyer, B.: 1998, 'Internal rotation, mixing and lithium abundances', in IAU Symp. 185, *New Eyes to See Inside the Sun and Stars*, in press.
Delyannis, C.P. and Pinsonneault, M.H: 1997, '110 Herculis: A Possible Prototype for Simultaneous Lithium and Beryllium Depletion, and Implications for Stellar Interiors' *ApJ* **488**, 836–840.
Herwig, F., Blöcker, T., Schönberner, D. and El Eid, M.: 1997, 'Stellar evolution of low and intermediate mass stars: IV. Hydrodynamically-based overshoot and nucleosynthesis in AGB stars', *A&A* **324**, L81–L84.
Freytag, B., Ludwig, H.-G. and Steffen, M.: 1996, 'Hydrodynamical models of stellar convection. The role of overshoot in DA white dwarfs, A-type stars, and the Sun', *A&A* **313**, 497–516.
Iglesias, C.A. and Rogers, F.J.: 1996, 'Updated Opal Opacities', *ApJ* **464**, 943–953.
Iglesias, C.A., Rogers, F.J., and Wilson, B.G.: 1992, 'Spin-orbit interaction effects on the Rosseland mean opacity', *ApJ* **397**, 717–728.
Jones, B.F., Fischer, D., Shetrone, M., Soderblom, D.R.: 1997, 'The evolution of the lithium abundances of solar-type stars', *AJ* **114**, 352–362.
Langer, N., Fricke, K.J., and El Eid, M.: 1985, 'Evolution of massive stars with semiconvective diffusion' *A&A* **145**, 179–191.
Martin, C.L. and Claret, A.: 1996 'Stellar models with rotation: an exploratory application to pre-main sequence Lithium Depletion', *A&A* **306**, 408–416.
Michaud, G.: 1986, 'The lithium abundance gap in the Hyades F stars - The signature of diffusion', *ApJ* **302**, 650–655.
Pinsonneault, M.H.: 1997 'Mixing in stars', *ARA&A* **35**, 557–605.
Pinsonneault, M.H., Delyannis, C.P., Demarque, P.: 1992 'Evolutionary models of halo stars with rotation: II.Effects of metallicity on lithium depletion, and possible implications for the primordial lithium abundance' *ApJS* **78**, 179–203.
Press, W.H.: 1981, 'Radiative and other effects from internal waves in solar and stellar interiors', *ApJ* **245**, 286–303.
Stephens, A., Boesgaard, A.M., King, J.R. and Deliyannis, C.P.: 1997, 'Beryllium in Lithium-deficient F and G Stars', *ApJ* **491**, 339–358.
Schramm, D.N., Steigmann, G., Dearborn, D.S.P.: 1992, 'Main-sequence mass loss and the lithium dip' *ApJ* **359**, L55–L58.
Zahn, J.-P.: 1992, 'Circulation and turbulence in rotating stars', *A&A* **265**, 115–132.

ON THE VELOCITY AND INTENSITY ASYMMETRIES OF SOLAR P-MODE LINES

M. GABRIEL
Institut d'Astrophysique, 5 Av. de Cointe, B-4000 Liège, Belgium

Abstract. We show that to explain the opposite line asymmetries shown by the velocity and intensity spectra, it is necessary to solve the full non-adiabatic problem which is at least of the fourth order.

Key words: solar oscillations

1. Introduction

Duvall *et al.* (1993) have discovered that lines of the solar acoustic spectrum show asymmetries and that these asymmetries have opposite signs for velocity and intensity spectra. The first one shows negative asymmetries (i.e. more power on the low frequency side of the line) while the second has positive asymmetries. This finding has been recently confirmed by MDI which is one of the SOHO experiments.

We show that if line asymmetries are connected to the properties and location of the source, these observations can be explained only through the solution of the full non-adiabatic problem and that any simplifications of the problem leading to a second order problem is due to fail. We then discuss the causes of asymmetries in the context of the non-adiabatic problem.

2. The Formal Solution to the Line Profile

To compute line profiles, we have to solve the following system:

$$\frac{d\mathbf{Y}}{dr} = A\mathbf{Y} + \mathbf{F} \tag{1}$$

The homogeneous system gives the usual stability equations and $\mathbf{F}$ is the driving force.

The solution of this system is given by:

$$\mathbf{Y}(r) = \int_0^R G(r, r') \cdot \mathbf{F}(r') dr' \tag{2}$$

where $G(r, r')$ is the Green function matrix.

Let $2N$ be the order of the system. The homogeneous problem has N independent solutions verifying the boundary conditions at the center $\mathbf{Y}_1, \mathbf{Y}_2, \ldots \mathbf{Y}_N$ and N independent solutions verifying the boundary conditions at the surface $\mathbf{Y}_{N+1}$,

$Y_{N+2}, \ldots Y_{2N}$. The corresponding fundamental matrix is $M = (Y_1, Y_2, \ldots, Y_N, Y_{N+1}, \ldots, Y_{2N})$.

The Green function matrix is given by:

$$G_{ij}(r,r') = -\sum_{k=1}^{N} M_{ik}(r) M_{kj}^{-1}(r') \qquad r < r'$$

$$= \sum_{k=N+1}^{2N} M_{ik}(r) M_{kj}^{-1}(r') \qquad r > r' \qquad (3)$$

Equation (2) can also be written as

$$\mathbf{Y}(r) = -\sum_{k=1}^{N} \mathbf{Y}_k(r) \int_r^R \frac{|M_k(r')|}{|M(r')|} dr'$$

$$+ \sum_{k=N+1}^{2N} \mathbf{Y}_k(r) \int_0^r \frac{|M_k(r')|}{|M(r')|} dr' \qquad (4)$$

$$= \sum_{k=1}^{k=2N} C_k(r) \mathbf{Y}_k(r)$$

M_k is obtained by replacing the kth column of the fundamental matrix by $\mathbf{F}$.

The last line shows the equivalence of the Green function method and that of variation of arbitrary constants.

If the excitation is caused by convection, $\mathbf{F}$ is different from zero below the observation point r and the integrals of the first line in Eq. (4) are equal to zero.

If the component Y_i is observed, the corresponding power spectrum is given by:

$$|Y_i|^2 = \sum_{j=N+1}^{2N} \sum_{k=N+1}^{2N} C_j(r) C_k^*(r) Y_{ij}(r) Y_{ik}^*(r)$$

or

$$|Y_i|^2 = \int\int \{G(r,r') \cdot [F(r')F(r'')] \cdot G(r,r'')\}_{i,i} dr' dr''$$

The expression for the statistical average of the matrix $[F(r')F(r'')]$ has been discussed in Gabriel (1993).

Any kind of simplification leading to a second order problem leaves only one constant $C_k(r)$ and the asymmetry related to the source term is the same for all components indeed.

For a fourth order problem, two constants $C_k(r)$ are present and will lead to different line asymmetries for intensity and velocity lines. Indeed numerical calculations are required to see whether they have opposite signs or not.

3. The Causes of Line Asymmetries

To simplify the discussion, let us assume (the generalization is straightforward):
1) the Cowling approximation,
2) a point source.
Three terms in Eq. (4) have to be discussed:
   1. the influence of $\mathbf{Y}_3(r)$ and $\mathbf{Y}_4(r)$,
   2. the variation of $|M(r')|$,
   3. the variation of the $|M_k(r')|$.

1) Because observation are done at small acoustic depth, the solutions regular at the surface vary slowly with ν. For narrow lines they may be considered as constant but if line profiles are considered over a frequency range of the order of the eigenvalue separation, these two solutions will also introduce some skewness different for each spectrum.

2) The denominator cancels for each eigenvalues $\sigma = \sigma_R + i\sigma_I$. Close to one of them

$$|M(r')| = f(r', \omega)(\omega - \sigma_R - i\sigma_I) \tag{5}$$

with $f(r', \omega) \neq 0$ and for ω real, $|M(r')|^2$ has minima close to the real part of the eigenvalues. Therefore it is nearly a periodic function (with a "period" equal to the frequency separation between two successive line centers) but not exactly, for two reasons:
   - the real parts of the eigenvalues are not exactly equidistant.
   - the extremal values of $|M(r')|^2$ show variations with frequency, especially close and above the cut-off frequency. This behaviour is already seen in simple idealized problems (Gabriel, 1992).

Therefore, close to the real part of an eigenvalue, $|M(r')|^2$ is symmetric (and as shown by Eq. (5) it gives a Lorentz profile if $f(r', \omega)$ is independent of ω) but over a wider frequency range, it produces some skewness of the line profiles. However the asymmetries introduced by this term will be the same for the velocity and for the intensity spectra.

3) The numerators $|M_k(r')|$ are also quasi-periodic functions which discard form a strict periodicity for the same reasons as $|M(r')|$. Since one column of $M(r')$ is replaced by $F(r')$ to get $M_k(r')$, the "periods" of the determinants in the numerator and the denominator of Eq. (4) are slightly different and they will generally be out of phase. For problems of degree higher than two, there are at least two $M_k(r')$ with different phases relative to $M(r')$.

Combining the influence of points 2 and 3 leads to asymmetric profiles.

To get a mean profile, one can think to $F(r')$ as a slowly varying function of frequency which is close to the average value of a white noise. If we are interested in the line profiles of one realization of the spectrum, then $F(r')$ must be taken as one realization of a function close to a white noise. It has a very complex frequency dependence and this gives the nasty behaviour of the line profiles in individual spectra.

References

Duvall, T.L.,Jr., Jefferies, S.M., Harvey, J.W., Osaki, Y. and Pomerantz, M.A.: 1993, 'Asymmetries of solar oscillation line profiles', *ApJ* **410**, 829.
Gabriel, M.: 1992, 'On the solar p-mode spectrum excited by convection', *A&A* **265**, 771.
Gabriel, M.: 1993, 'On the location of the excitation of solar p-modes', *A&A* **274**, 935.

Address for correspondence: Institut d'Astrophysique, 5 Av. de Cointe, B-4000 Liège, Belgium
mgabriel@ulg.ac.be

SENSITIVITY OF LOW FREQUENCY OSCILLATIONS TO UPDATED SOLAR MODELS

J. PROVOST, G. BERTHOMIEU and P. MOREL
Département Cassini, UMR CNRS 6529, OCA
BP 229, 06304 Nice CEDEX 4, France

Abstract. A large number of acoustic frequencies have already been detected, leading to a "seismic" model of the Sun rather close to the actual standard solar models. The core however is not yet well constrained by these observations and frequencies of low degree, low frequency modes which penetrate deeply into the solar core are needed. We present here a study on the sensitivity of low degree low frequency (50 – 900 μHz) modes to the structure of the solar interior, in order to help their detection and identification in the low frequency spectrum observed by SoHO experiments like VIRGO and GOLF. The frequencies of p and g modes have been computed for a set of solar models with updated physics (Morel *et al.*, 1997). We analyze their sensitivity to solar parameters like age and metallicity, and to various physical processes, like convective core overshoot and mass loss during the beginning of solar evolution

1. On the Solar Models

We consider a set of solar models with updated physics and with some physical processes which are likely to be present in the Sun, like microscopic diffusion, penetrative convection and mass loss (Morel *et al.*, 1997 (hereafter MPB); Morel *et al.*, 1998), in order to study the sensitivity of low degree low frequency modes to the solar structure. The global characteristics of these models and some properties of their core are given in Table 1. The notations are the same as in MPB and all details and references concerning the physics of the models and the observations can be found there. All the models include the microscopic diffusion of the chemical elements, except model S2. Models D1 is the actual reference model with CEFF equation of state, model D4 presents a mass loss of $\dot{M} = -5 \times 10^{-10} M_\odot$ yr^{-1} while the total mass is greater than one solar mass. Model D7 includes some overshoot of the convective elements at the boundary of the convective core which appears in the first stages of solar evolution. We assume an overshoot parameter of 0.2 min(H_p, r_{cv}) where H_p is the pressure scale height and r_{cv} the radius of the boundary of the convective core. All these models have been calibrated at the solar luminosity for a solar age 4.55Gy and a metallicity $Z/X = 0.0245$ (Grevesse and Noels, 1993), except D8 ($t_\odot = 4.65$ Gy). D9* and D10* are computed with OPAL equation of state and respectively $Z/X = 0.0245$ and 0.0265. The observed values of neutrino fluxes and the location of the lower boundary of the convection zone derived by inversion of the observations are given in the last column.

The sound speed of these models has been compared to that of the solar seismic model obtained by inversion of the observed frequencies of p-modes in the five minute frequency range (MPB). The agreement is within some 10^{-3} for the models

Table I

Global characteristics of solar models: r_{zc} is the location of the lower boundary of the convection zone; T_c, ρ_c and Y_c are the central temperature, density and helium content; Φ_{Ga}, Φ_{Cl} and Φ_{Ka} are the predicted capture rates for the three neutrino experiments; P_0 is the characteristic period of low degree gravity modes (see equation (3)). The protosolar mass is $1M_\odot$, except for the mass loss model D4 ($M_p = 1.1 M_\odot$); D7 is a model with a core overshoot; D9* and D10* are computed with OPAL equation of state and respectively $Z/X = 0.0245$ and 0.0265, with some numerical refinements compared to MPB.

	S2	D1	D4	D7	D8	D9*	D10*	$\odot$
$r_{zc}/R_\odot$	0.726	0.710	0.713	0.710	0.708	0.709	0.708	0.713±0.003
$T_c(10^6 K)$	1.539	1.557	1.558	1.557	1.560	1.561	1.571	
ρ_c(g cm^{-3})	147.1	151.5	152.0	150.8	153.1	149.8	150.4	
Y_c	0.619	0.637	0.639	0.634	0.642	0.639	0.647	
Φ_{Ga}(SNU)	121	128	136	128	129	127	131	69±8
Φ_{Cl}(SNU)	6.30	7.72	8.25	7.74	7.93	7.21	7.96	2.55±0.23
Φ_{Ka}(ev day)$^{-1}$	0.48	0.61	0.66	0.61	0.63	0.55	0.61	0.29±0.02
P_0(mn)	36.48	35.73	35.63	35.97	35.43	35.73	35.57	

including microscopic diffusion. The surface, a thin layer just below the convection zone and particularly the solar core are the regions which differ mostly from the models. The oscillations in the low frequency range depend both of the profiles of the sound speed c and of the Brunt-Väisälä frequency N by the relations:

$$c \sim \sqrt{\frac{\Gamma_1 RT}{\mu}}, \quad N^2 \sim g\left(\frac{1}{\mu}\frac{d\mu}{dr} + \frac{\delta}{T}\left(\left(\frac{dT}{dr}\right)_{ad} - \frac{dT}{dr}\right)\right) \tag{1}$$

where g is the gravity, T the temperature, r the radius, μ the mean molecular weight, $\delta = -\left(\frac{\partial \ln \rho}{\partial \ln T}\right)_P$ and Γ_1 the adiabatic exponent. Figure 1 gives the contributions of the temperature and mean molecular weight gradients to the derivative of the logarithm of the sound velocity $d \log c/dr$ and the square of the Brunt-Väisälä frequency N^2. It is seen that the mean molecular weight gradient contributes very significantly in the solar core.

Figures 2a and 2b show the relative differences of c and N as a function of the radius between the models of Table 1 and model D1, considered as the reference model. Both $\delta c/c$ and $\delta N/N$ are due to changes of gradient of μ in the nuclear core ($r < 0.2$) and to changes of gradient of temperature outside. It is shown that $\delta N/N$ is larger than $\delta c/c$ by an order of magnitude and that the differences in $\delta N/N$ are all concentrated in the core, except for the model S2 (without diffusion).

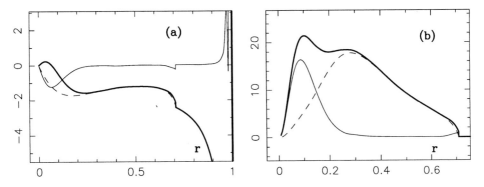

Figure 1. Variation of $d \log c/dr$ (a) and N^2 (b) as a function of the radius. The contributions of the temperature gradient and μ gradient are indicated respectively by dashed and thin full lines.

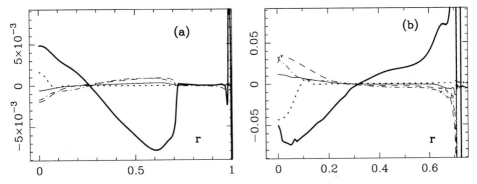

Figure 2. Relative difference of the sound speed c (a) and of the Brunt-Väisälä frequency N (b) between different solar models and the reference model D1. (S2 - D1 thick full (microscopic diffusion); D4 - D1 thin full (strong mass loss rate); D7 - D1 dotted (core overshoot); D8 - D1 dashed (age); D10* - D9* dotdash (metallicity).

2. Sensitivity of Low Frequency Modes to Solar Structure

The frequency differences between two models have been computed for gravity modes and pressure modes with degrees $\ell = 0$ to 3 and frequencies from 50 to 900 μHz. The general properties of solar oscillations can be found in Christensen-Dalsgaard and Berthomieu (1991). We have represented respectively the effect of microscopic diffusion, core overshoot, mass loss, metallicity and age in Figs. 3 and 6. In all these figures, three frequency domains may be distinguished. Above 400 μHz we have acoustic modes (p-modes) with frequencies depending principally on the sound velocity. In this frequency range, the frequency differences between two models are generally very small. Below 200 μHz, we are dealing with gravity modes (g-modes) which frequencies depend principally on N and which are already in the asymptotic range. In these two regions the differences are independent of the degree. In between are low radial order g-modes, f-modes (zero radial node) and p_1-modes (p-modes with one radial node).

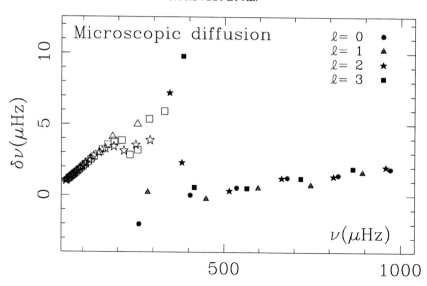

Figure 3. Difference of the frequencies of gravity modes (open symbols) and low frequency f and p-modes (full symbols) between models with (D1) and without (S2) microscopic diffusion, for modes of degree $\ell = 0, 1, 2$ and 3.

The behavior of these modes can be related to the properties of their eigenfunctions. The kinetic energy density of the modes of degree $\ell = 1$ in the intermediate frequency range 200 - 400 μHz is plotted in Figure 4. They appear to be very sensitive to the solar structure. The plots of the first line concern gravity modes which have their amplitude concentrated in the solar core ($r \leq 0.2 R_\odot$). In the two last graphs are represented two p-modes with amplitude in the solar envelop ($r \geq 0.75 R_\odot$). The mode p_1 has a mixed character, intermediate between p-and g-modes, with amplitude both in the envelope like a p-mode and in the center like a g-mode. The great sensitivity to the modifications introduced in the solar models of the p_1 mode is shown for $\ell = 1$ in Figure 5.

The frequency differences between two models vary almost linearly in g-modes range up to 200 μHz. This linear behavior at low frequencies can be explained in asymptotic approximation. For a frequency small enough, the first order asymptotic period is given by:

$$P_{n,\ell} = \frac{1}{\nu_{n,\ell}} \sim \frac{P_0}{\sqrt{\ell(\ell+1)}} (n + \ell/2 + \vartheta) , \qquad (2)$$

with $\qquad P_0 = 2\pi^2 / \int_0^{r_{zc}} (N/x) dx \qquad (3)$

ϑ is a phase factor sensitive to the properties below the convection zone (see Provost and Berthomieu, 1986, for more details). Thus the frequency of g-modes

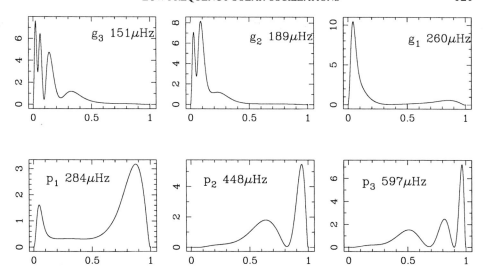

Figure 4. Kinetic energy density (i.e. normalized $\rho r^2 \delta \mathbf{r} \delta \mathbf{r}^*$ where $\delta \mathbf{r}$ is the displacement vector) as function of the radius for modes g_3, g_2, g_1, p_1, p_2 and p_3 of degree $\ell = 1$ computed for the reference model D1.

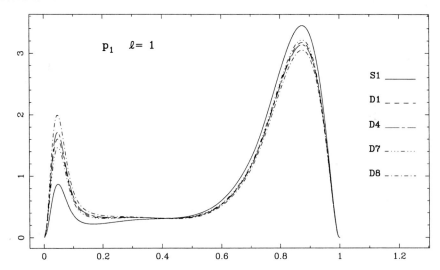

Figure 5. Kinetic energy density as function of the radius for the p_1 mode of degree $\ell = 1$ for different solar models.

is closely related to the Brunt-Väisälä frequency. Hence the relative frequency difference between models 1 and 2 is equal to:

$$\frac{(\nu_{n,\ell}^1 - \nu_{n,\ell}^2)}{\nu_{n,\ell}^1} \sim -\frac{(P_0^1 - P_0^2)}{P_0^1} . \quad (4)$$

The slopes measured on Figures 3 and 6 in the low frequency range correspond well to the relative differences of the characteristic periods P_0 of the gravity modes

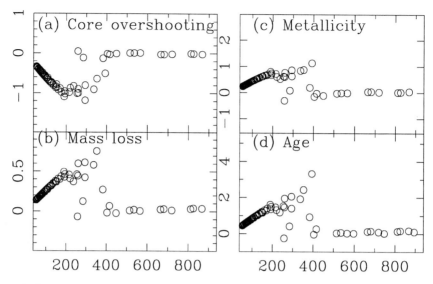

Figure 6. Frequency differences of low frequency modes between two solar models: **(a)** models with and without a convective core overshoot D7 - D1; **(b)** models with and without mass loss (D4 - D1); **(c)** : model $Z/X = 0.0265$ - model $Z/X = 0.0245$; **(d)** : model $t_\odot = 4.65$ Gy - model $t_\odot = 4.55$ Gy.

for the models (Table 1). Their signs are in agreement with the signs of the corresponding relative variations of $\delta N/N$ in the core plotted in Figure 2b.

The frequencies in the low frequency range are higher for a model with microscopic diffusion than for a standard one, with differences up to 10 μHz around 300 μHz and of order 2 μHz in p mode range (Fig. 3).

The g-mode frequencies are lower for models with convective core overshoot during the beginning of solar evolution. For a model with an overshoot parameter of $0.2 \min(H_p, r_{cv})$, the low frequencies are decreased by an amount up to 4 μHz, as seen in Figure 6a, while the p-mode ones except p_1 are almost not modified. We have studied the effect of a mass loss of 0.1 $M_\odot$ occurring during the first stage of evolution. The low frequencies are increased by an amount which depends on the mass loss rate. A strong mass loss rate (Fig. 6 b) results in a smaller change of the solar structure, hence of the solar oscillations frequencies. A mild mass loss rate not presented here would increase the frequencies up to 4 μHz but is not compatible with the p-modes properties (MPB).

The sensitivity of low frequency spectrum to some solar parameters, like the age and metallicity measured by the ratio of heavy elements to hydrogen contents Z/X at solar age is respectively represented in Fig. 6c and 6d. The frequencies increase with increasing metallicity Z/X and age. Work in progress by Grevesse and Sauval (1998) seems to indicate a smaller revised value of the solar metallicity.

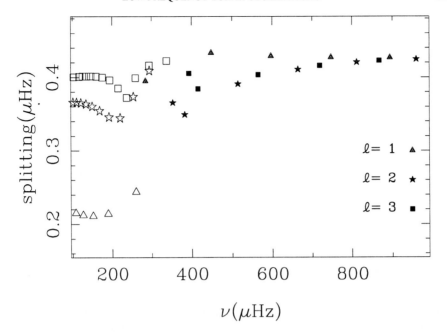

Figure 7. Splittings of low degree low frequency $\ell = m$ modes as a function of the frequency for the simplified rotation law described in the text with a central rotation of 0.433 μHz for the reference model D1. Open (full) symbols for g-(f- and p-) modes.

3. Rotational Splittings in Low Frequency Range

The solar rotation induces a splitting of the frequencies. To investigate the behavior of these splittings relatively to the frequency and the degree, we have used a simplified rotation law, as follows. For $r \geq 0.4$ we approach the rotation rate obtained by inversion from the p-mode data, i.e. rigid rotation below the convection zone and differential rotation inside (see for example Corbard *et al.*, 1997): for $0.7 < r < 1$, $\Omega = \Omega_0 + \Omega_1 \cos^2 \theta + \Omega_2 \cos^4 \theta$, with $\Omega_0 = 454$ nHz, $\Omega_1 = -54.6$ nHz and $\Omega_2 = -75.4$ nHz; for $0.4 < r < 0.7$, $\Omega = 433$ nHz. As the core solar rotation Ω_c remains more uncertain, we assume that it has a constant value equal to Ω_c. The results are given in Figure 7. They do not depend very much on the details of the rotation law. It can be shown for example that in the considered frequency range a rigid rotation of 0.433 μHz in the whole Sun will change the g-mode splittings by less than 0.5 nHz and the p-mode splittings by 2 to 7 nHz. At low frequency, the splittings of the modes of degree $\ell = 1$, 2 and 3 are very different according the degree ℓ with a behavior close to the asymptotic behavior of g modes, i.e. a splitting proportional to $1 - 1/\ell(\ell + 1)$ (see Berthomieu and Provost, 1991). At high frequencies the splittings have the asymptotic p-modes values, with a very weak dependence of the degree ℓ due to the differential rotation. In the frequency range 200 – 400 μHz the splittings depend on the degree and vary significantly with frequency. They are systematically larger for $\ell = 1$ p-modes.

4. Conclusion

In conclusion, the sensitivity of the low frequency modes to the physics and ingredients of the models we have considered is large below 400 μHz that is for gravity modes, f and p_1 modes principally. This shows the importance of these modes to constrain the solar model. However, they are difficult to detect. The properties we have studied here like the change of behavior between p-and g-modes splittings, specially for modes of degree $\ell = 1$ around 200 – 300 μHz have to be taken into account when searching to identify peaks in the low frequency spectrum.

References

Berthomieu, G., and Provost, J.: 1991, 'The asymptotic spectrum of gravity modes as a function of the solar structure: standard solar model', *Solar Phys.* **133**, 127–138.
Christensen-Dalsgaard, J. and Berthomieu, G.: 1991, 'Theory of solar oscillations', in *Solar interior and atmosphere*, eds. Cox A.N., Livingston W. C. and Matthews M., Space Science Series, University of Arizona Press, p. 401–478.
Corbard, Th., Berthomieu, G., Morel, P., Provost, J., Schou, J., and Tomczyk, S.: 1997, 'The solar rotation rate from LOWL data: A 2D regularized lest-squares inversion using B-splines', *Astron. Astrophys.* **324**, 298–310.
Grevesse, N., and Sauval, A.J.: 1998, 'Standard Solar Composition', *Space Sci. Rev.*, this volume.
Grevesse, N., and Noels, A.: 1993, 'Cosmic abundances of the elements', in *Origin and evolution of the elements* eds. Prantzos N., Vangioni-Flam E., and Cassé M., Cambridge University Press, p. 15.
Morel, P., Provost, J., and Berthomieu, G.: 1997, 'Updated Solar models', *Astron. Astrophys.* **327**, 349–360.
Morel, P., Provost, J., Berthomieu, G., and Audard, N.: 1998, 'Standard and Non Standard Solar models', in *Sounding Solar and Stellar Interiors* poster volume, eds. J. Provost and F.X. Schmider, OCA & UNSA, p. 109–110.
Provost, J., and Berthomieu, G.: 1986, 'Asymptotic properties of low degree solar gravity modes', *Astron. Astrophys.* **165**, 218–226.

COMPOSITION AND OPACITY IN THE SOLAR INTERIOR

SYLVAINE TURCK-CHIEZE
CEA/DSM/DAPNIA/Service d'Astrophysique, CE Saclay, 91191 Gif-sur-Yvette Cedex 01, France

Abstract. Detailed abundances of elements from hydrogen up to iron are necessary to perform a precise model of the solar structure. Most of them have been deduced from photospheric observed values, some others from the meteoritic composition. Nowadays, thanks to helioseismic constraints, they seem more and more under control.

1. Introduction

Hydrogen, helium and heavy element abundances are fundamental ingredients of stellar evolution for nuclear burning and radiative transport of energy. It is the reason why the knowledge of these ingredients has been a fundamental part of the solar study. The compilation of Grevesse and Noels on photospheric composition has resolved some inconsistencies between photospheric and meteoritic abundances, in particular regarding iron and gives better CNO determination (Grevesse and Noels, 1993; Grevesse and Sauval, 1998). This improvement has been complemented by some other progresses on the opacity coefficient calculations including nowadays 21 elements (Iglesias and Rogers, 1996; Rogers and Iglesias, 1998) and on the reaction rate determinations (Adelberger *et al.*, 1998). In parallel, the precise estimate of the position of the base of the convection zone and of the photospheric helium from helioseismology have pushed the modellers to include the microscopic diffusion of the elements.

Consequently, one rejects today the idea that the initial solar composition is strictly equal to the present photospheric one.

2. The Standard Solar Model and the Helium Content

All these improvements are included in our updated solar model (Brun, Turck-Chièze and Morel, 1998) and the results are summarized in figure 1a and the first column of table II. A comparison with table I, which summarizes the specific solar observations (see references in our paper), shows that our model fits reasonably well the helioseismic data for the base of the convective zone, the photospheric helium, the sound speed square difference between the Sun and the model, $\delta c^2/c^2 < 1\%$. This agreement will be even better by including turbulent mixing at the base of the convective zone to simulate the effect of the tachocline instability (Brun, Turck-Chièze and Zahn, 1999). The neutrino fluxes: 7.18 SNU for the chlorine experiment, 127.2 SNU for the gallium detector and 4.82×10^6 cm^{-2} s^{-1} for

Table I

Helioseismic Observations, Solar Neutrino Detections

$Y_{surf} = 0.249 \pm 0.003$
$(Z/X)_{surf} = 0.0245 \times (1 \pm 0.1)$
$R_{bcz}/R_{\odot} = 0.713 \pm 0.003$

$^{71}Ga = 76 \pm 8$ SNU (cal = 0.91 ± 0.08) for GALLEX
$^{71}Ga = 70 \pm 8$ SNU (cal = 0.95 ± 0.12) for SAGE
$^{37}Cl = 2.55 \pm 0.25$ SNU for Homestake
$^{8}B = 2.7 \pm 0.1 \times 10^{6}$ cm^{-2} s^{-1} for Kamiokande
$^{8}B = 2.44 \pm 0.26 \times 10^{6}$ cm^{-2} s^{-1} for Super Kamiokande

the ^{8}B neutrino flux, are slightly reduced in comparison with previous calculations (Bahcall and Pinsonneault, 1995) resulting from a partial compensating effect of the increase due to the microscopic diffusion and the reduction due to the updated reaction rates. However, the neutrino fluxes remain systematically higher than the observations.

In order to improve the theoretical predictions with the present helioseismic constraints obtained from ground networks and seismic measurements aboard SOHO, two other standard models have been calculated respecting the uncertainties of the physical ingredients: the model "Opac. inc." corresponds to a 1.5% increase of the opacity coefficients in the central part up to 4.5% at the base of the convective zone (see also next paragraph), the model "Min. nuc." introduces a 1 σ variation of the cross sections in order to reduce the neutrino fluxes. The results are illustrated by figure 2 and column 2 and 3 of table II.

We conclude that the standard model, inside its inherent uncertainties, is robust against the present acoustic mode detection except at the base of the convection zone, but neutrino fluxes are not yet well constrained: a 30% decrease is supported by the present knowledge of the nuclear cross sections.

The initial solar helium is one of the outputs of a solar model. It can be used to estimate the enrichment of helium in our galaxy since the primordial nucleosynthesis era. Some years ago, the initial helium was still poorly determined and values from 0.25 up to 0.29 were found. Such dispersion had two origins: the influence of the composition in heavy elements through the opacity coefficients (see Turck-Chièze and Lopes, 1993) and the coulomb term in the central equation of state, which was often absent in solar models (Christensen-Dalsgaard and Däppen, 1992). Nowadays, we note that the standard models which include the microscopic diffusion of the elements and a correct equation of state reasonably reproduce the low photospheric value and predict an initial helium of 0.277 ± 0.04 in respecting the sound speed profile suggested by seismic measurements.

COMPOSITION AND OPACITY IN THE SOLAR INTERIOR 127

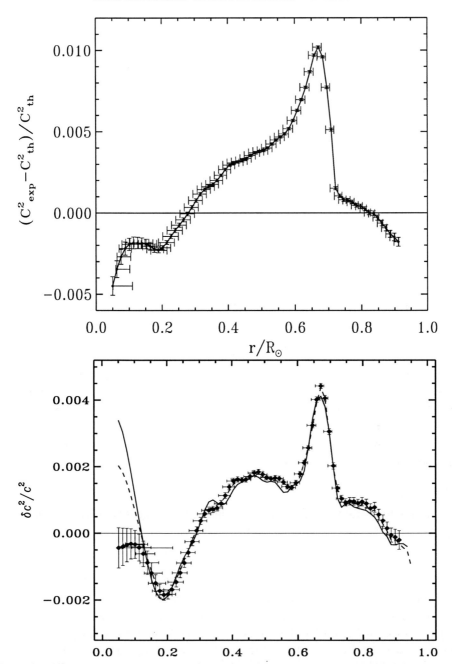

Figure 1. Difference between the square of the sound speed in the sun measured by GOLF+LOWL experiments (Lazrek *et al.*, 1997; Tomczyk *et al.*, 1995) and models. The upper figure corresponds to our model from Brun, Turck-Chièze and Morel (1998) the lower one to the model S of Christensen-Dalsgaard *et al.* (1996). For comparison, the solid line shows the results obtained by Basu *et al.* (1997) from analysis of a combination of BiSON and LOWL data, while the dashed line shows the results obtained with just 4 months of GOLF data, combined with LOWL data (from Turck-Chièze *et al.*, 1997).

Table II

Thermodynamical Quantities and Neutrino Predictions: α: mixing length parameter, Y_i, Z_i, $(Z/X)_i$: initial helium, initial heavy element and initial ratio of heavy elements over hydrogen in mass fraction, Y_S, Z_S, $(Z/X)_S$: idem for photospheric compositions, R_{bcz}, T_{bcz} are the radius and temperature at the base of the convective zone, Y_c, Z_c, T_c, ρ_c: central helium, heavy element contents, temperature and density; ^{37}Cl, ^{71}Ga, ^{8}B respective neutrino predictions for the chlorine, gallium and water detectors.

Parameters	Standard model	Opac. inc.	Min. nuc.
α	1.84	1.84	1.85
Y_i	0.273	0.277	0.273
Z_i	1.96×10^{-2}	1.95×10^{-2}	1.96×10^{-2}
$(Z/X)_i$	0.0277	0.0277	0.0283
Y_s	0.243	0.248	0.244
Z_s	1.810×10^{-2}	1.799×10^{-2}	1.809×10^{-2}
$(Z/X)_s$	0.0245	0.0245	0.0245
$R_{bcz}/R_\odot$	0.715	0.712	0.715
$T_{bcz} \times 10^6$ K	2.172	2.208	2.178
Y_c	0.635	0.640	0.631
Z_c	0.0208	0.0207	0.0208
$T_c \times 10^6$ K	15.67	15.72	15.61
ρ_c (g/cm^3)	151.85	152.63	150.64
^{37}Cl (SNU)	7.18	7.55	5.19
^{71}Ga (SNU)	127.2	129.0	119.0
^{8}B (10^6 cm^{-2} s^{-1})	4.82	5.10	3.21

3. The Determination of the Heavy Elements Composition Inside the Sun: a Dream or a Reality?

The present accuracy on seismic acoustic modes puts trust on a precise determination of the heavy element abundances. Three points may illustrate that:

– In figure 1, we present the difference between the squared solar sound speed and two theoretical results. The first model is our updated Saclay solar model (1998), the second one is the model S of Christensen-Dalsgaard et al. (1996). The main differences are attributed to the recent updated physics. Our previous works on the sound speed sensitivity to the nuclear reaction rates (Turck-Chièze and Lopes, 1993; Dzitko et al., 1995; Turck-Chièze et al., 1997) show that the effect of nuclear reaction rate modification is small except in the solar nuclear core; the sound speed increases by 0.1–0.2 % in the intermediate region between the core and the convective zone; so one can deduce that the main differences in

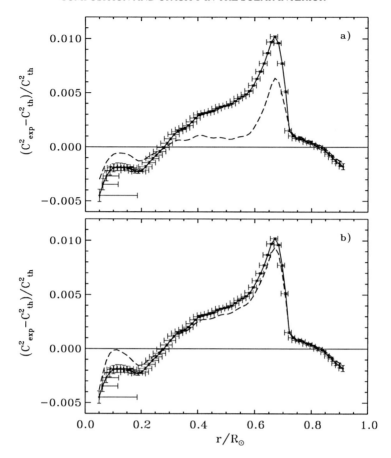

Figure 2. Sound speed square difference between the Sun seen by GOLF +LOWL (see Turck-Chièze et al. 97) and models. Dashed line: a) a model with an increase of opacity between 1.5% in the central region up to 5% at the bottom of the convective zone, b) a model taking into account the uncertainty on the nuclear reaction rates to define a "minimal nuclear" model.

this region arises from the update in the opacity coefficients and, more precisely, from the equation of state coupled with these calculations.

– The second model of table II seems to reproduce the solar profile (figure 2a) with an accuracy better than 0.1–0.2 %, except at the base of the convective zone. The variation of the opacity by 1.5 to 4.5 % has been chosen in order to mimic an uncertainty of the heavy element composition. Actually the present abundance uncertainties on the main heavy element contributors are still large ^{12}C: 12%, ^{14}N: 17%, ^{16}O: 17%, ^{20}Ne: 15%, ^{56}Fe: 9.5%. Figure 3 (from Courtaud *et al.*, 1990) recalls the radial dependence of the heavy element component in the opacity coefficients, following this radial dependence, mainly due to CNO and iron contributors. We used this shape to introduce the opacity variation simulating an increase of CNO elements by about 15%. This estimate has been confirmed by F. Rogers in his talk (Rogers and Iglesias, 1998). So one possible interpretation of the seismic

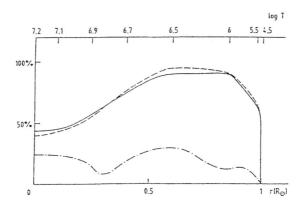

Figure 3. Radial contribution of the heavy elements to the total opacity (- . - for iron one.)

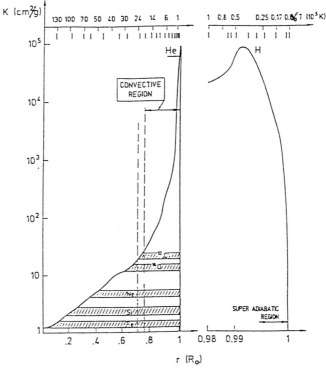

Figure 4. Radial dependence of the first bound bound heavy element contributions to the opacity coefficients (Turck-Chièze et al., 1993)

data could be that the central CNO abundances are underestimated by 15%. Mixing must be also questioned before giving final conclusions.

– Finally, we observe several small bumps, superimposed on the general broad deviation in the squared sound speed differences. If we can demonstrate that they

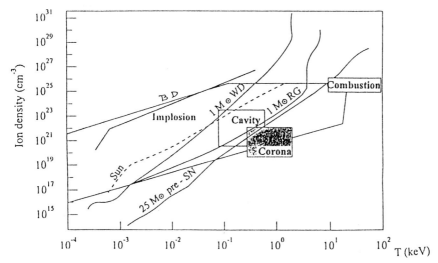

Figure 5. Large laser thermodynamical conditions compared to stellar cases (Chièze et al., 1998).

are not produced by the acoustic data or by the procedure of inversion which allows to determine the "solar sound speed", we will look for a physical process which may produce them, why not thinking to the heavy element partial ionisation? It is well known that bound-bound processes do not lead to continuous opacity, depending on the position of the lines in the vicinity of the maximum of the Rosseland coefficient ($h\nu/kT$ around 4). This remark has been pointed out by Turck-Chièze et al. (1993) and summarized in the illustrative figure 4, which shows where the bound-bound processes of the different heavy elements start in the solar interior. So one can imagine a reasonable sensitivity of the sound speed to these opacity coefficients. Therefore we cannot exclude that seismic data give us access to the iron abundance near the center or 0.2 $R_\odot$, the ^{3}He abundance at the edge of the nuclear core at 0.27 $R_\odot$, the silicon around 0.4 $R_\odot$, the iron again around 0.6 $R_\odot$, or the oxygen near the base of the convective zone. A partial answer to this question may be found in Rogers and Iglesias (1998) or Turcotte and Christensen-Dalsgaard (1998, figure 4) by the estimate of the opacity variations for 10 or 15% individual element variations. A 1% effect in the total opacity will lead to a 0.1% effect on the sound speed profile, which must be observable.

This could be a new constraint coming from the helioseismology, in complement to the photospheric helium determination.

In this difficult problem, the bound-bound opacity calculation must be tested. A confirmation of the Livermore calculations by laser experiments (Da Silva et al., 1992) at low density and temperature has been useful. So, we are studying the opportunity to measure opacities of individual elements or of a mixture near solar conditions on the large laser facilities in construction (LMJ in France and NIF in USA). They must be able to reproduce the thermodynamical conditions of stellar plasmas in local thermodynamic equilibrium (figure 5) and could help the interpretation of seismic measurements, as far as we have no direct access to the stellar temperature in seismology (Chièze et al., 1998).

I would like to thank S. Basu, S. Brun, J. Christensen-Dalsgaard and P. Morel with whom the most recent results have been obtained.

References

Basu, S., Chaplin, W. J., Christensen-Dalsgaard, J., Elsworth, Y., Isaak, G. R., New, R., Schou, J., Thompson, M. J., and Tomczyk, S.: 1997, 'Solar internal sound speed as inferred from combined BiSon and LOWL oscillation frequencies', *MNRAS* **292**, 243–251.
Bahcall, J., Pinsonneault, M. H.: 1995, 'Solar models with helium and heavy-element diffusion', *Rev. Mod. Phys.* **67**, pp 781-808.
Brun, S., Turck-Chièze, S. and Morel, P.: 1998, 'Standard solar model in the light of new helioseismic constraints. I The solar core', accepted by *ApJ*, vol **506**.
Brun, S., Turck-Chièze, S. and Zahn, J.P.: 1999, 'II The transition radiation-convection', to appear in *ApJ*.
Chièze, J.P., Arnaud, M., Teyssier, R., Turck-Chièze, S., Bouquet, S.: 1998, 'Plasmas Astrophysiques', *Rapport interne CEA/DSM*.
Christensen-Dalsgaard et al.: 1996, 'The current state of solar modeling', *Science* **272**, 1286–1292.
Christensen-Dalsgaard, J., Däppen, W.: 1992, 'Solar oscillations and equation of state', *Astron. Astrophs. Rev.* **4**, 267–361.
Courtaud, D., Dammame, G., Genot, E., Vuillemin, M., and Turck-Chièze, S.: 1990, 'Metallicity, opacity coefficients and the standard solar model', *Sol. Phys.* **128**, 49–60.
Da Silva et al.: 1992, 'Absorption measurements demonstrating the importance of $\Delta n = 0$ transitions in the opacity of iron', *Phys.Rev. Lett.* **64**, 438.
Dzitko, H., Turck-Chièze, S., Delbourgo-Salvador, P., and Lagrange, G.: 1995, 'The screened nuclear reaction rates and the solar neutrino puzzle', *ApJ* **447**, 428–442.
Grevesse, N. and Noels, A.: 1993, 'Cosmic abundances of the elements', *in Origin and Evolution of the Elements*, ed. N. Prantzos et al., Cambridge University Press, Cambridge, 15–25.
Grevesse, N. and Sauval, A. J.: 1998, 'Standard Solar Composition', *Space Sci. Rev.*, this volume.
Iglesias, C. A. and Rogers, F. J.: 1996, 'Updated OPAL opacities', *ApJ* **464**, 943.
Lazrek, M., Baudin, F., Bertello, L., Boumier, P., Charra, J., Fierry-Fraillon, D., Fossat, E., Gabriel, A. H., Garcia, R. A., Gelly, B., Gouiffes, C., Grec, G., Pérez Hernández, F., Régulo, C., Renaud, C., Robillot, J. M., Roca Cortés, T., Turck-Chièze, S. and Ulrich, R. K.: 1997, 'First results on p modes from GOLF experiment', *Sol. Phys.* **175**, 227–246.
Rogers, F. J. and Iglesias, C. A.: 1998, 'Opacity of Stellar Matter', *Space Sci. Rev.*, this volume.
Tomczyk, S., Streander, K., Card, G., Elmore, D., Hull, H., and Cacciani, A.: 1995, 'An instrument to observe low-degree solar oscillations', *Sol. Phys.* **159**, 1–21.
Turck-Chièze, S., Basu, S., Brun, A. S., Christensen-Dalsgaard, J., Eff-Darwich, A., Lopes, I., Pérez Hernández, F., Berthomieu, G., Provost, J., Ulrich, R. K., Baudin, F., Boumier, P., Charra, J., Gabriel, A. H., Garcia, R. A., Grec, G., Renaud, C., Robillot, and J. M., Roca Cortés, T.: 1997, 'First view of the solar core from GOLF acoustic modes', *Sol. Phys.* **175**, 247–265.
Turck-Chièze, S., and Brun, A. S.: 1997, 'Spatial seismic constraints on solar neutrino predictions', in *Proceedings of the fourth international solar neutrino conference*, ed. W. Hampel, MPI für Kernphysik, Heidelberg, 41–53.
Turck-Chièze, S., Däppen, W., Fossat, E., Provost, J., Schatzman, E. and Vignaud, D.: 1993, 'The solar interior', *Phys. Report.* **230 (2-4)**, 57–235.
Turck-Chièze, S. and Lopes, I.: 1993, 'Towards a unified classical model of the Sun: on the sensitivity of neutrinos and helioseismology to the microscopic physics', *ApJ* **408**, 347–367.
Turcotte, S. and Christensen-Dalsgaard, J.: 1998, 'Solar models with non-standard chemical composition', *Space Sci. Rev.*, this volume.

SOLAR MODELS WITH NON–STANDARD CHEMICAL COMPOSITION

S. TURCOTTE and J. CHRISTENSEN-DALSGAARD
Teoretisk Astrofysik Center, Danmarks Grundforskningsfond
Institut for Fysik og Astronomi, Aarhus Universitet, DK 8000, Aarhus C, Denmark

Abstract. The OPAL monochromatic opacity tables are used to evaluate the impact of a non-standard chemical composition on solar models.

A calibrated solar model with consistent diffusion including the effect of radiative forces and ionization on drift velocities is presented. It is shown that surface abundances are predicted to change slightly more than in traditional solar models where these refinements are not included. All elements included in the model settle at similar rates which is reflected in the relative variation in surface abundances ranging from 7.5% for calcium to 8.8% for argon. The structural difference between the consistent model and the traditional model is small, with a maximum effect of 0.3% for the isothermal sound speed at the base of the convection zone. The settling of CNO is only marginally affected.

Opacity profiles have also been calculated with varying abundances for volatile elements, for which the abundances are poorly known, and other selected elements. It is shown that if one allows a 10% variation of these elements individually one can expect a peak Rosseland mean opacity variation of 3% for oxygen, a little less 2% for Si and Ne, and around 1% for Mg and S in the radiative zone. Other light metals and volatile elements have no significant impact on the opacity.

Key words: Solar structure; Diffusion; Opacities

1. Introduction

In many solar models [Bahcall and Pinsonneault (1995); Christensen-Dalsgaard *et al.* (1996); and Morel *et al.* (1997) amongst others] the settling of heavy elements is treated very simply.

It is generally assumed that heavy elements are fully ionized, that they all settle at a common rate, and that the net outward radiative force is always negligible. In addition, thermal diffusion is often only coarsely taken into account, for example when using analytical expressions for diffusion velocities from either Proffitt and Michaud (1991) or Thoul *et al.* (1993). The errors caused by these approximations have been estimated to be small for solar models. However, the small but significant differences between the structure predicted by these models and the 'seismic' model (Basu *et al.*, 1996) require that all of their assumptions be tested. Some more recent models have begun to address some of those concerns. In particular, Richard *et al.* (1996) followed many elements and isotopes up to magnesium individually but keeping the assumption of full ionization. In the following we address mostly the cruder but more common models in which all assumptions are made; we label such models 'standard solar models' (SSMs).

The treatment of the settling of heavy elements in the SSMs ensures that the metallic composition remains constant outside of the nuclear burning region. In reality, individual elements settle at different rates because of differences in their profiles of ionization and radiative forces. The implications of the resulting departures from the solar mixture of heavy elements (Grevesse and Noels, 1993) can only be evaluated if the opacity can be reliably computed for an arbitrary composition, which in turn necessitates the use of monochromatic opacity tables for individual elements. The first such calculations have been done by Turcotte *et al.* (1998b; hereafter TRMIR98) using the OPAL monochromatic opacity tables [Iglesias and Rogers (1996); see also Richer *et al.* (1998) for a discussion of the opacity data and radiative forces].

In the following we revisit some of the results of TRMIR98 focusing on the evolution of abundances during the Sun's lifetime. Some additional calculations have been performed in order to understand what can be the implications of a revision of the abundance of the elements contributing the most to the opacity in the solar radiative region on the comparison of the predicted and 'seismic' structures.

2. Abundance Variations in the Sun

The abundance variation due to diffusion of a given element at a given point is mainly determined by the relative size of the opposite forces of gravity and radiation pressure. In the Sun, the luminosity is low enough and the mean ionization beneath the convection zone is high enough that the outward radiation pressure is smaller than gravity for all the elements at least up to Ni. Consequently, all the elements but hydrogen drift towards the center.

Figure 1 illustrates some results for the diffusion of He, O, Ti and Fe in a model with detailed diffusion (model H of TRMIR98). These elements are chosen because He, O and Fe are amongst the more important elements for the opacity in the solar interior, and Ti is the element most affected by radiative forces.

The left panel shows the ratio of radiative (g_{rad}) over gravitational (g) accelerations for these elements. The radiative acceleration on He is assumed to be zero and is indeed negligible under solar conditions. The only elements for which radiative forces are a significant fraction of gravity are those which still retain some electrons and whose spectra still feature a large number of spectral lines, which is true of the iron-peak elements in particular. Lighter elements such as oxygen are almost completely ionized in the radiative zone, and their g_{rad} is typically less than 5% of gravity*.

The ratio of the relative abundance variations after 4.6 Gyr for the model with detailed diffusion to the same for a model with a more standard settling procedure** is plotted in the right panel of Figure 1 for the same four elements. One can see

---

* One should realize that g_{rad}/g increases as the Sun ages on the main sequence so that the g_{rad} shown here at 4.6 Gr are larger than for earlier times in the Sun's life

** The most common prescription is followed so all elements heavier than oxygen settle at the same rate as Fe^{+26} and CNO are assumed completely ionized. Radiative forces are neglected.

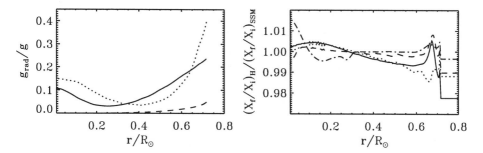

Figure 1. The left panel shows the ratio of radiative acceleration g_{rad} over gravitational acceleration g in the Sun at the solar age as function of the fractional radius. The curves are cut off at the base of the convection zone. The right panel shows the ratio between the solar (f) and initial (i) abundance profiles of model H with detailed diffusion over the ratio of the profiles for our 'standard' model in which it is assumed that heavy elements are fully ionized, not affected by radiation pressure and all elements heavier than oxygen settle at the same rate as Fe^{+26}. Elements shown are He (dot dash), O (dash), Ti (dots), and Fe (full), in both plots. The g_{rad} for He is zero.

that the largest differences are in the more superficial regions. Helium and oxygen for which g_{rad} is small and ionization is large show the least difference. Iron is the most affected. In the core, the difference for helium and oxygen can be attributed to differences in the nuclear reaction rates. The structure in the abundance ratios of Ti and Fe are related to the profiles of their respective radiative force.

The fractional surface abundance variations predicted by the consistent and standard models are shown in Figure 2. The model with detailed diffusion (filled circles) shows that all elements settle in a similar fashion, with a surface* depletion ranging from 7.5 to 8.8% for calcium and argon respectively. The SSM (diamonds) assuming the same settling rate for all elements heavier than oxygen, full ionization and no radiative forces predicts variations of the order of 7.5%. The squares show results for a consistent model where the g_{rad} were assumed to be zero. It is a blessing for SSMs that the elements for which the assumption of full ionization is the worst (those for which the difference between diamonds and squares is largest) are also the elements whose settling is slowed the most by the radiative forces (illustrated by the difference between filled circles and squares). The behavior of CNO does not change significantly from one model to another as they are almost completely ionized below the convection zone.

Unfortunately, both the absolute value of the surface abundance variation and the contrast between the abundance variation of different heavy elements are too small to be used as an observational test of diffusion because the uncertainties in the

---

* Here the surface abundances are actually the abundances in the convection zone. Since the latter extends up to the photosphere, the photospheric abundances are expected to reflect those in the convection zone.

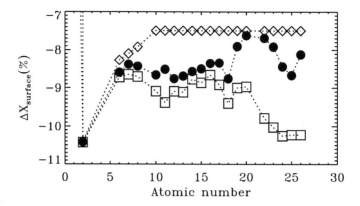

Figure 2. The fractional change of the surface abundance of all elements included in our models at the solar age as a function of atomic number. The models are: the SSM assuming fully ionized heavy elements and common settling rate for $Z > 8$ (diamonds); the model with detailed diffusion including radiative forces (filled circles); and the model with detailed diffusion but without radiative forces (squares).

observed photospheric abundances are of the order of 10% (Grevesse and Sauval, 1998).

In the solar core, diffusion is slowed considerably because of high densities and high degree of ionization of all elements. The abundance increases due to settling are close to 3% for all heavy elements with little difference from element to element (shown by TRMIR98).

These models neglect all mixing in the radiative zone. Such mixing would further reduce differences created by treating diffusion consistently, i.e., would result in lower contrast between elements and lower overall abundance variations.

3. Implications of a Non-standard Chemical Composition for Solar Models

Even though photospheric abundances cannot be determined with sufficient precision from direct observation, helioseismology can, in principle, be used to determine the abundance of at least some elements in the solar interior. This has been done for helium through its effect on the sound speed in regions of partial ionization. In the case of heavier elements, some information on abundances can only be expected for those which have a large impact on opacities.

For any abundance signature to be recognized, one must understand how sensitive the solar models are to variations in abundances. We first examine how the non-standard abundances produced by the detailed diffusion affect the structure of calibrated solar models (section 3.1). We then estimate how the opacity varies as the standard abundance for selected elements is varied (section 3.2).

3.1. CALIBRATED SOLAR MODELS WITH CONSISTENT DIFFUSION

The models were all calibrated to reproduce the observed solar radius and luminosity (6.9599×10^{10} cm and 3.86×10^{33} erg cm^{-2} s^{-1} at 4.6×10^9 years) to within a few parts in 10^5.

The relative difference of the isothermal sound speed squared ($u^2 = P/\rho$) for models with different assumptions for the diffusion of heavy elements is shown in Figure 3. The reference model in the following discussion is model H from TRMIR98 with detailed diffusion.

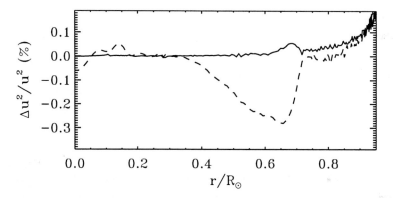

Figure 3. The relative difference ($\Delta u^2/u^2 = (u^2_{\text{reference}} - u^2)/u^2_{\text{reference}}$) of the isothermal sound speed squared ($u^2 = P/\rho$) with the fully consistent model is plotted for models with the complete treatment of diffusion except for radiative forces (full line) and our standard model (dashed line).

Neglecting radiative forces (full line), everything else being equal, has very little effect on the solar structure, with a maximum relative change of 0.03% at the base of the convection zone despite the large effect on surface abundances of the iron–peak elements as shown in Figure 2. The relative difference between the 'standard' treatment of diffusion assuming fully ionized heavy elements and the consistent treatment of diffusion reaches up to 0.3% which is at the limit of the current accuracy of solar models. However, every indication show that it is predominantly a numerical artifact due to a difference in interpolation schemes for the Rosseland mean and monochromatic opacity tables. Basically, the interpolation in the latter differ from interpolation in the former by a lower order of interpolation and a parametrization in terms of the electronic density rather than the total density of matter. Details are discussed by TRMIR98. Further work is required to quantify these effects.

Close to the convection zone, lighter elements, for which the assumptions for settling in the SSMs are essentially valid, dominate the opacity. In the warmer and denser regions where iron contributes more to the opacity, the chemical composition does not deviate greatly from the solar 'standard'.

3.2. UNCERTAINTIES IN THE SOLAR CHEMICAL COMPOSITION

The primordial abundance of heavy elements is thought to be identical to what is measured in meteorites. These measurements can be made with high precision for most elements and they agree with photospheric values within the error bars for the latter. There are some elements, the so–called volatile elements, that are not present in meteorites and for which there is no direct information on initial abundances. In addition, they are often very difficult to observe in the photosphere so that the abundance of some of these elements in the Sun is known only indirectly (Grevesse and Noels, 1993). As a matter of fact, preliminary results from Grevesse and Sauval (1998) show that the abundances of C, N and O may have been overestimated by up to 10% in the past. The abundance of neon and argon are also particularly uncertain.

One way to improve the agreement between the models and the 'seismic' solar structure is to increase the opacity in the interior by a few percents (Tripathy et al., 1998). For this to happen, either the OPAL opacity still underestimates the opacity of stellar plasmas or the abundance of strong absorbers is underestimated in the current 'standard' solar mix. Outside of the core, the elements which contribute the most to the opacity, apart from H and He, are the lighter elements (oxygen, neon and silicon for example), and iron.

The change in opacities due to a 10% variation in the abundance of most important light elements and of iron are shown in Figure 4. The abundance of a single element is varied at a time, and there is no simultaneous decrease in the abundance of other elements*. The abundance profiles are the same as for our model with detailed diffusion. The maximum influence of CNO is felt in the convection zone where opacity is unimportant but O is still the dominant element above 0.6 $R_\odot$ and is second only to Fe above 0.5 $R_\odot$. Silicon then dominates between 0.25 and 0.5 $R_\odot$. The opacity depends somewhat less on the abundance of Ne, Mg and S but the higher uncertainty in the Ne abundance suggests that it could play a role as important as oxygen in the fine-tuning of solar models.

The abundance of many, if not all, of these elements need to increase if one hopes to increase the opacity substantially in the interior. However, the uncertainty on the abundance of many of those elements is currently thought to be lower than 10%, especially Si and Mg which have been measured in meteorites. The uncertainty in the abundance of iron is probably too small for it to be considered a possibility and would as a side effect worsen the agreement in the core. The proposed reduction of the photospheric oxygen abundance by Grevesse and Sauval (1998) only aggravates this problem.

* For this reason and because the models have not been recalibrated with the corrected chemical composition, the sensitivity of the Rosseland mean opacity to abundance variations shown here is an upper limit.

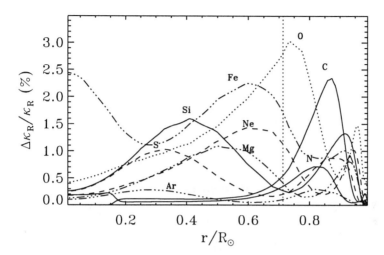

Figure 4. Ratio of the Rosseland mean opacity profiles for two chemical compositions. Both are identical with the exception of one element whose abundance has been increased by 10%. The abundance of other elements is not changed and the total mass fraction therefore exceeds unity. The boundary of the solar convection zone is identified by the vertical dotted line. Each curve is identified by the corresponding atomic symbol near their peak.

4. Conclusion

Solar models calculated with a fully consistent treatment of diffusion are discussed. These models are made possible by the new availability of monochromatic opacity tables from the OPAL group from which the radiative forces, the mean ionization for heavy elements and the Rosseland mean opacity for an arbitrary chemical composition are calculated. The models computed with a sophisticated treatment of diffusion are not very different from the much simpler standard solar models, with a maximum difference of 0.3%. Moreover, a significant fraction of that effect is attributable to purely numerical problems. The evolution of the abundance of individual elements is affected more strongly but the signature of diffusion on photospheric abundances is much lower than the current observational uncertainty.

All the models presented here neglect mixing in the radiative zone. There is strong evidence of such mixing both in the Sun (Richard *et al.*, 1996) and in other population I stars (Turcotte *et al.*, 1998a). Adding mixing to our models would lead to somewhat smaller abundance variations in and around the convection zone as well as smaller contrasts for the fractional surface abundance variation of different heavy elements. Mixing also improves the agreement between the models and the observed Sun.

The impact of varying the abundances of selected elements in the Sun is also examined. It is shown that the opacity is relatively sensitive to the variation in the abundance of oxygen, neon and silicon mainly. As the importance of each element varies from region to region in the solar interior, only a systematic increase of the abundance of these elements could help bridge the remaining gap between the predicted and the 'seismic' solar structure outside of the core.

Acknowledgements

This work was supported by the Danish National Research Foundation through its establishment of the Theoretical Astrophysics Center.

References

Bahcall, J. N. and Pinsonneault, M. H.: 1995, 'Solar models with helium and heavy element diffusion', *Rev. Mod. Phys.* **67**, 781–808.
Basu, S., Christensen-Dalsgaard, J., Schou, J., Thompson, M. J., and Tomczyk, S.: 1996, 'The Sun's hydrostatic structure from LOWL data', *Astrophys. J.* **460**, 1064–1070.
Burgers, J. M.: 1969, *Flow equations for composite gases*, (New–York: Academic Press).
Christensen–Dalsgaard, J., Däppen, W., Ajukov, S. V., *et al.*: 1996, 'The current state of solar modeling', *Science* **272**, 1286–1292.
Grevesse, N., and Sauval, A. J.: 1998, 'The standard solar chemical composition', *Space Sci. Rev.*, this volume.
Grevesse, N., and Noels, A.: 1993, 'Cosmic abundances of the elements', in *Origin and Evolution of the Elements*, eds N. Prantzos, E. Vangioni-Flam, M. Cassé (Cambridge: Cambridge Univ. Press), 15–25.
Morel, P., Provost, J., and Berthomieu, G.: 1997, 'Updated solar models', *Astron. Astrophys.* **327**, 349–360.
Iglesias, C. A., and Rogers, F. J.: 1996, 'Updated OPAL opacities', *Astrophys. J.* **464**, 943–953.
Proffitt, C. R., and Michaud, G.: 1991, 'Gravitational settling in solar models', *Astrophys. J.* **380**, 238–250.
Richard, O., Vauclair, S., Charbonnel, C., and Dziembowski, W. A.: 1996, 'New solar models including helioseismological constraints and light-element depletion', *Astron. Astrophys.* **312**, 1000–1011.
Richer, J., Michaud, G., Rogers, F. J., Iglesias, C. A., Turcotte, S. and LeBlanc, F.: 1998, 'Radiative accelerations for evolutionary model calculations', *Astrophys. J.* **492**, 833–842.
Thoul, A. A., Bahcall, J. N., and Loeb, A.: 1993, 'Element diffusion in the solar interior', *Astrophys. J.* **421**, 828–842.
Tripathy, S. C., Basu, S., and Christensen–Dalsgaard, J.: 1998, 'Helioseismic determination of opacity corrections', in *Poster Volume; Proc. IAU Symp. 181: Sounding solar and stellar interiors*, eds J. Provost and F. X. Schmider (Nice: Université de Nice), p. 129–130.
Turcotte, S., Richer, J., and Michaud, G.: 1998a, 'Consistent evolution of F stars: diffusion, radiative accelerations, and abundance anomalies', *Astrophys. J.*, in the press.
Turcotte, S., Richer, J., Michaud, G., Iglesias, C. A., and Rogers, F. J.: 1998b, 'Consistent solar models including diffusion and radiative acceleration effects', *Astrophys. J.*, in the press.

Address for correspondence: Teoretisk Astrofysik Center, Institute of Fysik og Astronomi, Aarhus Universitet, DK 8000, Århus C, Danmark

ON THE COMPOSITION OF THE SOLAR INTERIOR

Rapporteur Paper I

DOUGLAS GOUGH
*Institute of Astronomy
and Department of Applied Mathematics and Theoretical Physics
University of Cambridge, UK*

Abstract. Standard solar models, although they are free from the influence of much of the fluid motion that is bound to be present in the Sun, have been shown by helioseismology to represent the spherically averaged structure of the Sun amazingly well. This state of affairs has come about after painstaking refinements by a great many people of the pertinent microphysics, including that which controls the equation of state, the opacity, the nuclear reaction rates and the diffusion that inhibits gravitational segregation of chemical elements. It has instilled confidence in the modellers in being able to predict the composition of the solar interior. But there are consequences of the flow, related particularly to redistribution of chemical species, that can be difficult to identify observationally, yet which may degrade any inferences we might make. Their potential presence must at least be acknowledged by anyone who tries to asses the reliability of the models. This report summarizes the discussions in the preceding pages of this volume of the current theoretical and observational status of the subject, pointing to many of the caveats that have been raised, and attempting at the same time to put them into a seemingly coherent discourse in the context of our present understanding of the workings of the solar interior.

Key words: equation of state, opacity, helium abundance, lithium abundance, tachocline, solar neutrinos, helioseismology

1. Introduction

In the last decade or so, our view of the Sun – its structure, its composition and its dynamics – has crystallized substantially. This is due principally to the advance of helioseismology. It has transpired that the spherically averaged stratification of the solar interior is quite close to appropriately calibrated theoretical models. Indeed, it is really quite amazing that the so-called standard solar model, which is conceptually about the simplest model one can envisage that incorporates all the microphysics that we believe to be pertinent, is so close to reality. Many of the ideas that were rife three decades ago to resolve the solar neutrino problem appear not to be correct, and we seem to be left with a relatively simple picture of the star. Perhaps we should rejoice in that, for it appears to afford us the opportunity of understanding the Sun really well – understand it well enough that we can use it as a physics laboratory to study matter under conditions that cannot be achieved on Earth. But we do know that the Sun is actually not as simple as the hydrostatic spherically symmetrical models are. The Sun rotates, and associated with that rotation there must necessarily be a flow of material with a component circulating in meridional planes. Waves are generated in the convection zone, and possibly in the core, which propagate through the star, transporting momentum and some energy, and also contributing to the redistribution of the chemical elements. Does such motion modify the underlying hydrostatic model in a substantial way? We do not yet have a complete answer. But we are gathering fascinating evidence, much of which is discussed in the lectures and posters of the first day of this workshop.

2. The Standard Solar Model

The stage is admirably set by Jørgen Christensen-Dalsgaard (1998), who explains the simplifying assumptions that are adopted for constructing a standard model, and how the properties of that model depend on certain of the values of the free parameters that prescribe it. He explains also how the interior of the Sun is probed by seismic waves, how we deduce from the frequencies of those waves the hydrostatic stratification, and how we have tried to interpret the inferred differences between the Sun and the model. Those differences have motivated further advances in both modelling and in the theory of the microphysics required for calculating the opacity and the equation of state. The outcome is always an improved model, from the point of view of its physics; and after each advance in the physics, the model often, though not always, agrees better with observation. Thus, even though it might be said that the theory has not been formally calibrated by the helioseismic data, helioseismology has certainly played an important role in motivating the improvements.

Perhaps one of the most prominent additions to the physics in the last decade, and one which is central to this workshop, has been the incorporation of the gravitational settling of heavy elements. If each chemical constituent were thought of as a gas which does not interact with the other constituents, save only to equalize the temperatures of all the constituents at each point in space, then in a steady state each constituent would have its own hydrostatic structure, the local scale height being proportional to the molecular mass of the constituent. Therefore, the chemical composition would vary from place to place. But in reality the constituents do interact, and the consequent particle diffusion acts in such a way as to reduce chemical inhomogeneity; the steady state is one of a balance between the smoothing tendency of diffusion and the segregating action of gravity and other phenomena such as the interaction of the particles with the anisotropic radiation field. However, the Sun is far too young for equilibrium to have been achieved: the equilibration timescale is comparable with, even somewhat greater than the age of its Universe. Slow gravitational settling from a presumed homogeneous state, moderated by the diffusion process, has led to only a ten-per-cent reduction of the helium abundance in the convection zone, and typically a somewhat lesser reduction of the abundances of heavier elements. The segregation diminishes with depth, because settling is fastest in the outermost quiescent layers of the radiative interior, where the diffusion coefficients are least. In the convection zone material is redistributed by macroscopic turbulent motion; the effective diffusion time there is only about a year, and the composition is therefore essentially homogeneous.

An obvious consequence of gravitational settling is that the chemical composition of the convection zone today is not the same as it was when the Sun first formed. It follows that to infer the initial abundances from any direct abundance measurements that we might make today requires the application of the theory of gravitational settling. One cannot make the measurement directly.

One way to infer the initial helium abundance Y_0 is simply to calibrate the solar model by adjusting Y_0, and the mixing-length parameter α, such as to reproduce the observed current luminosity and radius at the age one presumes the Sun to be, usually between 4.5×10^9y and 4.6×10^9y. It is also necessary to determine the initial total heavy-element abundance Z_0, which is achieved by insisting that the current ratio Z/X of heavy-element to hydrogen abundances in the convection zone, after taking settling into account, agrees with observation. Strictly speaking, one should take into account the settling of each heavy element individually, and, of course, take account of the consequent influence on opacity, nuclear reaction rates and the equation of state. But usually some simplifying approximation is made at some stage (or several stages) by considering several chemical species as one. The most detailed calculation that has yet been performed is presented in this volume by Sylvain Turcotte and Jørgen Christensen-Dalsgaard (1998). The full calculation has yet to be performed. Nevertheless, the current models represent the physics very well compared with the earlier models in which gravitational setting was ignored. And as Christensen-Dalsgaard, Sylvaine Turck-Chièze and Sylvie Vauclair all demonstrate, the agreement with observation has been dramatically improved.

The sound speed of the current standard models differs from the spherically averaged sound speed in the Sun by typically about a part in 10^3. Broadly speaking, the discrepancy that remains can be considered to be composed of a smoothly varying component plus a relatively rapidly varying component, the latter being dominated by a hump immediately beneath the convection zone, centred at $r/R \simeq 0.68$. I shall discuss these later in connexion with the tachocline.

3. Direct Measurement of the Helium Abundance

In the convection zone the helium abundance Y is uniform. As Werner Däppen (1998) explains, ionization of helium causes the adiabatic exponent γ_1 to decrease, which influences the adiabatic sound speed directly. In the zone of second ionization of helium the turbulent motion is sufficiently weak for Reynolds stresses to be negligible, because the heat capacity of the fluid is so great that the fluid can easily transport at very low speed the heat that the Sun demands. The buoyancy force that drives the flow must therefore be low too. This implies that the stratification is close to being adiabatic. These conditions, together with the hydrostatic constraint, give us enough information about the stratification of the convection zone that we can infer from a helioseismic determination of the sound speed a purely thermodynamic function, Θ, which depends on γ_1 and its partial derivative with respect to pressure on an isentropic surface. With the help of the equation of state, this can be converted into a value for Y.

Various determinations of Y in the He II ionization zone have been made, the results of the most recent of which are discussed here by Christensen-Dalsgaard (1998), by Wojciech Dziembowski (1998), by Turck-Chièze (1998) and by Däppen (1998). Some of the methods use the thermodynamic function Θ overtly, others do not. But they all depend on the special constraints in the adiabatically stratified body of the convection zone that lead to the relation between Θ and the sound speed. Moreover, they all give more-or-less the same result, which is in quite good agreement with the values obtained by calibrating the standard solar model. Of course, there is some variation amongst the values of Y obtained, and that arises both because different equations of state were employed in the analysis and because different methods were used to analyse the seismic data.

One might wonder why it is that the result depends on the method of analysing the seismic data. Surely, one might think, all good seismological procedures should determine almost the same value of γ_1, and converting that into a helium abundance is an issue related solely to the equation of state. But actually, that is not quite the case. The reason is that the application of most of the methods requires the use of the equation of state to interpret adiabatic stratification in seismic terms. When the equation of state is not exactly correct, which is bound to be the case, erroneous quantities are created in the numerical manipulations, and different manipulations incorporate them differently to corrupt the final outcome. Tests have been performed with artificial seismic data: if the procedures determining Y from the data use the same equation of state as was used in the computation of the data, then they give more-or-less the (same) correct result, whereas if they use a different equation of state, they produce different erroneous results (Kosovichev *et al.*, 1992). It is interesting that Christensen-Dalsgaard picks out an asymptotic method that appears to be the most robust under changes of equation of state. Some effort should be expended in trying to find out why that is so, for then one might have more confidence in the outcome of the method.

4. The Equation of State

It goes without saying that we need to understand the microphysics of the equation of state, the opacity and the nuclear reaction rates, irrespective of whether the standard solar model is correct. Werner Däppen's (1998) succinct account of the two fundamentally different approaches to the problem is an excellent introduction to the difficulties that need to be faced. Both approaches have their drawbacks. The chemical approach has the immediate appeal of being able to accommodate quite complicated situations, with many different species of particle interacting weakly, and with a free energy composed of many easily identifiable parts. Once one has constructed the free energy, the mathematical manipulations required to generate the equation of state should be straightforward. However, as Däppen explains, it is not easy to incorporate some of the necessary interactions between neighbouring particles – some of them quite obviously necessary because without taking them into account the partition function is formally infinite. Various almost *ad hoc* procedures have been adopted for describing those interactions, and it is only by comparing their consequences with observation that one can test them. The physical approach does not suffer these drawbacks, for it deals with 'elementary' particles, rather than composite atoms and ions. It is executed by a formal activity expansion which, if taken far enough (and if it were to converge) should account automatically for bound states and their interactions with their neighbours. However, the expansions are complicated, especially for a realistic gas of a mixture of many chemical species, and the fear is that they have not been taken far enough, particularly in the outer layers of the Sun. Indeed, as Däppen reports, seismic estimates of γ_1 suggest that in the outer layers better results are obtained with the chemical approach, whereas the physical approach is better at greater depths. However, it should be noted that we are discussing discrepancies in γ_1 which are only a few tenths per cent; these are really quite tiny compared with the uncertainties that were being discussed before helioseismology. This is a demonstration of how the Sun is being used as a laboratory for the physics of dense plasmas.

The discrepancies in γ_1 translate into a much greater variation in our inferences of chemical composition. The range of values of Y that are currently being obtained is about 7 per cent, which must represent a lower bound to our uncertainty. Evidently, it is important to look more carefully at the seismological measures of the equation of state, in order to assess and subsequently improve the approximations that are being used.

One of the interesting problems arising in the chemical approach concerns how to treat the finite size of positive ions. Baturin *et al.* (1994) have introduced a simple formalism depending on an effective radius which is a constant multiple λ of the Bohr radius associated with the charge of the ion, and have calibrated λ seismologically using an asymptotic estimate of the thermodynamic function Θ in the convection zone. Not only did they find a not obviously implausible value ($\lambda \simeq 1.7$), but with that value they were able to construct a model convection zone that was in good seismic agreement with the Sun, at least beneath the main body of the He II ionization zone. This result is an illustration of how ionic properties are accessible seismologically. However, it would be wrong to assume that ionic dimensions have actually been measured, because there may be other aspects of the plasma, not taken into account by Baturin *et al.*, that mimic the property that they did take into account.

Another interesting phenomenon that Däppen discusses concerns the effect of a density-dependent formalism for excited-state occupation probabilities that accounts for interactions between neighbours. The outcome is a characteristic 'wiggle' in thermodynamic quantities, particularly in the He II ionization zone. This is indicative of a region of exceptional sensitivity of the thermodynamics to the physical conditions. It appears that there is some seismological verification of the phenomenon (Basu et al., 1998), based on inversions of recent data. But there may be older evidence too, from the temporal solar-cycle variation in the frequencies measured by Libbrecht and Woodard (1990). These variations show a frequency-dependent oscillation with respect to mode frequency (Goldreich et al., 1991; Balmforth et al., 1996), which is indicative of there being a localized variation of wave propagation speed (probably sound speed) in the vicinity of the He II ionization zone.

With precise seismological measurements of Θ throughout the convection zone we are approaching the point where we might measure the abundances of some of the more abundant heavy elements, through the variations in Θ they induce in regions of partial ionization. When this was first suggested more than a decade ago, the community considered the idea to be ludicrous. Now it is within sight of becoming reality.

5. Opacity

The determination of the helium abundance in the Sun's radiative zone is based on a calibration of standard solar models. That calibration depends critically on the opacity. Opacity calculations are extremely complicated, and only few of the many astrophysicists who use the tables kindly supplied by the experts have much idea of what goes into them. Forrest Rogers' contribution to this workshop (Rogers and Iglesias, 1998) gives us an inkling of some of the issues that have been addressed in recent years. The outcome has led to marked improvements to our knowledge of the opacity in stellar interiors, judged both by the greater thoroughness of the calculations and by the various astrophysical issues – almost all of them associated with stellar pulsation, because dynamics depends more sensitively on the functional form of the opacity than statics does – that have recently been resolved by using the most modern opacity tables. On the whole, the history of the subject has been one of general increase in the values of the opacity, due either to the discovery of new sources of opacity or to an increase in the frequency resolution in the computations, thereby capturing the absorption in spectrum lines more faithfully. (Inadequately resolved numerical integrals tend to underestimate the contribution from the lines.)

The computation of radiative transitions depends on the states of the atoms and ions, which are influenced by interactions with their neighbours. These are the interactions that were discussed in connexion with the equation of state, and indeed the computation of the equation of state forms part of the computation of opacity.

It has from time to time been argued that collective effects, extending well beyond nearest neighbours, causes substantial diminution of the opacity. In addition, Tsytovich *el al.* (1996) have recently produced a list of corrections that should be made, all negative, and which together amount to a reduction in the modern values by about 14 per cent. Rogers argues against them in this volume, partly on the grounds that some of the effects discussed by Tsytovich *et al.* are included already. A modification to the opacity of the magnitude suggested by Tsytovich *et al.* would destroy the impressive agreement with the seismic constraints, and would therefore demand some compensating phenomenon in the theoretical models of the Sun in order to restore it.

The ionization of particularly the more abundant heavy elements causes sharp variation in the opacity, which might be detectable in the seismologically determined sound speed. One might thus have a means of measuring heavy-element abundances in the radiative zone. The opacity variation is greater than the variation of γ_1, but on the other hand the influence on the sound speed is only indirect. Whether the sensitivity of the sound speed is more or less than it is in the convection zone is yet to be determined. In any case, owing to gravitational settling the two measurements would be complementary. Sylvaine Turck-Chièze suggests that the undulations evident in the current sound-speed inversions in the radiative zone might be the result of such opacity variation. It would be very exciting if that were to be so. But we must not jump too hastily to believe it yet, for there are a host of possible sources of error correlations that may be the true cause of the undulations.

It is hardly necessary for me to point out that the importance of the opacity in standard solar models is through the relation between the heat flux and the temperature gradient in the radiative zone. But that relation depends on the assumption that radiative transport provides the sole contribution to the heat flux. There are other possibilities, not considered in the standard model, resulting from macroscopic fluid motion, either oscillatory or direct, and it is important not to forget this. For example, the cores of standard solar models are seats of linear instability driven by the burning of ^{3}He; the nonlinear development of this instability is not yet understood, but one possibility is a shallow layer of direct convection, which Ghosal and Spiegel (1991) called shellular convection. A careful realistic nonlinear calculation of this motion has never been carried out, but it is interesting to note that, aside from the dominant hump immediately beneath the convection zone, the greatest undulation in the sound-speed difference between the Sun and a standard model, evident for example in Christensen-Dalsgaard's Figure 3, is precisely where the ^{3}He abundance is greatest. The coincidence may, of course, be fortuitous.

6. The Solar Tachocline

Advection by macroscopic flow is a potentially important agent for redistributing chemical species in the Sun. Jean-Paul Zahn summarizes some of the pertinent processes. Central to much of his discussion is the variation of angular velocity Ω. Although it influences neither the radial nor the latitudinal distribution of elements directly – in discussions of the global distributions of chemical elements the Sun is presumed to be axisymmetric – associated with it is necessarily a meridional flow which does. There are basically two kinds of meridional flow that are considered: small-scale turbulence and large-scale circulation. These flows redistribute not only the chemical elements but also the angular momentum. It is important for us to consider the latter, therefore, not just for its own sake, to provide understanding of the global dynamics of the Sun and of the other stars, in itself both interesting and important to stellar physics, but also because the chemical elements are redistributed by the very same processes that redistribute angular momentum, which can thus be investigated seismologically through their effect on Ω.

Roughly speaking, helioseismology has told us that the latitudinal variation of Ω observed at the surface is more-or-less maintained throughout the convection zone, whereas at least the outer parts of the radiative interior rotate almost uniformly, with an angular velocity Ω_c which is intermediate between the polar and equatorial values in the convection zone. The transition between the two is very sharp – too sharp to be resolved seismologically. It occurs as a thin shear layer situated immediately beneath the base of the convection zone, which Spiegel and Zahn (1992) have called the tachocline. Two questions are immediately raised: why should the radiative zone, which is normally assumed to be quiescent, rotate uniformly in the face of both the tachocline shear and the spinning down of the Sun by the solar-wind torque, and why does it rotate at the rate it does? One presumes that the anisotropic turbulence in the convection zone, which can adjust the angular-momentum distribution within the convection zone essentially instantaneously (on a timescale of about a year), is oblivious of the details of the rotational flow in the radiative zone (but is no doubt dependent on the value of Ω_c), and so can be regarded as posing a separate, albeit interesting and difficult problem.

Spiegel and Zahn (1992) have addressed the two questions. They pointed out that the shear in the tachocline would generate an Ekman-like circulation. However, they estimated that in the absence of an appropriate stress the circulation would have penetrated throughout much of the radiative zone, causing it to rotate nonuniformly. To inhibit the penetration, they invoked the existence of weak turbulence in the tachocline, which they presumed to act in the manner of a viscosity and thereby reduce shear. Because the tachocline is highly stably stratified (the mean buoyancy period in the tachocline is about 2 hours, to be compared with the timescale of differential rotational, which is several months) the turbulence is essentially layerwise two-dimensional, in horizontal surfaces, and therefore exerts little stress between neighbouring horizontal surfaces. However, it does lead, in the model, to latitudinally independent rotation at the base of the tachocline, where the viscous stress is presumed to dominate over advection by the circulation. In a strictly steady state, that in turn would lead, as required, to uniform rotation in the quiescent region beneath, leaving no substantial baroclinicity to drive a circulation.

What is left unexplained, however, is how the turbulence is driven to an amplitude high enough to suppress the shear, and by what mechanism the angular velocity of the radiative zone immediately beneath the tachocline is caused to be independent of radius in the more realistic unsteady state in which angular momentum is being extracted from the Sun by the solar-wind torque. Perhaps a magnetic instability is operative, as Gilman and Fox (1997) advocate. There is also the very important issue of whether the turbulence really does act in such as way as to suppress shear, or whether instead, as seems more likely of two-dimensional flow, it is potential vorticity rather than angular velocity that is transported, and presumably somewhat smoothed by the turbulence. That would not lead to uniform rotation, but instead to an angular velocity that decreases with latitude. There is substantial evidence from meteorological studies of the Earth's stratosphere to suggest that this would be the case (McIntyre, 1994). However, given that the source of the turbulence in the Sun is not the same as that in the stratosphere, the analogy is not perfect, and there must remain some doubt. Indeed, Zahn (1998) believes the analogy to be so poor that he does not even mention the objection it raises to his turbulent-tachocline theory.

Whilst I am on this subject, it behoves me also to discuss the role of angular-momentum transport by internal gravity waves. Both Kumar and Quataert (1997) and Zahn *et al.* (1997) have simultaneously invoked differential dissipation of gravity waves generated at the interface between the convection zone and the radiative interior to spin down the radiative interior of the Sun. In his contribution to the workshop Zahn (1998) admits that the respective roles of prograde and retrograde waves were not treated correctly: to be more explicit, in both papers the sign of the angular-momentum transport was wrong, so the initial response would actually be to spin the interior up, not down. This property of gravity-wave dissipation had already been pointed out in connexion with possible critical-layer absorption in the Sun (Gough, 1977). But, if gravity-wave transport is important, what might really be its consequences? One's first thought might be that a quasi-periodic oscillation in the angular velocity is generated in the tachocline, in a manner similar to that believed to generate the quasi-bienniel oscillation of the Earth's atmosphere (e.g. McIntyre, 1994), and demonstrated in the laboratory by Plumb and McEwan (1978). Unlike the terrestrial case, it is probably only the gravity waves in the high-frequency low-degree tail of the spectrum that penetrate deeply enough to have a significant effect over the scale of the tachocline; however, there is substantial uncertainty in the value of the resonant frequency at which the spectrum is expected to peak, partly because of the uncertainty in the coupling of the gravity waves with the convection, but mainly because of the uncertainty in the properties of the convection itself. A plausible detailed study of the entire process has yet to be undertaken. It would be interesting if the outcome were an eleven-year oscillation of the tachocline, leaving the magnetic field to play only a secondary role in the dynamics controlling the solar cycle.

Another consequence of the presence of gravity waves is the transport of material by the rectified Stokes drift, a large-scale wave-induced circulation which mixes material between the convection zone and the radiative interior, thereby modifying the variation of chemical composition. In earlier discussions (e.g. Gough, 1988) the effect was estimated to be negligible, but perhaps the issue should be revisited in the light of the more recent estimates of the possible amplitudes of the waves (cf. García López and Spruit, 1991). The wave enhancement of the material diffusion that results from the increase in gradients produced by differential advection by the waves may be a significant and a more important transport process than the Stokes drift (Knobloch and Merryfield, 1992).

If neither turbulence nor gravity waves can hold the radiative interior rigid, what can? It seems that Lorentz forces are all that remain. It takes only a weak field, of order 10^{-9}T, to maintain Ω uniform against the solar-wind torque (Mestel, 1953; Cowling, 1957). A rather stronger field is needed to provide the rigidity required to prevent the latitudinal shear in the convection zone from penetrating far into the radiative zone. Gough and McIntyre (1998) have shown how the intensity of that field depends on the mean thickness Δ of the tachocline. If their simplified description is correct, one might thus use seismological inferences to estimate the large-scale magnetic field in the Sun's radiative interior.

Whatever the details of the tachocline dynamics, baroclinicity induced by the shear is bound to drive a circulation in the tachocline, which advects helium that has settled under gravity back into the convection zone, replacing it with relatively hydrogen-rich material and thereby augmenting the adiabatic sound speed $c = (\gamma_1 p/\rho)^{1/2}$ to produce the characteristic hump in $\delta c^2/c^2$ (where δc^2 is the difference between c^2 in the Sun and c^2 in a reference theoretical model) evident in nearly all of the seismic inversions illustrated in this volume. Of course there is also the possibility of some material mixing by convective penetration and overshoot, but Zahn's discussion suggests that that is likely to be insignificant compared with transport by the laminar tachocline circulation. One can therefore calibrate against the inversions a sequence of theoretical solar models with mixed regions of different thicknesses. In so doing it is necessary to separate the slowly varying discrepancies, which are presumed to have arisen from errors in either the microphysics or the global parameters, such as radius, adopted for the reference model, from the rapidly varying tachocline anomaly. Provided the slowly varying discrepancies are small, it is probably safe to fit any smooth curve through them. The outcome, using model S of Christensen-Dalsgaard *et al.* (1996) as a reference, is $\Delta \simeq 2 \times 10^{-2} R_\odot$ (Elliott and Gough, 1998). However, I must caution that without an understanding of the global discrepancies, the removal using an arbitrary smooth curve leaves room for doubt. This is evident in the illustrations of $\delta c^2/c^2$ that Turck-Chièze (1998) presents in this volume, in some of which the tachocline anomaly appears to be substantially larger than that inferred from model S. Admittedly, the global discrepancies are larger too, which might seduce one into naively believing that Turck-Chièze's theoretical reference models are not as good; however, it could be that model S merely appears to be better as a result of having errors of similar magnitude which happen nearly to cancel. Indeed, as Christensen-Dalsgaard points out, the tachocline anomaly could be reproduced by introducing an appropriately designed, relatively rapidly varying component to the opacity. Although it is unlikely that such an extreme suggestion might provide the explanation of the entire anomaly, such an opacity component could upset the tachocline calibration.

It should be recognized that other forms of material mixing might be the major cause of the hump in the sound-speed difference. Wave transport, which I mentioned earlier, is one such possibility. Another is suggested by Maurice Gabriel (1998), who reports an investigation of mixing by presumed rotationally induced turbulence. Although he is able to reduce the magnitude of the sound-speed discrepancy, his diffusion process produces too broad a modification to remove the hump entirely.

It is interesting to note that the sound-speed calibration of the tachocline thickness Δ yields a smaller value than the characteristic thickness of the shear layer measured from inversion of rotational-splitting data, estimates of which lie between $0.04R_\odot$ (e.g. Basu, 1997) and $0.09R_\odot$ (Kosovichev, 1996). There are several possible reasons for this disparity. First, one or both of the seismic calibrations might be in error: for example, the global sound-speed discrepancy may have been removed incorrectly, or the resolution of the splitting data may have been inadequate to infer so thin a transition in Ω. Alternatively, the disparity may be real, with the shear penetrating into the convection zone because the Reynolds stresses are not strong enough (or are of the wrong form) to maintain the latitudinal variation throughout the zone.

One of the implications of the tachocline circulation is that the decline of the helium abundance Y in the convection zone is less than it would otherwise have been, by about 10 per cent. A second implication concerns the photospheric lithium abundance. Because the Lorentz force which confines the tachocline by deflecting sideways the downwelling segment of the circulation (which, in turn, acts to keep the magnetic field buried in the radiative zone) must vanish somewhere – for example, at the poles if the field is axisymmetric and aligned with the axis of rotation – there must be regions where the tachocline is thicker, possibly substantially thicker, than average; the penetration may even be great enough for sufficient destruction of lithium by nuclear reactions to account for the spectroscopically observed depletion in the photosphere.

Finally, it is interesting to speculate on the upwelling segment of the circulation. There, the tendency of the flow is to drag the magnetic field upwards through the tachocline into the convection zone. The tendency is resisted by the Lorentz force. Whether or not the equilibrium balance is one in which there is a crossing of the tachocline by the field over a substantial range of latitude is yet to be determined. But if so, one might expect there to be a modification to the angular-velocity, possibly in such a way as to nullify the rotational shear throughout that latitude range. A new look at the rotational splitting data is called for to assess whether that is likely to be the case. It is interesting to note that the centres of upwelling, which are located at the latitudes where the tachocline shear vanishes (or at the centres of the regions of no shear), namely at about ±30°, are situated at the boundaries of the sunspot belt. This coincidence gives pause to wonder whether the source of the magnetic field that eventually emerges through the photosphere, after amplification by the convection, is primordial rather than having been generated by a putative solar dynamo.

7. Neutrinos

It seems to be impossible to reconcile the predictions of standard solar models with the whole set of neutrino measurements. As Yoichiro Suzuki (1998) points out, the main problem seems to be that in order to bring theory in line with observation a severe reduction of the medium-energy ^{7}Be neutrino flux $F_{\nu 7}$ relative to the high-energy ^{8}B flux $F_{\nu 8}$ is required. Most discussions of theoretical neutrino flux changes are in terms of global changes to the core temperature T_c, to which $F_{\nu 8}$ is the more sensitive; therefore, reducing T_c reduces $F_{\nu 8}$ more than it reduces $F_{\nu 7}$. (The low-energy p-p flux is closely linked with the total rate of thermal energy production, which for standard models is fixed by the observed solar luminosity, and so it cannot be modified substantially.) Consequently, some have argued that the neutrino flux measurements taken together demonstrate that neutrino transitions must occur, irrespective of astrophysical considerations.

That argument has been shown to be fallacious in an inventive discussion by Cumming and Haxton (1996). By introducing, admittedly arbitrarily, an appropriately configured macroscopic flow into the solar core, nuclear reactions can be thrown out of equilibrium. In particular, ^{3}He is destroyed at locations different from where it is produced. The outcome is a preferential reduction of $F_{\nu 7}$, as required by the neutrino flux measurements. Cumming and Haxton emphasize that the details of their model are not meant to be believed; and indeed, Bahcall et al. (1997) have pointed out that a literal interpretation of their model appears to be contradicted in its detail by helioseismology. However, that does not invalidate Cumming and Haxton's demonstration that arguments for neutrino transitions based on the solar neutrino flux measurements must rely on astrophysics.

The SuperKamiokande measurements of zenith-angle variation of atmospheric neutrinos discussed by Suzuki (1998) does provide strong evidence for neutrino oscillations. These oscillations are between μ and τ neutrinos, and require both $|\nu_\mu\rangle$ and $|\nu_\tau\rangle$ to be associated with eigenstates of nonzero mass. This being so, it is highly likely that electron neutrinos oscillate too, and might provide the 'solution' to the solar neutrino problem. The current conjecture is that the atmospheric neutrino measurements, with some support given by the hint from SuperKamiokande that the energy spectrum of the solar electron neutrinos is anomalous, are compatible with there being vacuum oscillations alone. But particle physics suggests that if vacuum oscillations can occur, MSW transitions are likely too; there is much work to do to sort out the values of the mass differences and mixing angles that characterize them all. As Suzuki points out, the new generation of neutrino flux observations, both from SuperKamiokande and from the imminent operation of the Sudbury Neutrino Observatory, to detect the transitions from measurements of the distortion of energy spectra, from possible day-night, seasonal and solar-cycle variations, and from neutral-current measurements, will revolutionize neutrino physics. Although it is Suzuki's hope that all the parameters of the theory will be determined independently of solar models by such measurements, it is my suspicion that that will not be possible, at least in the near future, particularly if the transition length associated with electron neutrinos is much greater than terrestrial dimensions. It is likely that particle physicists will continue to rely on the Sun. Our responsibility as astrophysicists, therefore, is to do our utmost to characterize as accurately as we can the solar source of neutrinos.

With this thought in mind, I return briefly to the idea of Cumming and Haxton. That there should be macroscopic flow in the solar core which unbalances the nuclear reactions is to be expected, for standard solar models are unstable, the source of the instability being the energy released in the ^{3}He-destroying reactions in the core (e.g. Christensen-Dalsgaard et al., 1974). The initial linear growth of the instability is well understood. But the subsequent nonlinear development is not. The hypothesis of Cumming and Haxton is similar to the suggestion by Ghosal and Spiegel (1991): that the outcome is slow, possibly steady, diffusively controlled convective motion. If that were to be the case, then both Reynolds stresses and buoyancy forces would perforce be very small, implying that the Eulerian fluctuations in both pressure and density are also very small. Consequently, so too would be the sound-speed fluctuations, because in the core the adiabatic exponent γ_1 is essentially constant ($\simeq 5/3$). The only seismologically accessible signature of the flow would be the global, approximately spherically symmetrical reaction of the Sun to the modified nuclear reaction rates and the additional energy transport by convection. (The flow might have the tendency to make the temporally averaged angular momentum density more nearly uniform, so a hint of its existence would also be provided if seismology were to reveal that the angular velocity in the core increases with latitude at constant radius.) It would be interesting to determine whether a flow configuration exists that is both compatible with the seismic observations and which has a substantial influence on the fluxes of neutrinos. Of course, the problem of whether it can be sustained dynamically would also need to be addressed.

8. Mild Mixing

In order to explain the low photospheric abundance of ^{7}Li in the Sun it is almost universally agreed that some form of material mixing beneath the convection zone must have taken place. There are many physical mechanisms that could have caused it, as Zahn discusses. Amongst them are meridional circulation, diffusive mixing resulting from rotational shear instability, material transfer by gravity waves, and convective penetration. All these forms of mixing are likely to be weak, in the sense that, although they transfer a significant quantity of material on the timescale of the solar age, they transfer only a small, if not negligible, amount of heat. Since these processes are difficult to incorporate in detail into stellar evolution calculations, they are usually parametrized in some simple way, often in terms of an effective diffusion coefficient. An example is the interesting recent investigation of convective penetration by Freytag *et al.* (1996), which can be characterized by a diffusion process with a diffusion coefficient which declines exponentially with distance from the convection zone (cf. Turner, 1973). As Hartmut Holweger and his collaborators discuss in their contribution to the workshop (Blöcker *et al.*, 1998), a not implausible value of the diffusion efficiency can reproduce the solar ^{7}Li and ^{3}He abundances. Richard *et al.* (1996) have carried out a similar investigation, except that in their calculation the diffusion process was assumed to result from shear instability.

A common aim of such studies is to extend the calculations to other stars, trying to reconcile the theory with the observed surface abundances. Thus, Holweger *et al.* report at this workshop that their convective penetration calculations for A stars leads to a value of the effective diffusion coefficient that is similar to that needed to account for the solar photospheric ^{7}Li abundance (Blöcker *et al.*, 1998). They cannot calculate the penetration explicitly for the Sun because, as Zahn also emphasizes, the Péclet number is about 10^6, too high for direct numerical simulation to be possible, and very much greater than the O(1) values characteristic of A stars. Indeed, one might therefore expect the degree of penetration to be very different in the Sun from that in A stars; is it not surprising that a very similar diffusion coefficient works for both?

Although it is an interesting and often instructive exercise to try to compare processes in different stars, one must beware of the consequences of assuming that the dominant processes are always the same, or even that a single process dominates. For example, in the Sun a (horizontally) large-scale circulation in the tachocline is inevitable, and, whether it is essentially inviscid except in a lower diffusive boundary layer (Gough and McIntyre, 1998) or diffusive overall as a consequence of small-scale anisotropic turbulence (Spiegel and Zahn, 1992), (irrespective of whether that turbulence acts in such a way as to render the angular velocity or the potential vorticity more uniform), the circulation timescale is only about 10^6y, which is much less than the gravitational settling timescale. I have already suggested that such a circulation might account for the solar tachocline sound-speed anomaly, and perhaps even the photospheric lithium abundance. But alternatively, it might be the case that the penetration of the circulation is actually much too shallow to account for the photospheric abundances in the Sun, and that some other process actually dominates the overall material transport. In the more rapidly rotating A stars one might expect rotational instability to be a more likely transport process. We cannot tell how that might compete with large-scale circulation, particular because we have little evidence relating to whether other stars even have tachoclines. We should also keep in mind that significant abundance changes can be produced by accretion or mass loss.

An important contribution to our endeavours is modelling of the kind discussed at this workshop by Sylvie Vauclair (1998), to determine what kind of material redistribution is necessary to account for the observations. We thus learn that the transport must be greatest in the outer layers of the radiative zone, in order to provide the required depletion of lithium by a factor of about 10^2 yet not to dredge up so much ^{3}He from the edge of the energy-generating core as to augment the photospheric abundance by more than 20–50 per cent. The detailed modelling of gravitational setting of heavy elements of the kind that Turcotte and Christensen-Dalsgaard report is also of extreme importance, and might be testable by helioseismology in the future through its influence on the opacity, as Turck-Chièze now dreams. The results of these settling calculations must of course be modified in the light of whatever macroscopic transport processes are taking place.

9. Concluding Remarks

Solar interior modelling is entering a new era of sophistication. Transport of the chemical elements, both by microscopic gravitational settling and macroscopic material motion, is now being considered very seriously. Indeed, it is now a worthwhile task to do so, because the helioseismological inferences concerning the spherically averaged hydrostatic stratification and the variation of angular velocity throughout much of the solar interior have reached a level of detail that enables some comparison with the predictions of the transport theories. In the recent past, the greatest changes to the models have resulted from the correction of the treatment of spin-orbit coupling and the improved spectral resolution in the opacity calculations, which has a direct influence on the transport of radiation and therefore on the overall thermal stratification, and from the treatment of gravitational settling of helium and heavy elements, which influences both the equation of state and the opacity. Both changes have led to a marked improvement of the agreement between the theoretical models and observation. More detailed calculations are now being undertaken, particularly of the microphysical processes. As Christensen-Dalsgaard reports, not all of these bring the seismic structure of the theoretical model closer to observation. There is evidently further improvement to make; the end of interior solar modelling is certainly not yet in sight. Of course, a great uncertainty lies in the macroscopic processes, most of which are too weak to be detected directly. They must be determined through their global consequences, but at present the state of the art cannot easily distinguish the consequences of one process from another.

We can expect substantial improvements in our ability to make helioseismic diagnosis. Indeed, so gross a parameter as the radius of the Sun is in the process of revision, as both Christensen-Dalsgaard and Dziembowski discuss. Previously, most modellers had accepted the value in Allen's book on Astrophysical Quantities (Allen, 1973), without questioning seriously whether that value has the same meaning as that adopted by the modellers. In the old days, when the models and the seismic diagnostics were first brought face to face, the issue was considered, and any difference was judged to be insignificant. But with the increase in precision of the models and particularly the seismic diagnostics, the meaning of an insignificant difference changes. It is important to reassess the tenets of the subject continually.

In order to test many of the theoretical ideas that have been addressed at this workshop, either extension or improvement of the seismic data is required, preferably both. Janine Provost and her collaborators emphasize the importance of the low-order p and g modes of low degree, which have mixed p-mode and g-mode character and which are very sensitive to conditions in the solar core (Provost *et al.*, 1998). Such modes would be of particular diagnostic importance for investigating the consequences of the ^{3}He-driven core instability, which could, perhaps, be a convective circulation of the genre suggested by Cumming and Haxton (1996), and which could influence both the mean hydrostatic stratification and the variation of angular velocity. However, the mixed seismic modes, and more especially the g modes of moderate and high order, have relatively low surface amplitudes, and are difficult to detect. We must be prepared for the possibility that measurements of the frequencies of such modes will not be available in the near future. Therefore, we must think seriously about how to determine more accurately the properties of the higher-order low-degree p modes that we can detect.

One of the difficulties encountered in measuring the properties of the modes, particularly the frequencies of free oscillation which are the most commonly used diagnostic quantities, arises because the free oscillation of the modes is masked by the excitation and damping processes that modulate the motion. Of course, they are masked also by other phenomena not directly related to the oscillations, which can be regarded as noise. It is normal to work with the Fourier spectra of the data, either the amplitude spectra or the power spectra, because techniques exist for ready analysis in these terms. However, the oscillatory motion is not purely sinusoidal, so projecting onto sinusoids, which is what Fourier decomposition is, is not the most appropriate procedure to adopt. Unfortunately, we do not know what is. There have been attempts at various wavelet transforms, but these have not yet yielded a convincing improvement. This may be simply because the subject is in its infancy. Such methods should certainly be pursued further.

A disturbing property of the Fourier power spectra is that the lines produced by the modes are asymmetric. Evidently, this would not matter if the asymmetry were understood, because it could be duly taken into account by the data-analysis procedures. And indeed, there have been various studies of the asymmetry, which are summarized in Maurice Gabriel's (1998) contribution to this workshop. The principal ideas, which relate to the spatial nonuniformity of the excitation and damping processes, are certainly plausible, and I have no doubt that much of the asymmetry is accounted for in this way. Another possible source of asymmetry is frequency ber of elements. Just above the convection zone, the photosphere is a well mixed region (see however Solanki, 1998) whereas the outer solar layers show a very heterogeneous and changing structure. The structure and the physical processes of the photosphere are also rather well known allowing to reach good accuracies. It is also the layer that has been studied quite a long time before the other layers for obvious reasons: the solar photospheric spectrum has been recorded since quite a long time. For all these reasons, photospheric abundances will be adopted as a reference for all the other solar data (see also section 4).

3. Interest of Solar Abundances

The chemical composition of the Sun is a key data for modelling the Sun, the interior as well as the atmosphere. The role of the opacities and the crucial contributions of elements like Fe in the central layers and of O and Ne at the bottom of the convective zone has been stressed by Rogers (1998) and Turck-Chièze (1998).

Standard chemical composition is also the basic data that has to be reproduced by nucleosynthesis theories (see also section 5) and it plays also a key role in the chemical evolution of galaxies (see e.g. Pagel, 1997).

The Sun, being the best known star, has always been considered as the typical star, the reference to which the abundance analyses of other stars are compared (see section 5). It has been suggested that the Sun might be somewhat anomalous (see Grevesse *et al.*, 1996, for the references) but Gustafsson (1998) has convinced us at this Workshop that *the Sun is like many other stars* of the same age in our galaxy and that the claimed slight metal richness of our Sun is well within the real cosmic scatter.

When studying solar abundances we have access to important tracers of the structure and of the physical processes in the outer solar layers.

The *Sun* is also *unique* because *chemical composition* data *can be acquired for other types of matter in different objects of the solar system like the Earth, Moon, planets, comets, meteorites*. Few data come from planets; for the terrestrial planets, including the Earth, elements have either evaporated or fractionated. Very few reliable data are available for comets. This is one of the main goals of a future comet rendez-vous mission. *A very rare class of meteorites, the so-called CI carbonaceous chondrites, is of particular interest*. These meteorites have preserved the bulk composition of their parent bodies (planetesimals) and have thus retained most of the elements present in the primitive matter of the solar nebula, except for the few most volatile elements (see section 5).

4. Solar Abundances

For many reasons given above, the solar chemical composition, to which the results for the other layers will be compared, is the composition derived from the analysis of the solar photospheric spectrum.

Much progress has been made during the last decades. Solar photospheric spectra with very high resolution and very high signal over noise ratio, obtained from the ground and from space, are now available for quite a large wavelength range, from the UV to the far IR (see Kurucz, 1995 for a recent review). Empirical modelling of the photosphere has now reached a high degree of accuracy [see also section 4.5 and the reviews in this volume by Solanki (1998) and Rutten (1998)]. And, last but not least, accurate atomic and molecular data, in particular transition probabilities, have progressively been obtained for transitions of solar interest; these data play a key role in solar spectroscopy.

We have recently reviewed in detail this *key role* and, in particular, the role *of transition probabilities in solar spectroscopy*, not only in improving the abundance results but also as tracers of the physical conditions and processes in the solar photosphere (Grevesse and Noels, 1993; Grevesse et al., 1995). Most of the progress in our knowledge of the solar photospheric chemical composition during the last decades has been essentially, if not uniquely, due to the use of more accurate transition probabilities as seen in the examples given in the hereabove mentioned papers. Large discrepancies previously found between the Sun and CI meteorites (see section 5) have progressively disappeared as the accuracy of the transition probabilities has been increased. Actually, the dispersion of solar photospheric abundance results reflects the internal accuracy of the transition probabilities used to derive the abundances. The Sun is rarely (never?) at fault but unfortunately, older sets of transition probabilities were too often at fault! Hopefully, the techniques now allow to measure transition probabilities with high accuracy, even for rather faint lines. It has to be mentioned that many analyses of solar abundances have resulted from *close collaborations between atomic spectroscopists and solar spectroscopists*. Too rare groups, however, work to fill the gaps in the many data still needed by the astronomers.

Two recent papers on atomic transition probabilities and the solar abundance of Lu better illustrate the role of atomic data on solar abundances (Den Hartog et al., 1998; Bord et al., 1998). The solar photospheric abundance of Lu was 4 times larger than the very accurate meteoritic value. New accurate measurements of atomic transition probabilities for Lu II lines of solar interest have allowed to get the photospheric result down to the meteoritic value (see Table I).

In Table I, we give the best solar photospheric abundances taken from our latest review (Grevesse et al., 1996). Values are given in the logarithmic scale usually adopted by astronomers, $A_{e\ell} = \log N_{e\ell}/N_H + 12.0$, where $N_{e\ell}$ is the abundance by number. We now comment on a few elements.

4.1. HELIUM

In 1868, a new element is discovered in the solar spectrum obtained during an eclipse. The name of the Sun was given to the new element, helium. Helium is only discovered on Earth in 1895. Nowadays, its primordial abundance is known

Table I
Element Abundances in the Solar photosphere and in Meteorites

El.	Photosphere*	Meteorites	Ph-Met	El.	Photosphere*	Meteorites	Ph-Met
01 H	12.00	–	–	42 Mo	1.92 ±0.05	1.97 ±0.02	−0.05
02 He	[10.93 ±0.004]	–	–	44 Ru	1.84 ±0.07	1.83 ±0.04	+0.01
03 Li	1.10 ±0.10	3.31 ±0.04	−2.21	45 Rh	1.12 ±0.12	1.10 ±0.04	+0.02
04 Be	1.40 ±0.09	1.42 ±0.04	−0.02	46 Pd	1.69 ±0.04	1.70 ±0.04	−0.01
05 B	(2.55 ±0.30)	2.79 ±0.05	(−0.24)	47 Ag	(0.94 ±0.25)	1.24 ±0.04	(−0.30)
06 C	8.52 ±0.06	–	–	48 Cd	1.77 ±0.11	1.76 ±0.04	+0.01
07 N	7.92 ±0.06	–	–	49 In	(1.66 ±0.15)	0.82 ±0.04	(+0.84)
08 O	8.83 ±0.06	–	–	50 Sn	2.0 ±(0.3)	2.14 ±0.04	−0.14
09 F	[4.56 ±0.3]	4.48 ±0.06	+0.08	51 Sb	1.0 ±(0.3)	1.03 ±0.07	−0.03
10 Ne	[8.08 ±0.06]	–	–	52 Te	–	2.24 ±0.04	–
11 Na	6.33 ±0.03	6.32 ±0.02	+0.01	53 I	–	1.51 ±0.08	–
12 Mg	7.58 ±0.05	7.58 ±0.01	0.00	54 Xe	–	2.17 ±0.08	–
13 Al	6.47 ±0.07	6.49 ±0.01	−0.02	55 Cs	–	1.13 ±0.02	–
14 Si	7.55 ±0.05	7.56 ±0.01	−0.01	56 Ba	2.13 ±0.05	2.22 ±0.02	−0.09
15 P	5.45 ±(0.04)	5.56 ±0.06	−0.11	57 La	1.17 ±0.07	1.22 ±0.02	−0.05
16 S	7.33 ±0.11	7.20 ±0.06	+0.13	58 Ce	1.58 ±0.09	1.63 ±0.02	−0.05
17 Cl	[5.5 ±0.3]	5.28 ±0.06	0.22	59 Pr	0.71 ±0.08	0.80 ±0.02	−0.09
18 Ar	[6.40 ±0.06]	–	–	60 Nd	1.50 ±0.06	1.49 ±0.02	+0.01
19 K	5.12 ±0.13	5.13 ±0.02	−0.01	62 Sm	1.01 ±0.06	0.98 ±0.02	+0.03
20 Ca	6.36 ±0.02	6.35 ±0.01	+0.01	63 Eu	0.51 ±0.08	0.55 ±0.02	−0.04
21 Sc	3.17 ±0.10	3.10 ±0.01	+0.07	64 Gd	1.12 ±0.04	1.09 ±0.02	+0.03
22 Ti	5.02 ±0.06	4.94 ±0.02	+0.08	65 Tb	(−0.1 ±0.3)	0.35 ±0.02	(−0.45)
23 V	4.00 ±0.02	4.02 ±0.02	−0.02	66 Dy	1.14 ±0.08	1.17 ±0.02	−0.03
24 Cr	5.67 ±0.03	5.69 ±0.01	−0.02	67 Ho	(0.26 ±0.16)	0.51 ±0.02	(−0.25)
25 Mn	5.39 ±0.03	5.53 ±0.01	−0.14	68 Er	0.93 ±0.06	0.97 ±0.02	−0.04
26 Fe	7.50 ±0.05	7.50 ±0.01	0.00	69 Tm	(0.00 ±0.15)	0.15 ±0.02	(−0.15)
27 Co	4.92 ±0.04	4.91 ±0.01	+0.01	70 Yb	1.08 ±(0.15)	0.96 ±0.02	+0.12
28 Ni	6.25 ±0.04	6.25 ±0.01	0.00	71 Lu	0.06 ±0.10	0.13 ±0.02	−0.07
29 Cu	4.21 ±0.04	4.29 ±0.04	−0.08	72 Hf	0.88 ±(0.08)	0.75 ±0.02	+0.13
30 Zn	4.60 ±0.08	4.67 ±0.04	−0.07	73 Ta	–	−0.13±0.02	–
31 Ga	2.88 ±(0.10)	3.13 ±0.02	−0.25	74 W	(1.11 ±0.15)	0.69 ±0.03	(+0.42)
32 Ge	3.41 ±0.14	3.63 ±0.04	−0.22	75 Re	–	0.28 ±0.03	–
33 As	–	2.37 ±0.02	–	76 Os	1.45 ±0.10	1.39 ±0.02	+0.06
34 Se	–	3.41 ±0.03	–	77 Ir	1.35 ±(0.10)	1.37 ±0.02	−0.02
35 Br	–	2.63 ±0.04	–	78 Pt	1.8 ±0.3	1.69 ±0.04	+0.11
36 Kr	–	3.31 ±0.08	–	79 Au	(1.01 ±0.15)	0.85 ±0.04	(+0.16)
37 Rb	2.60 ±(0.15)	2.41 ±0.02	+0.19	80 Hg	–	1.13 ±0.08	–
38 Sr	2.97 ±0.07	2.92 ±0.02	+0.05	81 Tl	(0.9 ±0.2)	0.83 ±0.04	(+0.07)
39 Y	2.24 ±0.03	2.23 ±0.02	+0.01	82 Pb	1.95 ±0.08	2.06 ±0.04	−0.11
40 Zr	2.60 ±0.02	2.61 ±0.02	−0.01	83 Bi	–	0.71 ±0.04	–
41 Nb	1.42 ±0.06	1.40 ±0.02	+0.02	90 Th	–	0.09 ±0.02	–
				92 U	(< −0.47)	−0.50 ±0.04	–

Values between square brackets are not derived from the photosphere, but from sunspots, solar corona and solar wind particles – Values between parentheses are less accurate results – For He, see section 4.1; for Th, see Grevesse et al. (1996).

to a high degree of accuracy (see e.g. Pagel, 1997). As many data concerning solar He are discussed at length in many of the review papers presented in this volume, we shall only give a brief summary of the present state-of-the-art.

Despite its name and its very high abundance, He is not present in the photospheric spectrum and is largely lost by the meteorites. Solar wind and solar energetic particles show a very variable and rather low value (i.e. low when compared to values observed in hot stars and in the interstellar medium from H II regions around us). Coronal values derived from spectroscopy have large uncertainties: N_{He}/N_H = 7.9 ± 1.1 % (Gabriel et al., 1995) and 8.5 ± 1.3 % (Feldman, 1998). Giant planets, as observed by the Voyager spacecraft, do not allow to settle the question: Jupiter and Saturn show anomalously low values whereas higher values (9.2 ± 1.7 %) are found for Uranus and Neptune. Note that the recent Galileo spacecraft has recently measured an intermediate value on Jupiter, Y = 0.234, or N_{He}/N_H = 7.85 % (von Zahn and Hunten, 1996).

Progress in our knowledge of *the solar He content* has recently come *from solar standard models* as well as non standard models and the inversion of helioseismic data. While the *calibration of the standard models leads to an abundance of He* by mass of Y = 0.27 ± 0.01 (N_{He}/N_H = 9.5 %) *in the protosolar cloud* (Christensen-Dalsgaard, 1998), non standard models (i.e. taking element migration, for example, into account) start with an helium abundance of Y = 0.275 (Gabriel, 1997). *Inversion of helioseismic data leads to* a very accurate, but smaller, value, Y = 0.248 ± 0.002 (i.e. 8.5 %) as the value of *the present solar abundance of He in the outer convection zone* (Dziembowski, 1998). *The difference of 10 percent between these two values is* now interpreted as *due to element migration at the basis of the convection zone* during the solar lifetime (see e.g. Vauclair, 1998; Turcotte and Christensen-Dalsgaard, 1998).

In Table I, we give the present value in the outer layers, Y = 0.248 ± 0.002, or N_{He}/N_H = 8.5 %, or A_{He} = 10.93 ± 0.004. The value at birth of the Sun is Y = 0.275 ± 0.01, or N_{He}/N_H = 9.8 %, or A_{He} = 10.99 ± 0.02.

4.2. LITHIUM, BERYLLIUM, BORON

Since our latest review (Grevesse et al., 1996), things have changed for Be. Very recently, Balachandran and Bell (1998) have shown that *the solar abundance of Be* should be increased because an extra opacity source has to be introduced in the near UV region of the Be II lines and their result *is now in perfect agreement with the meteoritic value* (see Table I).

The Li Be B problem is now reduced to explaining how the Sun can deplete Li by a factor 160 whereas Be and B are not destroyed. Although conventional models fail to do that, mixing just below the bottom of the convection zone seems to be successful (see Blöcker et al., 1998; Vauclair, 1998; Zahn, 1998).

STRUCTURE OF THE SOLAR PHOTOSPHERE

SAMI K. SOLANKI
Institute of Astronomy, ETH, CH-8092 Zürich, Switzerland

Abstract. The majority of measured solar abundances refer to the solar photosphere. In general, when determining photospheric abundances a plane-parallel atmosphere and LTE are assumed. However, the photosphere is structured by granulation, magnetic fields and p-modes. They change line profiles by the thermal inhomogeneities and wavelength shifts they introduce. A brief description of the first two of these phenomena is given and some of the ways in which they influence abundances are pointed out. Departures from LTE also occur. The magnitude of the errors introduced into elemental abundances by neglecting such departures is also briefly discussed.

Key words: Solar Abundances, Solar Granulation, Solar Magnetic Fields, Radiative Transfer

1. Introduction

The photosphere of the sun is a comparatively thin layer (by solar standards), being only a few hundred km thick. Nevertheless, it is the most massive layer of the directly observable parts of the sun and contains some of the coolest solar gas. The most striking characteristic of the photosphere, however, is that it encompasses the visible surface of the sun, i.e. it is the layer in which the visible (and large parts of the UV and IR) solar spectrum is formed.

The richness of the photospheric spectrum and the relative simplicity with which it can be modelled are mainly responsible for our knowledge of such a large number of photospheric elemental abundances. Indeed, the photosphere is the solar layer for which the largest number of elemental abundances is known, so that in many cases the term "solar abundance" is synonymous with "photospheric abundance". The current status of photospheric abundance determinations is reviewed by Grevesse and Sauval (1998).

The solar photosphere is also special in another respect. It is the layer for which we possess the most sensitive diagnostics of temperature, velocity and magnetic field. Once again, this is largely due to the richness of the visible solar spectrum and the relative simplicity of line-formation physics in this layer. This means that we can actually test the assumptions underlying the determination of photospheric abundances. The availability of relatively realistic simulations of photospheric features and processes is also a significant asset.

The most important assumptions about the sun made when deriving abundances are:

1. Neglect of NLTE effects (i.e., departures from thermodynamic equilibrium),
2. Neglect of the horizontal structure of the photosphere.

Further simplifications are the description of solar velocity fields in terms of micro- and macroturbulence and the use of fudge factors to enhance the Van der Waals damping of spectral lines.

The aim of the present paper is to illustrate the validity (or otherwise) of these assumptions on the basis of examples taken from the literature. In Sect. 2 the influence of NLTE is discussed (while remaining within the framework of plane-parallel atmospheres). In Sects. 3 and 4 the influence of convection and magnetism, respectively, is assessed under the assumption of LTE.

2. NLTE Effects

The influence of NLTE on photospheric spectral lines has been studied in greatest detail for iron, so that I mainly discuss its behaviour here. One of the basic ingredients of every photospheric abundance analysis is the model atmosphere employed. Not only does the derived abundance directly (i.e. even in LTE) depend on the choice of atmosphere, but, as shown by Rutten and Kostik (1982), Steenbock (1985) and Holweger (1988), the magnitude of NLTE effects is also dictated by the model atmosphere.

In this context the empirical photospheric models can be divided into three classes, examples of each of which are plotted in Fig. 1.

1. LTE-based models: The most prominent example of this type of model is that of Holweger and Müller (1974, abbreviated here as HOLMUL), which is based on the model of Holweger (1967). Such models have a gentle $T(z)$ gradient in the photosphere and no chromospheric temperature rise, since the temperature must continue to follow the approximate source function of the stronger lines. These models look very similar to the radiative equilibrium models of Bell et al. (1976), Anderson (1989), Kurucz (1992a,b), etc. NLTE effects in such LTE models are very small in the photospheric layers, i.e. LTE provides an excellent description of the spectral lines when such HOLMUL-like models are used together with LTE; an effect referred to as LTE masking by Rutten and Kostik (1982).
2. Old NLTE-based models: The HSRA (Gingerich et al., 1971) and the VAL models (Vernazza et al., 1976, 1981) belong to this type. They are distinguished by a larger temperature gradient in the photosphere and a cool temperature minimum. Such models produce large departures from LTE.
3. New NLTE-based models: This is basically the model presented by Avrett (1985), as well as its derivates the MACKKL (Maltby et al., 1986) and FAL-C (Fontenla et al., 1993). These models are characterized by a high minimum temperature and a photospheric temperature stratification which is very similar to that of the HOLMUL model. As with that model, photospheric NLTE effects are relatively unimportant.

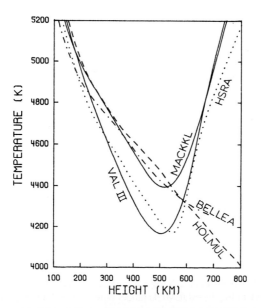

Figure 1. Temperature vs. height of 4 empirical and 1 radiative equilibrium model atmospheres. HOLMUL: Empirical, LTE-based model of Holweger and Müller (1974), HSRA: Empirical, NLTE-based model by Gingerich et al. (1971), VAL III: similar model by Vernazza et al. (1981), MACKKL: newer NLTE-based model due to Maltby et al. (1986), BELLEA: LTE-based radiative equilibrium model of Bell et al. (1976). Figure from Rutten (1988) by permission.

The residual NLTE effects for FE I and II (for the HOLMUL atmosphere) are plotted in Fig. 2. Note that for almost all lines the logarithmic abundance correction

$$\Delta \log \epsilon = \log \epsilon_{NLTE} - \log \epsilon_{LTE} < 0.05 \,,$$

where ϵ is the derived elemental abundance relative to hydrogen. Departures of this magnitude are only exhibited by low excitation Fe I lines. For the majority ion, Fe II, $\Delta \log \epsilon < 0.005$ and is thus completely negligible.

In summary: Abundance determinations based on photospheric and 1-D models are only little affected by neglecting departures from LTE, if the plane-parallel assumption is correct.

3. Structuring the Photosphere: Convection

3.1. Properties of Solar Granulation

Convection cells of different sizes are observed in the solar photosphere. They are generally divided into 3 classes. The smallest, the granules, are typically 1500 km

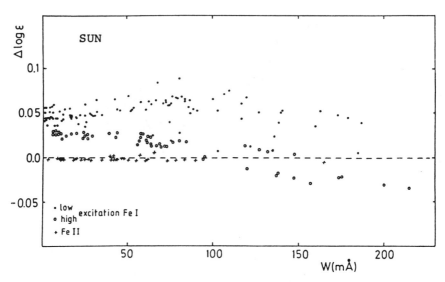

Figure 2. NLTE abundance correction $\Delta \log \epsilon = \log \epsilon_{\text{NLTE}} - \log \epsilon_{\text{LTE}}$ for Fe I and Fe II lines plotted vs. equivalent width (W_λ) for the HOLMUL model atmosphere. Each symbol represents a spectral line. Lines with different excitations or belonging to different ionization stages are distinguished by different types of symbols (from Holweger, 1988, by permission).

across, mesogranules are roughly 7000 km, while supergranules possess diameters of 20000–30000 km. There is, as for any highly dynamic fluid phenomenon, a large scatter around these values. As far as their influence on abundances is concerned granules are the most important and the further discussion is restricted to them.

Granules have mean lifetimes of 6–8 minutes, rms vertical velocities on the order of 1 km/s and horizontal rms velocities that are somewhat larger. Peak velocities are considerably larger and there is evidence for supersonic horizontal flows. Granulation is clearly visible in continuum images, exhibiting contrasts $(\delta I/I)_{\text{rms}}$ in the visible that can exceed 10% in the best images (here I is the intensity). Note that the contrast value is a strong function of spatial resolution and wavelength. Granulation also possesses a distinctive surface topology, being composed of isolated, bright granules, that correlate with upflows, and multiply connected, dark intergranular lanes, correlating with downflows. The narrow lanes cover a smaller part of the solar surface than the broad granules. The downflow velocities are also correspondingly higher than the upflows.

These and further observed properties of granulation are reviewed by Muller (1989), Title *et al.* (1990), Karpinsky (1990) and Spruit *et al.* (1990). References to the vast literature on observations of solar granulation may be found therein.

There is also a growing body of work on granule simulations, i.e. time-dependent solutions of the radiation-hydrodynamic equations on a two- or three-dimensional spatial grid. Modern 2-D (Steffen, 1990; Steffen and Freytag, 1991;

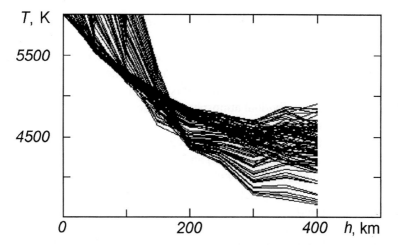

Figure 3. Temperature stratification (T vs. height h) along vertical rays passing through a 3-D simulation of granular convection. Note the large scatter. A careful inspection reveals a systematic correlation between the temperature in the lower photosphere and the temperature gradient (adapted from Atroshchenko and Gadun, 1994; figure kindly provided by A. Gadun).

Gadun *et al.*, 1997, 1998; Solanki *et al.*, 1996a) and in particular 3-D simulations (Nordlund, 1984, 1985; Stein and Nordlund, 1989; Lites *et al.*, 1989; Malagoli *et al.*, 1990; Atroshchenko and Gadun, 1994; Rast, 1995) reproduce a large number of observations, often with considerable quantitative accuracy. The remaining differences between the best simulations and the observations are partly caused by the need to introduce a numerical viscosity into the simulations and the consequently limited Reynolds numbers, but in part are also due to uncertainties in the magnitude and exact form of spatial distortions produced by the Earth's atmosphere.

3.2. GRANULATION AND CONVECTION

The main diagnostic of elemental abundances is spectral line equivalent width. The equivalent width is also strongly dependent on the temperature stratification and, to a lesser degree, on velocity.

Consider first the temperature. Figure 3 shows $T(z)$ at different horizontal positions of a set of the 3-D simulations of Atroshchenko and Gadun (1994). The scatter in temperatures is greater than 500 K in both the upper and lower photosphere. The 3-D models of Nordlund (1985) exhibit an even larger scatter (over 1000 K in the upper photosphere). Such a difference in temperature non-linearly affects line strengths of minor-ion lines (such as those of Fe I).

It is not just this scatter that is important, but also the correlation between the temperature near the continuum-forming layer (i.e. $h \approx 0$ in the figure) and the temperature gradient. Although perhaps not so well visible in the figure, regions that are hot in the lower layers are cooler in the upper layers (cf. Stein and Nordlund, 1989), in qualitative agreement with the observations, although these also indicate the importance of wave motions in determining the structure of the upper photosphere (e.g., Komm *et al.*, 1990).

This structure implies, however, that spectral lines are particularly deep over granules and particularly shallow over intergranular lanes. Since this dependence on temperature stratification is non-linear it is unlikely that the average of the profiles resulting from the individual atmospheres possesses the same W_λ as the profile resulting from the averaged atmosphere.

The influence of granulation on Fe abundances has been studied by Nordlund (1984) and Gadun and Pavlenko (1998) and on Li by Kiselman (1997), Uitenbroek (1998) and Gadun and Pavlenko (1998). The spate of work on Li followed an initial suggestion by Kurucz (1995) that convection could falsify the Li abundance by up to an order of magnitude.

The detailed calculations of Uitenbroek (2-D NLTE radiative transfer), Kiselman (1.5-D NLTE) and Gadun and Pavlenko (1.5-D LTE) showed that these fears are unfounded for the solar case. Of interest is the difference in A, the logarithmic abundance relative to hydrogen, between the multi-dimensional and the 1-D case. In the former case the radiative transfer is carried out along multiple rays passing through the simulation and the emerging profiles are then averaged. This average profile is compared with the observations (which have low spatial and temporal, but high spectral resolution and high S/N ratio). In the latter case in principle a plane-parallel reference model (e.g. HOLMUL) can be used, but in order to isolate the influence of spatial inhomogeneities it is better to use an atmosphere constructed by forming a horizontal average of the temperature stratifications of the granulation simulation. The average $T(\tau)$ looks rather similar to HOLMUL or RE models, although the seemingly small differences have a significant effect (see below).

For Li the abundance derived from 2-D and 3-D models relative to the plane-parallel case differs by 0.1–0.2 dex (Gadun and Pavlenko, 1998) and by 0.1 dex or less (Kiselman, 1997; Uitenbroek, 1998).

In the case of iron the difference is also small: 0.1 dex for Fe I and only 0.05 dex for Fe II (Gadun and Pavlenko, 1998). Earlier, Nordlund (1984) found a greater difference (0.3–0.4 dex), but ascribed that to shortcomings in his old, incompressible models. Gadun and Pavlenko (1997) did find an additional and more disturbing deviation from the HOLMUL results, however. Whereas for HOLMUL A(Fe II) = A(Fe I) the simulations imply a difference between the abundance derived from the two ions: A(Fe II) = A(Fe I) + 0.3.

The reason for this discrepancy is unclear. Due to the sensitivity of NLTE on the detailed temperature stratification, it is conceivable that NLTE effects need to be taken into account when using 2-D simulations to determine abundances. On the other hand, this discrepancy may be an artifact caused by (on average) somewhat incorrect temperature gradients in the granulation simulations due to a not sufficiently exact description of the UV radiation field (the opacities of Kurucz, 1979, are used, which do not yet include sufficient lines in the UV).

An empirical approach was taken by Kiselman (1994), who considered the variation of spectral line parameters across granules. He also concluded that the influence of convection on abundances should be small (cf. Holweger *et al.*, 1990).

Granular velocity fields also influence line profiles. Traditionally, spectral lines have been broadened using a mixture of micro- and macroturbulence. In reality, convective and wave-like or oscillatory motions dominate the solar velocity field; the former in the lower and mid photosphere, the latter in the higher layers. Note that in the above investigations these motions were consistently taken into account. They have strong horizontal and vertical gradients. The former is clearly visible in the wiggly line shapes along the slit in high resolution spectra. The latter can be deduced from line asymmetries in line profiles observed with high spatial resolution; e.g., Bonnacini (1989, see his Fig. 6). As illustrated by Fig. 4 the sum of these motions broadens the line profile sufficiently to match the observations without requiring additional micro- and macroturbulence (e.g., Nordlund, 1984; Lites *et al.*, 1989; Steffen and Freytag, 1991; Gadun *et al.*, 1998). Note that at disc centre horizontal gradients of the velocity produce no change in W_λ, i.e. to first order act like macroturbulence, while vertical gradients produce an enhancement of W_λ, i.e. act like microturbulence. The mixture of vertical and horizontal gradient, however, need not have the same influence on W_λ as the mixture of micro- and macroturbulence used to reproduce the line profile shape. Hence the velocity structure also affects the derived abundances, although the influence is expected to be smaller than of the thermal inhomogeneities (at least for not too strong spectral lines).

Finally, the wings of strong lines appear to be enhanced by granulation (Bruls and Rutten, 1992), which may provide an alternative to the generally used fudge-factors to the Van der Waals damping constants. How far this has an effect on the abundance is unclear, although the too strong wings produced by the simulations (Bruls and Rutten, 1992) might point to the need for lower abundances of the elements in question (Na and K), or alternatively to shortcomings of the simulations, although Stuik *et al.* (1997) question such a need.

4. Structuring the Photosphere: Magnetism

The main structuring agent of the solar photosphere besides convection is the magnetic field. Magnetic features range from the smallest magnetic elements (estimated diameters of approximately 50 km) to sunspots (diameters of up to 50000 km). Over the 6 orders of magnitude of flux per magnetic feature thus spanned the intrinsic field strength in the lower photospheric layers remains remarkably constant, lying between 1000 and 2000 G (averaged over the magnetic feature). A simulated magnetic element is illustrated in Fig. 5 (Steiner *et al.*, 1996). Due to the requirement of horizontal force balance (which for smaller magnetic features is basically horizontal pressure balance) and exponential decrease of gas pressure with height

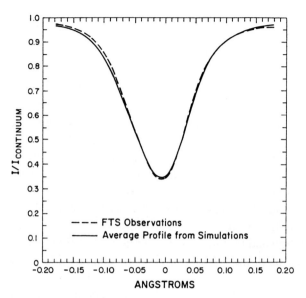

Figure 4. Spatially and temporally averaged profile of Fe I 630.25 nm observed at solar disc centre with an Fourier transform spectrometer (dashed curve) and synthesized from 3-D granular simulations (solid curve). Note the similarity between the two profiles (from Lites *et al.*, 1989, by permission).

the field strength also decreases rapidly with height. Magnetic flux conservation then requires that the magnetic field flares out with height.

Thus, although magnetic features cover at the most 1% of the solar surface in the lower photosphere they fill ever larger portions of the atmosphere with increasing height. In and above the middle-chromosphere the field occupies all available space.

In contrast to the field strength the thermal structure of magnetic features depends strongly on their size. Magnetic elements are brighter than the quiet sun, while sunspots are darker. The properties of magnetic elements and sunspots are reviewed in greater detail by Stenflo (1989), Spruit *et al.* (1990), Solanki (1993) and Schüssler (1993).

Sunspots are clearly visible and can be easily avoided when observing. Consequently, they mainly play a role for abundance determinations when molecular species provide the most reliable abundances (but see, e.g., Ritzenhoff *et al.*, 1997, who determine the Li abundance from atomic lines in sunspot spectra). Magnetic elements, on the other hand, are distributed all over the sun, including the quiet sun (at the supergranule boundaries). In the middle and upper layers of the photosphere they are considerably hotter than the average quiet sun (see Fig. 6), which leads to a considerable weakening of spectral lines, in particular of minor ions. If the enhanced temperature is not taken into account it can lead to over a factor of

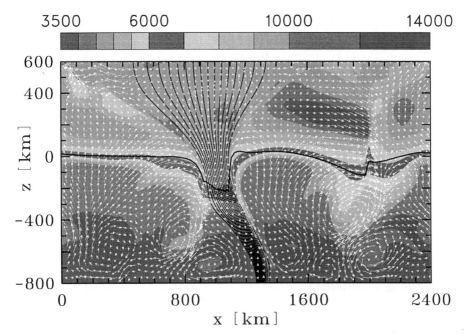

Figure 5. Snapshot of a 2-D simulation of a magnetic element (magnetic flux slab) lasting over a total period of 18.5 minutes real time. The nearly vertical solid lines near the centre of the frame are representative field lines (note their spreading with height). The thick nearly horizontal black curve marks unit continuum optical depth ($\tau_{5000} = 1$) for vertically incident lines of sight. The arrows indicate the strength and direction of the flow field, while the double grey scales indicate the temperature according to the bar at the top of the figure (from Steiner et al., 1996, by permission).

2 error in the elemental abundance within the magnetic element. However, since outside active regions magnetic elements cover only 1% of the surface they have no practical implications for photospheric abundances.*

In addition to the highly visible kG fields discussed above, there is at least as much magnetic flux in weak fields, i.e. those with intrinsic field strengths of a few G to a few hundred G (Faurobert-Scholl et al., 1995; Solanki et al., 1996b; Meunier et al., 1998). Only recently have new observing techniques made these fields amenable to study.

Due to their intrinsic weakness they cover a much larger portion of the solar surface than magnetic elements. Could they significantly affect abundance determinations?

Unfortunately, the thermal structure of these weak fields is unknown. However, theoretical arguments indicate that the temperature is not very different from that of

---

* Velocities and in particular magnetic fields have a far smaller effect even than the temperature. Thus for most studies of the quiet sun it is much more important to select spectral lines according to their temperature sensitivity rather than to follow the common practice of choosing $g = 0$ lines in order to avoid undue influencing by magnetic features.

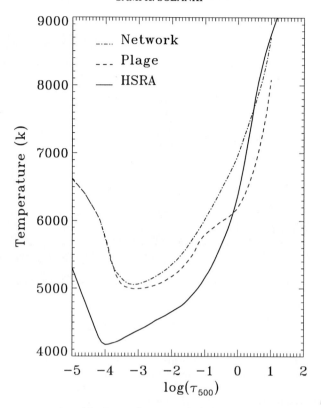

Figure 6. Temperature vs. logarithmic continuum optical depth (log τ_{5000}) of 3 empirical model atmospheres. Solid curve: Quiet sun model HSRA, dashed curve: model describing the magnetic elements found in active region plage, dot-dashed curve: magnetic elements forming the network (from Briand and Solanki, 1995).

the surrounding gas. Hence they probably play a smaller role than the granulation in affecting abundance determinations.

Can we therefore completely forget magnetic features when considering solar abundances? Although this may well be true it appears somewhat premature to draw this conclusion in its full generality. Recall that although magnetic elements cover less than 1% of the solar surface, their magnetic fields fill much of the chromosphere and all the corona and heliosphere. Hence the gas in which First Ionisation Potential (FIP) fractionation takes place (see Geiss, 1998; von Steiger, 1998) is connected to the magnetic features and not to the bulk of the photospheric material to which the photospheric abundances apply. In principle it is possible that abundances in magnetic elements are considerably different form those in the "field free" photosphere (e.g., photospheric magnetic elements could exhibit the FIP effect), without this being noticed, since no rigorous abundance determinations have been carried out so far. Although it appears unlikely on theoretical grounds, abundances in the photospheric layers of magnetic elements do provide a boundary condition for FIP theories and should be determined.

Acknowledgements

This paper is dedicated to H. Holweger, who has devoted a large part of his life to the topics covered here and who has been right more often than he admits. C. Briand, A. Gadun, R. Rutten, M. Schüssler and O. Steiner kindly provided original figures.

References

Anderson, L.S.: 1989, *Astrophys. J.* **339**, 558.
Avrett, E.H.: 1985, in *Chromospheric Diagnostics and Modelling*, B.W. Lites (Ed.), National Solar Obs., Sunspot, NM, p. 67.
Atroshchenko, I.N. and Gadun, A.S.: 1994, *Astron. Astrophys.* **291**, 635.
Bell, R.A., Eriksson, K., Gustafsson, B., and Nordlund, A.: 1976, *Astron. Astrophys. Suppl. Ser.* **23**, 37.
Bonnacini, D.: 1989, in *High Spatial Resolution Solar Observations*, O. von der Lühe (Ed.), National Solar Observatory, Sunspot, NM, p. 241.
Briand, C. and Solanki, S.K.: 1995, *Astron. Astrophys.* **299**, 596.
Bruls, J.H.M.J. and Rutten, R.J.: 1992, *Astron. Astrophys.* **265**, 257.
Faurobert-Scholl, M., Feautrier, N., Machefert, F., Petrovay, K., and Spielfiedel, A.: 1995, *Astron. Astrophys.* **298**, 289.
Fontenla, J.M., Avrett, E.H., and Loeser, R.: 1993, *Astrophys. J.* **406**, 319.
Gadun, A. and Pavlenko, YA. V.: 1997, *Astron. Astrophys.* **324**, 281.
Gadun, A.S., Hanslmeier, A., and Pikalov, K.N.: 1997, *Astron. Astrophys.* **320**, 1001.
Gadun, A., Solanki, S.K., and Johannesson, A.: 1998, *Astron. Astrophys.* , submitted.
Geiss, J.: 1998, *Space Sci. Rev.*, this volume.
Grevesse, N., and Sauval, A.J.: 1998, *Space Sci. Rev.*, this volume.
Holweger, H.: 1967, *Z. Astrophys.* **65**, 365.
Holweger, H.: 1988, in *The Impact of Very High S/N Spectroscopy on Stellar Physics*, G. Cayrel de Strobel, M. Spite (Eds.), Kluwer, Dordrecht *IAU Symp.* **132**, 411.
Holweger, H. and Müller, E.A.: 1974, *Solar Phys.* **39**, 19.
Holweger, H., Heise, C., and Kock, M.: 1990, *Astron. Astrophys.* **232**, 510.
Karpinsky, V.N.: 1990, in *Solar Photosphere: Structure, Convection and Magnetic Fields*, J.O. Stenflo (Ed.), *IAU Symp.* **138**, 67.
Kiselman, D.: 1994, *Astron. Astrophys. Suppl. Ser.* **104**, 23.
Kiselman, D.: 1997, *Astrophys. J.* **489**, L107,
Komm, R., Mattig, W., and Nesis, A.: 1990, *Astron. Astrophys.* **239**, 340.
Kurucz, R.L.: 1991, in *Precision Photometry: Astrophysics of the Galaxy*, A.G. Davis Philip, A.R. Upgren, K.A. Janes (Eds.), L. Davis Press, Schenectady.
Kurucz, R.L.: 1992, *Rev. Mexicana Astron. Astrof.* **23**, 181.
Kurucz, R.L.: 1995, *Astrophys. J.* **452**, 102.
Lites, B.W., Nordlund, Å., and Scharmer, G.B.: 1989, in *Solar and Stellar Granulation*, R.J. Rutten, G. Severino (Eds.), Reidel, Dordrecht, p. 349.
Malagoli, A., Cattaneo, F., and Brumell, N.H.: 1990, *Astrophys. J.* **361**, L33.
Maltby, P., et al.: 1986, *Astrophys. J.* **306**, 284.
Meunier, N., Solanki, S.K., and Livingston, W.C.: 1998, *Astron. Astrophys.* **331**, 771.
Muller, R.: 1989, in *Solar and Stellar Granulation*, R.J. Rutten, G. Severino (Eds.), Reidel, Dordrecht, p. 101.
Nordlund, Å.: 1984, in *Small–Scale Dynamical Processes in Quiet Stellar Atmospheres*, S.L. Keil (Ed.), National Solar Obs., Sunspot, NM, p. 181.
Nordlund, Å.: 1985, in *Theoretical Problems in High Resolution Solar Physics*, H.U. Schmidt (Ed.), Max-Planck-Inst. f. Astrophys., Munich, p. 1.

Rast, M.P., Nordlund, Å., Stein, R.F., and Toomre, J.: 1993, *Astrophys. J.* **408**, L53.
Rast, M.P.: 1995, *Astrophys. J.* **443**, 863.
Ritzenhoff, S., Schroter, E. H., and Schmidt, W.: 1997, *Astron. Astrophys.* **328**, 695.
Rutten, R.J.: 1988, in *Physics of Formation of Fe II Lines Outside LTE*, R. Viotti (Ed.), Reidel, Dordrecht *IAU Coll.* **94**, 185.
Rutten, R.J. and Kostik, R.I.: 1982, *Astron. Astrophys.* **115**, 104.
Schüssler, M.: 1992, in *The Sun — a Laboratory for Astrophysics*, J.T. Schmelz, J.C. Brown (Eds.), Kluwer, Dordrecht, p. 191.
Solanki, S.K.: 1993, *Space Science Rev.* **61**, 1.
Solanki, S.K., Rüedi, I., Bianda, M., and Steffen, M.: 1996a, *Astron. Astrophys.* **308**, 623.
Solanki, S.K., Zuffrey, D., Lin, H., Rüedi, I., and Kuhn, J.: 1996b, *Astron. Astrophys.* **310**, L33.
Spruit, H.C., Nordlund, Å., and Title, A.M.: 1990, *Ann. Rev. Astron. Astrophys.* **28**, 263.
Spruit, H.C., Schüssler, M., and Solanki, S.K.: 1992, in *Solar Interior and Atmosphere*, A.N. Cox et al. (Eds.), University of Arizona press, Tucson, AZ, p. 890.
Steenbock, W.: 1985, in *Cool Stars with Excesses of Heavy Elements*, M. Jaschek, P.C. Keenan (Eds.), Reidel, Dordrecht, p. 231.
Steffen, M.: 1991, in *Stellar Atmospheres: Beyond Classical Models*, L. Crivellari et al. (Eds.), Kluwer, Dordrecht, p. 247.
Steffen, M. and Freytag, B.: 1991, *Rev. Mod. Astron.* **4**, 43.
Stein, R.F. and Nordlund Å.: 1989, *Astrophys. J.* **342**, L95.
Steiner, O., Grossmann-Doerth, U., Knölker, M., and Schüssler, M.: 1996, *Solar Phys.* **164**, 223.
Stenflo, J.O.: 1989, *Astron. Astrophys. Rev.* **1**, 3.
Stuik R., Bruls J.H.M.J., Rutten R.J., 1997, *Astron. Astrophys.* **322** 911
Title, A.M., et al.: 1990, in *Solar Photosphere: Structure, Convection, Magnetic Fields*, J.O. Stenflo (Ed.), Kluwer, Dordrecht *IAU Symp.* **138**, 49.
Vernazza, J.E., Avrett, E.H., and Loeser, R.: 1976, *Astrophys. J. Suppl. Ser.* **30**, 1.
Vernazza, J.E., Avrett, E.H., and Loeser, R.: 1981, *Astrophys. J. Suppl. Ser.* **45**, 635.
von Steiger, R.: 1998, *Space Sci. Rev.*, this volume.

THE STRUCTURE OF THE CHROMOSPHERE
Properties Pertaining to Element Fractionation

P. G. JUDGE and H. PETER
High Altitude Observatory
NCAR* PO Box 3000, Boulder CO 80307

Abstract. We review the structure and dynamics of the solar chromosphere with emphasis on the quiet Sun and properties that are relevant to element fractionation mechanisms. Attention is given to the chromospheric magnetic field, its connections to the photosphere, and to the dynamical evolution of the chromosphere. While some profound advances have been made in the "unmagnetized" chromosphere, our knowledge of the magnetically controlled chromosphere, more relevant for the discussion of element fractionation, is limited. Given the dynamic nature of the chromosphere and the poorly understood magnetic linkage to the corona, it is unlikely that we will soon know the detailed processes leading to FIP fractionation.

Key words: Sun:chromosphere; abundance patterns

Abbreviations: FIP: first ionization potential; FIT: first ionization time

1. Introduction

At a meeting devoted to solar abundances from core to corona and into the heliosphere, it is daunting to discuss one of the least understood regions of the Sun: the chromosphere. Our lack of understanding, arising from some difficult physical regimes encountered in the chromosphere, contrasts with a need for understanding driven in part by the subject of this workshop: observed abundance patterns. Geiss (1982) was first to point out that the well-established first ionization potential, or "FIP" effect observed in the Sun's wind and corona, almost certainly arises from processes in the chromosphere. The FIP effect is such that the abundances of elements with low FIP values are enhanced by a factor of four relative to high FIP elements in the slow wind and solar energetic particles, compared with the photosphere. The hypothesis that the FIP effect has its origin in the chromosphere will be adopted as a focusing point for the rest of this paper, which is organized as follows. First, a brief review of the multi-fluid equations is given to see what kind of physical forces and conditions can influence element fractionation in the chromosphere. Next, a review of the properties of the chromosphere is given, from the static 1D models through to a 3D time-dependent picture in which the chromosphere is driven by photospheric motions and mediated by magnetic fields. Finally, we discuss some existing fractionation models in the context of these properties.

* The National Center for Atmospheric Research is sponsored by the National Science Foundation

2. Forces Influencing Element Fractionation in the Chromosphere

To address element fractionation, a multi-fluid description has to be applied. The simplest physically meaningful approach is to start from the Boltzmann equation and use the five-moment approximation, yielding for each ion the equations of continuity, momentum and energy, see e.g. Schunk (1975).

$$\partial_t n_i + \nabla \cdot (n_i \boldsymbol{u}_i) = \delta N_i/\delta t, \tag{1}$$

$$n_i m_i (\partial_t + \boldsymbol{u}_i \cdot \nabla) \boldsymbol{u}_i + \nabla p_i - n_i m_i \boldsymbol{g}$$
$$- n_i Z_i e (\boldsymbol{E} + \frac{\boldsymbol{u}_i}{c} \times \boldsymbol{B}) = \delta \boldsymbol{M}_i/\delta t + n_i m_i \boldsymbol{D}_i, \tag{2}$$

$$(\partial_t + \boldsymbol{u}_i \cdot \nabla)(3/2 p_i) + (5/3) p_i \nabla \cdot \boldsymbol{u}_i + \nabla \cdot \boldsymbol{q} = \delta E_i/\delta t + H_i - L_i. \tag{3}$$

The external forces are those due to gravity and electro-magnetic fields, $\boldsymbol{F} = m_j \boldsymbol{g} + Z_i e(\boldsymbol{E} + \frac{\boldsymbol{u}_i}{c} \times \boldsymbol{B})$. All "others" (e.g., due to radiation pressure gradients) are included schematically in the term $n_i m_i \boldsymbol{D}_i$, which in some work can also represent wave motions not included on the LHS of equations (2). Similarly, sources and sinks of energy are written schematically on the RHS of equation (3) in the terms H_i and L_i respectively. Other symbols have their usual meanings. The sources and sinks for the population densities of the different species are described by

$$\delta N_i/\delta t = \sum_k \left(n_k \gamma_{ki} - n_i \gamma_{ik} \right), \tag{4}$$

where γ_{ik} denotes rates for the reactive processes like ionization/recombination or excitation/deexcitation.

The momentum transport due to collisions, reactive processes and thermal forces is described by

$$\delta \boldsymbol{M}_i/\delta t = \sum_k n_i m_i \nu_{ik} (\boldsymbol{u}_k - \boldsymbol{u}_i) + \sum_k n_k m_k \gamma_{ki} (\boldsymbol{u}_k - \boldsymbol{u}_i)$$
$$+ \sum_k z_{ik} \nu_{ik} \mu_{ik}/(k_B T_{ik}) \left(\boldsymbol{q}_i - (\rho_i/\rho_k) \boldsymbol{q}_k \right). \tag{5}$$

Here ν_{ik} is the elastic collision frequency, which is density and temperature dependent, e.g., Schunk (1975). At densities and temperatures characteristic of the upper chromosphere, typical values for collisional frequencies are $10\,\text{sec}^{-1}$ (neutral–neutral, ion–neutral) and $10^4\,\text{sec}^{-1}$ (ion–ion), with coefficients z_{ik} in the thermal force of $-1/5$, 0 and $3/5$ respectively. Equations (1)–(3) are most simply closed by assuming the (isotropic) ideal gas pressure, $p_i = n_i k_B T_i$, and a heat flux following Spitzer's law, $\boldsymbol{q}_i = \kappa_0 T_i^{5/2} \nabla T_i$. The electric field in the momentum equation (2) can be re-written in terms of ∇p_e, ∇T_e using the electron momentum equation (e.g., Hansteen et al. 1997).

Element fractionation must occur through different net forces on different ions. The observed FIP or FIT ("first ionization time") patterns indicate that the most important forces must depend critically on whether a given atom is charged or not, and less critically on, for instance, the atom's mass. The momentum equation shows that charge dependent forces fall into two groups: the electromagnetic and collisional forces. The former (along with pressure gradients, gravity, thermal forces) can "drive" the system, in the sense of producing velocity differences between different species. In contrast, the latter tend to "relax" the system to a state that approaches a single fluid (i.e. fully collisionally coupled).

It is interesting that, owing to the complexity of calculating some driving terms, notably the Lorentz force, models of the FIP effect have often adopted a kinematic model in which a flow field is assumed *a priori* and the fractionation occurs owing to incomplete collisional coupling (e.g., Peter 1998). The exact process or processes which dominate the fractionation processes have not yet been identified, although many possibilities have been proposed (e.g., Hénoux 1998).

To assess the relative merits of the proposed fractionation processes we now review thermal properties of the gas, configuration(s) of the magnetic field, and their observable dependence on time.

3. Quasi-Static Structure of the Chromosphere

3.1. THERMAL PROPERTIES — 1D ATMOSPHERES

To get a feel for physical conditions in the chromosphere we can look at the VAL-III C model (Vernazza *et al.* 1981). This kind of model remains an essential starting point: it can help to give an idea of the physical regimes in which chromospheric plasma exists; the density stratification is probably reasonable (outside of the enigmatic spicules); therefore it can give us indications of where a variety of observed spectral features are formed.

Table I lists some physical properties derived in part from this model. The model was derived by extending ideas from the classical theory of stellar atmospheres (e.g., Mihalas, 1978). The model is a static, single fluid model with constant elemental abundances. Furthermore, "microturbulence" was included (we will return to this below). The main conceptual difference in the way the chromosphere was handled involves the energy equation. Classically, the energy equation for an atmosphere in radiative and convective equilibrium takes the form $\frac{d}{dz}(F_R + F_C) = 0$, where F_R, F_C are known or calculable flux densities of radiative and convective energy, and z is a height variable. With accompanying standard equations, the structure of the atmosphere can be calculated. In the chromosphere the energy equation is instead assumed to take the form $\frac{d}{dz}(F_R + F_M) = 0$, where F_M is an *unknown* flux density of mechanical energy whose divergence is assumed to heat the chromosphere. Because F_M is not known *a priori*, no energy equation was solved in VAL-III, instead temperature T_e as a function of height z was adjusted to match observed intensities with those computed from the model in a non-LTE treatment.

Table I
Some Relevant Physical Properties

Physical quantity	Description	Photosph.	Lower Chromosph.	Upper Chromosph.
T_e	electron temperature K	5840	5030	8180
N_e	electron density cm^{-3}	2×10^{13}	8×10^{10}	4×10^{10}
N_H	hydrogen density cm^{-3}	9×10^{17}	3×10^{14}	7×10^{10}
c_S [km s^{-1}]	sound speed km s^{-1}	8	7	9
ξ [km s^{-1}]	micro-turbulence km s^{-1}	1	1	9
$|B|$	typical field strength G (in flux tubes)	1000	100	10
c_A	Alfvén speed km s^{-1}	22	40	260
β	$\frac{P_{gas}}{(B^2/8\pi)}$	2	0.6	0.05
ω_e	electron gyro freq. Hz	2×10^{10}	2×10^9	2×10^8
τ_{ee}	electron-electron collision time sec.	8×10^{-10}	10^{-7}	5×10^{-7}
ω_i	proton gyro freq. Hz	10^7	10^6	10^5
τ_{pp}	proton-proton collision time sec.	10^{-7}	2×10^{-5}	3×10^{-5}
τ_{pn}	proton-neutral collision time sec.	9×10^{-9}	3×10^{-6}	3×10^{-2}
$I/k_B T_e$	low-FIP /thermal energy	10	12	7
$I/k_B T_e$	high-FIP/thermal energy	30	35	21

Some properties of the VAL-III C model are shown in Fig. 1. Although $T_e(z)$ is a free parameter in such models, the derived structure makes physical sense. The temperature steadfastly refuses to rise above 10^4K until hydrogen becomes fully ionized. This is because of two micro-physical effects (ionization of hydrogen, leading to a high specific heat, and radiative losses due to inelastic collisions of electrons with trace elements) which turn the entire chromosphere into a giant "thermostat". Since this behavior results from basic gas thermodynamics, more complex chromospheric models can be expected to share this characteristic.

Several initial conclusions can be drawn concerning element fractionation using this model. First, particle densities at the base and top of the chromosphere differ by four orders of magnitude. The collisional (frictional) coupling between different species is correspondingly larger at the base of the chromosphere, requiring correspondingly higher "driving" forces to fractionate elements there. With the exception of the Lorentz force, all the forces in the momentum equation that compete with frictional forces scale with density in the same way, and thus are of similar relative magnitudes at the base and top of the chromosphere (pressure gradients, gravity)*. Unless there are enormous Lorentz forces operating at the chromospheric base the most natural site for ion fractionation is the upper chromosphere. Most fractionation models have therefore focussed on this region. All fractionation mod-

---

* Here we assume the thermal force, depending on the temperature gradient, is negligible in the chromosphere, a reasonable approximation.

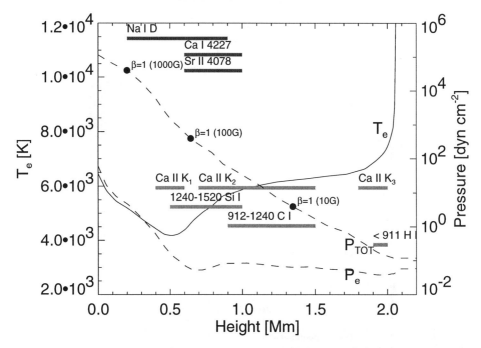

Figure 1. Properties of the VAL III C model of the chromosphere (Vernazza, Avrett, and Loeser, 1981). Formation heights (relative to $\tau_{5000\text{Å}} = 1$) of several spectral features discussed in the text (light shading) and of lines that have been used to diagnose magnetic fields (dark shading) are shown. Heights where gas and magnetic pressures are equal are marked as $\beta = 1$ for field strengths of 1000, 100 and 10 Gauss.

els depend on the magnetic structure they are embedded in, e.g. involving particle transport across the field, or use emerging magnetic field to "lift" the ions, or use current systems or simply magnetic fields to channel a flow. It is then of crucial importance to think about the magnetic structure and its dynamics.

3.2. MAGNETIC STRUCTURE OF THE CHROMOSPHERE

Measurement of chromospheric magnetic fields is difficult, for several physical reasons, including the fact that the fields are intrinsically weak and have a correspondingly weak signature in the emitted spectrum. Fig. 1 includes several chromospheric lines that have been used to determine magnetic field strengths and configurations (see below). These measurements are, to date, relatively crude and are limited to regions low in the chromosphere, so first we examine simple physical arguments based upon extrapolations of much stronger fields measured more accurately in the photosphere.

Chromospheric densities follow closely an exponential stratification, resulting from the assumption of hydrostatic equilibrium and a gas temperature that is nearly constant. The solar photosphere is permeated by finely structured and dynamic

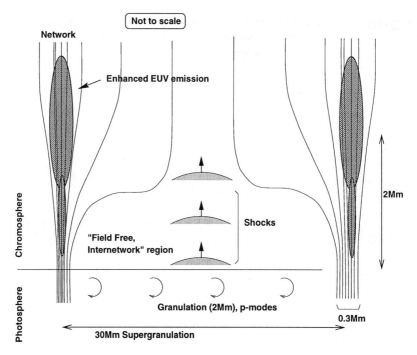

Figure 2. A cartoon showing a vertical 2D slice through the solar atmosphere in a region representative of the quiet Sun's network. Photospheric flux tubes are all assumed to be of the same polarity. Indicated are field lines which, for reasons determined by the unknown heating mechanism(s), have more hot plasma than others- statistically these lie vertically above regions of strong field in the photosphere.

magnetic structures that extend through the chromosphere into the corona. These structures, clumped into $\approx$ 0.3 Mm or less diameter "flux tubes", are formed by convective collapse. The tubes are subject to hydrodynamic forces in the photosphere — small length scale (2 – 3 Mm) and time scale (2 – 15 min.) granular motions and the larger ($\sim$ 30 Mm), more stable (tens of hours) convective pattern called supergranulation. Flux tubes are buoyant, like corks in water, and similarly they are forced to clump around places of convergent fluid flow, i.e. boundaries of supergranule cells. At a spatial resolution in excess of an arcsecond or so ($\sim$ 0.725 Mm on the Sun's surface), the magnetic flux is seen to be concentrated into a supergranular *network* pattern, built up of many unresolved flux tubes spending most of their time bobbing around in the network (Simon and Leighton 1964).

The effect of these fields must be included in the chromosphere and higher layers. The Lorentz force, in the MHD limit, can be decomposed into a gradient of an isotropic pressure term, and tension force that acts along field lines. The ratio between the scalar thermal pressure and magnetic pressure, is $\beta = \frac{P_{gas}}{(B^2/8\pi)}$. β is $>$ 1 outside of the photospheric flux tubes, so that the photospheric gas dominates the force balance. The exponential decay of gas pressure with height

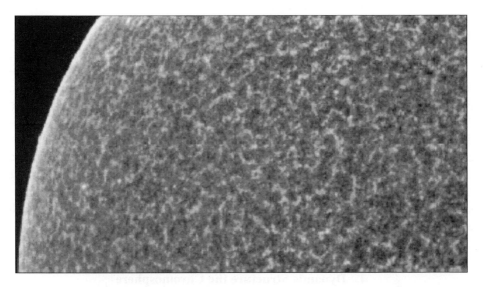

Figure 3. An example of the chromospheric network pattern, showing the intensity of the Si I emission line and continuum near 1256Å as a function of position on the solar disk (the abscissa encompasses 1 solar radius, or ∼700 Mm). Conditions are typical of the quiet Sun. These SUMER data were acquired on 7 June 1996 by D. Hassler.

in the chromosphere is not matched by a similar drop-off in magnetic pressure, whose structure is controlled by the MHD equations. The magnetic fields therefore spread out laterally with height through the chromosphere, forming a magnetic "canopy". The details of the spreading depends primarily on the distribution of magnetic "charges" in the photosphere — if they are essentially unipolar (neighboring field lines have the same polarities) then the fields are expected to be like those of Solanki and Steiner (1990), and references therein. A cartoon showing the expected structure is shown in Fig. 2.

A remarkable observation is that the network as seen in chromospheric emission features does not fill the available volume — for instance Skumanich *et al.* 1975 determined that 39% of the Sun is covered by this network pattern, in Ca II K line core emission. This pattern is persistent in plasmas at much lower pressures, up to and including the upper transition region (e.g., Feldman, 1987). In contrast, MHD models imply that the magnetic fields must fill the available volume at gas pressures ≤ 1 dyn cm^{-2}, or heights ≥ 1.5 Mm for the model of Fig. 1. The conclusion is simply that, for some reason, only those field lines overlying the photospheric fields are bright. Fig. 2 shows a schematic picture of the relationship between photospheric and chromospheric magnetic fields, and regions where the high intensity "network" chromospheric and transition region emission is formed. Fig. 3 shows a typical chromospheric network pattern seen in a spectral feature, probably formed at heights near 1 Mm.

These physical arguments are supported to some extent by measurement, although there may exist some important problems. For example, the chromospheric "canopy" appears to be seen substantially lower than 1.5 Mm (e.g., Giovanelli, 1980). Recent sensitive polarization work using the Hanle effect has yielded useful field strength measurements which agree broadly with those expected (e.g., Faurobert-Scholl 1994, Bianda *et al.* 1998). The "canopy" picture appears to have considerable support, but very recent work has suggested further problems (Landi, 1998). These discrepant measurements have yet to be understood and reconciled, but problems with complex line formation mechanisms in 3D atmospheric structures have not been properly addressed yet, which may yield insight. Further progress therefore appears to depend on the development of physical models constrained by photospheric observations (e.g., Steiner *et al.*, 1998), to analyze such effects, as much as on direct measurements of the chromospheric magnetic field.

4. Dynamic Structure the Chromosphere

The chromosphere is a partially ionized layer overlying the photosphere. In the network it is coupled via strong, localized magnetic fields. The photosphere is a very dynamic region, driven by a response to the need to transport the solar luminosity. This dynamic behavior is probably the driver for all the phenomena seen in the chromosphere, corona and wind. The picture shown in Fig. 2 must therefore be interpreted as just a "snapshot" of what must be a very dynamic region.

Photospheric granulation evolves continuously on short timescales (typically 5 minutes and less). Detailed numerical studies of granulation show spectacular evolution of small scale structures. These may serve as sources of *acoustic waves* in regions between the magnetic fields shown in Fig. 2 (R. Skartlien, in preparation, 1998). Numerical studies that include flux tubes in the photosphere have only recently appeared (e.g., Steiner *et al.*, 1998, and references therein) — these too show remarkable dynamic responses on similar timescales as the flux tubes are buffeted by the granular motions. These must serve as sources of *magnetic free energy* that is injected into the chromosphere and higher layers — both in the form of waves (at "high" frequencies, say $> 10^{-2}$ hz) and of secular changes (current systems) to the magnetic fields there. The chromosphere is therefore a region which is not only in a rather difficult physical regime, but is also strongly driven by magnetic and acoustic perturbations from below.

4.1. "Non-Magnetic" Chromosphere

Recently there has been substantial progress in the area of understanding the chromosphere as a system driven from below, in regions believed to be free of the influence of strong magnetic fields- the "internetwork" regions. A set of observations of network and internetwork regions were obtained by Lites *et al.* (1993), in a spectral region including the chromospheric Ca II H line. In the wing of this line is a line of neutral iron, formed much lower in the atmosphere, near 260 km. These data were used in an important study by Carlsson and Stein (1994), who applied a "forward" modeling approach. They solved the 1D equations of non-LTE radiation hydrodynamics. The photosphere and chromosphere were modeled as a single fluid, in the absence of magnetic fields. The equations solved were those of continuity, momentum, energy, coupled with a set of atomic rate equations for hydrogen and calcium and the equation for radiative transfer. These equations were solved on a numerical grid (the grid itself was simultaneously solved so as to provide adequate resolution to treat shocks), with appropriate boundary conditions. Starting from an atmosphere in radiative equilibrium, the lower boundary's velocity was specified as a function of time, and the response of the atmosphere was computed. A transmitting upper boundary condition was used, and all radiation at this boundary was assumed to be optically thin with no important incoming radiation. The spectrum of the observed Fe I absorption line was calculated, and the lower boundary's behavior was then modified to attempt to match the observed Fe I line's dependence on time. A second calculation using the modified boundary condition was performed. This calculation matched well the observed behavior of the Fe I line, indicating that the photospheric driver was adequately modeled. The computed Ca II H spectrum was then compared with the observed one. The comparison between observed and computed data is shown in Fig. 4. The results surely mark one of the most important steps taken in our understanding of the chromosphere.

While the above results could be interpreted as simply providing a match between a chromospheric line's computed and observed behavior (Fig. 4), the additional implications of this work are more profound: (1) It is the first model that can account for the detailed *time-dependent behavior of the Ca II H line*, formed in the chromosphere. (2) The model is (almost!) of an *ab-initio* nature — the only prescribed feature of the model is the vertical photospheric velocity at the base of the model, several pressure scale heights below the Ca II H line. Particularly noteworthy is the fact that *no additional arbitrary mechanical heating is present*. In the absence of the vertical driving, the "chromosphere" would return to radiative equilibrium. (3) Points (1) and (2) provide a compelling argument in favor of the model's believability: *the model appears to capture the basic physics important for the formation of the Ca II H line*. To date, no similar breakthrough in our understanding of the *magnetized* chromosphere has been obtained.

Assuming that the model captures the essential physics, we can examine it for more insight. First, the upward propagating waves steepen into shocks near ~ 0.6 Mm. The underlying density stratification of the atmosphere filters out oscillations with periods in excess of the the acoustic cutoff period of 3 minutes. Since the motions inferred from the Fe I observations are typical of the photosphere, showing a broad power peak near 5 minutes, the chromosphere "oscillates" mostly at 3 minutes. Thus, a "snapshot" of the thermal structure reveals a chromosphere consisting mostly of fairly cool material with temperature spikes associated with

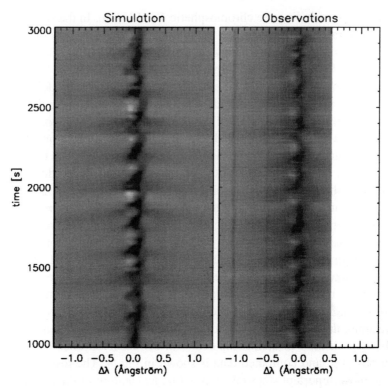

Figure 4. A comparison of observed and computed Ca II H line profiles as a function of time for one region of the Sun's internetwork, from Carlsson and Stein (1994).

shocks propagating upwards, separated by periods close to 3 minutes. Since they propagate at ~ 10 km s^{-1}, they are 2 Mm apart, so typically one shock dominates the chromosphere at any time. Second, if we consider time-averaged quantities, a surprising result arises. The *time averaged electron temperature shows no temperature rise*.

Furthermore the *time-averaged spectrum that arises from the model depends on what frequency ν you look at*, through the ratio $u = \frac{h\nu}{k_B T_e}$. Lines and continua have source functions which scale as $1/(e^u - 1)$. In UV continua and lines, $u \gg 1$, so the high temperature regions of the atmosphere (i.e. post-shock regions) completely dominate the source functions and emergent intensity. When averaged over time, the UV intensities are much higher than one would compute from an atmosphere with no temperature rise, such as the time-averaged temperature structure. In the infrared region, $u \leq 1$, the source functions scale more linearly with temperature, as u^{-1}. The shocks contribute little to the spectrum, and so the atmosphere looks like it has no temperature rise at all!

There are then several additional consequences of this kind of model: (4) The "non-magnetic" chromosphere is in its very essence *dynamic*, controlled by upward propagation of acoustic shocks of ± 10 km s^{-1} amplitude, driven by photospheric motions. (5) Interpretation of chromospheric data in terms of a *static* picture can be very misleading. The static approach can lead to *erroneous* interpretations (Carlsson and Stein, 1995). Relaxing the static picture may also resolve a long-standing controversy concerning the interpretation of infrared bands of the CO molecule *versus* Ca II emission data, without the need to invoke (time independent) spatial inhomogeneities. (6) One must re-examine previous element fractionation models, since most assume that rapid time evolution of the plasmas can be neglected.

4.2. FURTHER STUDIES OF THE CARLSSON AND STEIN PICTURE

Ground-based polarimetric data obtained recently by Lites, Rutten and Berger (1998) provide evidence that magnetic fields are very weak and dynamically insignificant in typical internetwork regions. This at least gives observational support to one critical fundamental assumption behind the model.

We are fortunate to work in the era of the SOHO spacecraft. Several groups have reported dynamic evolution of chromospheric spectra, mostly from the SUMER instrument on SOHO (Wilhelm *et al.* 1995), between 700 and 1560 Å. While radiation hydrodynamic calculations are needed to make detailed comparisons, we can nevertheless first compare UV continua of varying opacity, formed at various heights in the chromosphere. We expect UV continua to be formed roughly at heights indicated in Fig. 1. All SUMER data obtained away from the network that we have seen at wavelengths > 911 Å qualitatively confirm the Carlsson and Stein picture (e.g. Carlsson *et al.* 1997). Three minute oscillations are present in continua and lines. Clear phase differences between continuum variations (leading) and lines (following) exist, indicating upward wave propagation (the lines must be formed above the layers where the neighboring continua form). The data also show an important difference with preliminary computed data: observations show a pervasive emission, but calculations show periods of effectively zero intensity, corresponding to periods when shocks are absent from the formation heights of the spectra.

The Lyman continuum reveals another important difference. The 3 minute period shocks have apparently disappeared at heights where the continuum is formed. Instead, relatively low amplitude intensity variations are seen. This result is also not consistent with the models of Carlsson and Stein (Carlsson, 1997, private communication). It is probably no coincidence that the magnetic field is expected to begin playing an important role below the 2 Mm or so formation height of this continuum (cf. Figs. 1 and 2). This observation suggests that the gas dynamics changes radically in character, as the upward propagating shock waves interact with the overlying magnetic field. This is an area where almost nothing is presently known.

We conclude that the Carlsson-Stein picture is reasonable for heights below 1.5 Mm for the quiet chromosphere outside of network regions. While a limited region, it nevertheless represents the only area of the chromosphere for which we have even a basic understanding.

5. Issues and Speculations Concerning Chromospheric Element Fractionation

Thus far we have reviewed thermodynamic properties of the chromosphere, magnetic properties (away from active regions), and properties of regions of the quiet Sun's chromosphere that lie between the bright network elements in the quiet Sun. We have argued that a viable physical model successfully accounts for the essential properties of this "internetwork" chromosphere, at least in layers below the magnetic canopy ($z < 1.5$ Mm or so). In contrast, very little is known about the basic gas dynamics of the network and all regions above 1.5 Mm, and no physical model is available that can be used as a basis for understanding these components of the chromosphere. This situation leads us to make several comments concerning element fractionation models that have been proposed.

5.1. THE CHROMOSPHERE/CORONA CONNECTION, ROLE OF SPICULES

Element fractionation was first identified through measurements of particles in the solar wind. Later, EUV observations of the corona (mostly in closed field regions) showed the same effect. Thus we come to a problem. To understand element fraction as seen in photospheric and corona/wind data, we must understand the *network* chromosphere since the *internetwork* chromosphere is not so directly connected to the corona (Fig. 2).

An important component of the chromosphere has not been discussed at all: *spicules*, which have been known for more than a century. The reason for not discussing them is that they are not understood at all. These are structures first recognized in limb H-α data which are part of the chromospheric network, seen also in EUV network emission. At the limb they extend *much* higher into the atmosphere (typically by several Mm), and they are very dynamic. They may be crucial to the element fractionation discussion since it is known from kinematic arguments that they can supply the corona with 100 times the mass flux needed to account for the solar wind!* Furthermore, spicules are likely to exist at lower densities than the rest of the chromosphere suggesting that collisional coupling is less effective.

These problems arise because of our lack of knowledge of conditions in the *magnetized* chromosphere itself. To address the FIP effect as seen in the solar wind particle measurements, there remains the additional problem of knowing the *source* of the slow solar wind in relation to the observed chromosphere. The solar wind clearly connects somewhere to the network regions — magnetic field lines in the wind have to connect back to field lines (either steadily, or intermittently by reconnection with closed loop structures) at the top of the chromosphere.

---

* It is not clear that observed spicular motions are mass motions or if they correspond to some kind of wave. The above statement is based upon the assumption that mass is flowing into the corona.

5.2. IMPLICIT ASSUMPTIONS: ROLE OF MIXING

After the workshop, it became clear that a basic and important element was missing from our discussions. This arose from an apparent discrepancy between different models attempting to explain element fractionation. A multi-fluid (H-He), dynamic model of the chromosphere and solar wind by Hansteen et al. (1997) revealed an important problem. To get even a little helium in the solar wind, collision cross sections between neutral helium and protons had to be increased by a factor of 15 to avoid strong gravitational settlement. Taken at face value, their results suggest it is difficult to get *any* elements heavier than hydrogen into the corona! This clearly is contradicted by observations, and it stands in apparent contradiction to other (kinematic) models, in which element fractionation is difficult to achieve in standard (i.e. VAL-like) models.

The resolution of this apparent contradiction appears to rest in assumptions concerning unresolved plasma motions, or "turbulence". In the work of Hansteen et al. (1997), no turbulence was included, but in all fractionation calculations based upon the VAL (or similar) models, turbulence is included implicitly. Non-thermal motions are required by observations (see, e.g. Vernazza, Avrett, and Loeser, 1981). Turbulent eddies (and in some cases waves) have the crucial effect of coupling all species together to produce the same pressure scale height for all species, and thus no strong gravitational settling. In essence, the pressure scale height for a species i in a plasma with turbulence characterized by r.m.s. speed ξ is $\frac{k_B T_i + \frac{1}{2} n_i m_i \xi^2}{m_i g}$. Thus, turbulence serves to "lift" the heavier elements to higher layers of the atmosphere in a radically different way than can be achieved in its absence.

This observation should raise a "red flag" for those seeking to construct fractionation models. The results depend critically on something that cannot be computed in a rigorous manner, although various schemes are well known in the atmospheric community. For example Lilli (1989) includes the gradient of a stress tensor to the right hand side of the momentum equation.

5.3. EFFECT OF RAPID DYNAMIC EVOLUTION OF CHROMOSPHERIC MODELS ON THE FIP EFFECT

We consider one final question: if the chromosphere evolves in the manner indicated in the Carlsson and Stein work, what qualitative effect will this have on existing fractionation models in which the chromosphere was assumed to be in a steady state? The FIP effect indicates that the fractionation process(es) is determined primarily by the fractional ionization of a given element in the upper layers of the chromosphere. We will therefore calculate the time average of the ionized fraction of elements in a fluid element subject to changes similar to those computed by Carlsson and Stein (1994).

Let the quantities x and y represent the fractional number densities (i.e. $x + y = 1$) of neutral and (singly) ionized ions of a given element (we ignore higher ionization stages). Consider a volume of fluid, at time $t < 0$ in a specific state, in which the thermodynamic variables are changed abruptly at $t = 0$, and held constant thereafter. This can be used as a first approximation for a volume of fluid through which a strong shock travels. In the fluid rest frame, the time dependence of, for example, y, can be written $\frac{D}{Dt}y = -\frac{y}{\tau_R} + \frac{x}{\tau_I}$, where τ_R and τ_I are the recombination and ionization times (i.e. the inverse of coefficients γ_{ki} in eq. (4)) respectively. Since $x + y = 1$,

$$\frac{D}{Dt}y = -\frac{y}{\tau_M} + \frac{1}{\tau_I}, \qquad (6)$$

Where $\tau_M^{-1} = (\tau_R^{-1} + \tau_I^{-1})$. This equation, for fixed ionization and recombination times, has two solutions, an equilibrium solution

$$y^* = \frac{\tau_M}{\tau_I} \qquad (7)$$

and a time dependent solution which satisfies

$$y(t) - y^* = (y_0 - y^*)e^{-t/\tau_M}, \qquad (8)$$

where y_0 is the ion density at $t = 0$. Previous models of the FIP effect have concentrated on the stationary solution $y = y^*$. The important atomic timescale is τ_M, which is roughly the smaller of τ_I and τ_R. For all elements τ_R is typically $\sim 30(\frac{10^{11}}{n_e})(\frac{T_e}{10^4})^{0.7}$ sec, for electron temperatures and densities T_e (K) and n_e (cm^{-3}) respectively. In contrast the quantity τ_I is an (exponential-like) function of the first ionization potential, varying from 0.15 sec for Mg to 220 sec for He in the calculations of Geiss and Bochsler (1985), for example. This simple calculation indicates that ionization equilibrium is a reasonable approximation for changes on times t substantially greater than τ_M which is approximately τ_R for high FIP elements, and τ_I for low FIP elements.

Now consider the situation where a shock hits the fluid element periodically. Following Carlsson and Stein's results, let the time between the shocks be t_B (≈ 3 min), and the duration of shock be t_S (≈ 10 s). Thermodynamic variables are assumed constant between and during the shocks. For simplicity we examine the extreme case in which there is no ionization between the shocks ($\tau_I^{(B)} \to \infty$), and choose the recombination rate to be the same in and between the shock. After a few shocks have traversed the fluid element the time averaged ionization fraction $<y>$ tends to the value

$$<y_\infty> = \left\{ \frac{(1-Q)(1-R)}{1-QR} \frac{\tau_R - \tau_M}{t_S + t_B} + \frac{t_S}{t_S + t_B} \right\} y^* \qquad (9)$$

where $Q = \exp(-t_B/\tau_R)$ and $R = \exp(-t_S/\tau_M)$ arise from equation (8) for periods between and during the shocks, and y^* denotes the equilibrium solution during the shocks. One can show that the bracketed coefficient fulfills $0 \leq \{\cdots\} \leq 1$. This solution has two important properties: first, *even in a dynamically evolving plasma the fraction of ionized particles varies in proportion to the equilibrium value y^**. Thus the results for the ionization time etc. as obtained e.g. by Geiss and Bochsler (1985) for a steady state situation might still be applicable in a dynamic situation. Second, a parameter study of the coefficient $\{\cdots\}$ in (9) shows that the dynamics will *enhance* the differences between the low- and high-FIPs: for reasonable values for the timescales of the shocks $\{\cdots\}$ is a step-like function of the first ionization time with the values for the low-FIPs clearly higher than for the high-FIPs.

We conclude that existing models of the FIP effect based on ionization equilibrium calculations contain some (but not all) of the important physical processes determining the mean ionization fractions, if the chromosphere is indeed as dynamic as implied by recent work. However, it is worrisome that information on higher frequency variations (in both space and time) is fundamentally limited (e.g. Judge *et al.* 1997), and "micro-turbulence" is a ubiquitous feature of chromospheric observations. Thus, higher frequency variations than those resolved as shocks in the internetwork cannot be ruled out, and should perhaps be expected.

6. Conclusions

The chromosphere is a highly structured region driven by photospheric motions and magnetic fields that are only beginning to be understood. It is a region believed to be the seat of the physical processes leading to elemental abundance differences measured in the interior, photosphere and corona (the helium abundance and FIP effect). The chromosphere is possibly the most difficult region of the Sun to model accurately owing to a variety of difficult physical conditions. Semi-empirical models, used as a basis for modeling the FIP effect, are demonstrably poor in modeling the temperature structure of the non-magnetic regions of the chromosphere, because it is strongly time dependent, but their stratification may be reasonable. Magnetized regions of the chromosphere, connected more directly to the overlying corona (and with solar wind) are not even basically understood. Open questions therefore remain as to the role of dynamic evolution, turbulent mixing, and the long standing problem of spicules, in the attempts to understand the physical origin of element fractionation between photosphere and corona.

References

Bianda, M., Solanki, S. K. and Stenflo, J. O.: 1998, 'Hanle Depolarisation in the Solar Chromosphere', *Astron. Astrophys.* **331**, 6760–770.

Carlsson, M., Judge, P. G. and Wilhelm, K.: 1997, 'SUMER Observations Confirm the Dynamic Nature of the Quiet Solar Outer Atmosphere: The Internetwork Chromosphere', *Astrophys. J. Lett.* **486**, L63–L67.

Carlsson, M. and Stein, R. F.: 1994, 'Non-LTE Radiation Shock Dynamics in the Solar Chromosphere', *in* M. Carlsson (ed.), *Chromospheric Dynamics*, Institute of Theoretical Astrophysics, Oslo, pp. 47–77.

Carlsson, M. and Stein, R. F.: 1995, 'Does a Nonmagnetic Solar Chromosphere Exist?', *Astrophys. J. Lett.* **440**, L29–L32.

Faurobert-Scholl, M.: 1994, 'Hanle Effect of Magnetic Canopies in the Solar Chromosphere', *Astron. Astrophys.* **285**, 655–662.

Feldman, U.: 1987, 'On the unresolved fine structures of the solar atmosphere. II - The temperature region 200,000 - 500,000 K', *Astrophys. J.* **320**, 426–429.

Geiss, J.: 1982, 'Processes Affecting Abundances in the Solar Wind', *Space Science Reviews* **33**, 201–217.

Geiss, J. and Bochsler, P.: 1985, 'Ion Conposition in the Solar Wind in Relation to Solar Abundances', *Rapports Isotopiques dans le Systeme Solaire*, Cepadudues-Editions, p. 213.

Giovanelli, R. G.: 1980, 'An exploratory two-dimensional study of the coarse structure of network magnetic fields', *Solar Phys.* **68**, 49–69.

Hansteen, V. H., Leer, E. and Holzer, T. E.: 1997, 'The Role of Helium in the Outer Solar Atmosphere', *Astrophys. J.* **482**, 498–509.

Hénoux, J.-C.: 1998, *Space Sci. Rev.*, this volume.

Judge, P. G., Carlsson, M. and Wilhelm, K.: 1997, 'SUMER Observations of the Quiet Solar Atmosphere: The Network Chromosphere and Lower Transition Region', *Astrophys. J. Lett.* **490**, L195–L198.

Landi Degl'Innocenti, E.: 1998, 'Evidence Against Turbulent and Canopy-Like Fields in the Solar Chromosphere', *Nature* **392**, 256–258.

Lilli, D.: 1989, in J. R. Herring and J. C. McWilliams (eds), *Lecture Notes on Turbulence*, World Scientific, Singapore, pp. 171–218.

Lites, B. W., Rutten, R. J. and Berger, T.: 1998, in preparation.

Lites, B. W., Rutten, R. J. and Kalkofen, W.: 1993, 'Dynamics of the Solar Chromosphere. I. Long-Period Network Oscillations', *Astrophys. J.* **414**, 345–356.

Mihalas, D.: 1978, *Stellar Atmospheres*, W. H. Freeman and Co., San Francisco (second edition).

Peter, H.: 1998, 'Element Fractionation in the Solar Chromosphere Driven by Ionization-Diffusion Processes', *Astron. Astrophys.* **335**, 691–702.

Schunk, R. W.: 1975, 'Transport Equations for Aeronomy', *Planet. Space Sci.* **23**, 437–485.

Simon, G. W. and Leighton, R. B.: 1964, 'Velocity fields in the chromosphere. III Large scale motions, the chromospheric network, and magnetic fields', *Astrophys. J.* **140**, 1120–1147.

Skumanich, A., Smythe, C. and Frazier, E. N.: 1975, 'On the statistical description of inhomogeneities in the quiet solar atmosphere. I. Linear regression analysis and absolute calibration of multichannel observations of the Ca^+ emission network', *Astrophys. J.* **200**, 747–764.

Solanki, S. K. and Steiner, O.: 1990, 'How magnetic is the solar chromosphere?', *Astron. Astrophys.* **234**, 519–529.

Steiner, O., Grossmann-Doerth, U., Knoelker, M. and Schuessler, M.: 1998, 'Dynamical Interaction of Solar Magnetic Elements and Granular Convection: Results of a Numerical Simulation', *Astrophys. J.* **495**, 468–484.

Vernazza, J. E., Avrett, E. H., and Loeser, R.: 1981, 'Structure of the Chromosphere. III. Models of the EUV Brightness Components of the Quiet Sun', *Astrophys. J. Suppl. Ser.* **45**, 635–725.

Wilhelm, K., and 15 co-authors: 1995, 'SUMER: Solar Ultraviolet Measurements of Emitted Radiation', *Solar Phys.* **162**, 189–231.

Address for correspondence: Philip Judge and Hardi Peter, High Altitude Observatory, PO Box 3000, Boulder CO 80307.

THE SOLAR QUIET CHROMOSPHERE–CORONA TRANSITION REGION

L. S. ANDERSON–HUANG
The University of Toledo
Toledo OH 43606, U.S.A.

Abstract. The chromosphere–corona transition region of the Sun enjoys both simplicities of character and complexities of character which result from its very thin geometrical extent. The simplicities derive from the reasonably clear view of the energy balance (both observationally and theoretically), while the complexities derive from both the proximity of the not–so–clearly viewed regions below and above, and the almost certain convolutions and perhaps discontinuities in the three dimensional geometry of the transition sheet. While observational resolution and spectral information has improved greatly in recent years, the problems associated with a single vantage point, the Earth and its environs, have not gone away. To understand the transition region we must resolve structures radially and temporally as well as in the plane of the sky.

Key words: Sun: transition region

1. Introduction

The solar chromosphere–corona transition region is what its name describes; the (relatively spatially confined) *region* in the Solar atmosphere between the classical chromosphere and corona. I emphasize the word 'region' because, while on average material must in some manner *make* a transition from chromospheric (or in some cases photospheric) to coronal conditions, we are a long way from understanding how that transition is made. For the purposes of this review, I will define the region by its radiative emission characteristics. It may be that to refer to any temperature, whether it be an ionization temperature or an electron temperature, is misleading. Any spectroheliogram of the Sun in the radiation of any ion stripped of three to six electrons shows an image of some part of the transition region. I will not discuss flares, spicules, prominences or structures associated with active regions of the Sun, although certainly much of the physics in these phenomena is related, and perhaps much material enters the corona via spicules. The transition region is characterized by very steep ionization gradients, unresolved (and unresolvable) structures, rapidly time dependent phenomena, velocity fields, and nonequilibrium physical processes. Despite those daunting characteristics, it does have certain properties which make it tractable for study. The transition region is effectively optically thin; that is, every photon created by collisional excitation and radiative decay escapes (although it may scatter many times before doing so). Even so, the transfer problem is nontrivial, because the three dimensional structure will influence the escape direction. Line radiation from transition region ions is produced only in the transition region, and to the extent (possibly very weak) that the ionization is in equilibrium, comes from very narrow temperature ranges associated with the particular ion. Finally, if bulk velocities are subsonic and the sound travel time across structures is less than the lifetimes of those structures, the narrow geometrical extent means that the gas pressure $P \simeq 2N_e kT$ is approximately constant throughout the region.

As in all other astronomical endeavor where the subject is not directly retrievable, the study of the transition region ranges from purely *a priori* physical modeling to semi–empirical conclusions drawn more directly from observational data. I have used the expression 'semi–empirical' because *any* conclusions concerning the state of transition region material must rely on physical models and implicit assumptions. Only the photons are directly observable. At this time, there are two rather different proposals for explaining the transition region and its processes. One school of thought suggests that the region is energetically connected to its chromospheric and coronal boundaries; that is, its physical state derives from processes (e.g. waves, advection, and/or conduction) which are continuous across the boundaries. The other school suggests that the region is energetically disconnected; that is, its physical state derives from processes which occur *in situ* (e.g. short range electric currents and/or magnetic reconnection). I will try not to choose between the two; however, at this point I do believe that a more probable choice may be established via William of Ockham's razor (c1320; however, here is a lesson from history: original authors are usually misquoted, and have eloquent predecessors–see the reference).

The rest of this paper is organized as follows. I will discuss in turn the geometrical structure, energy balance, dynamics, and statistical equilibrium of the region, and conclude with comments about observational interpretation. Most of the discussion is based on easily derivable length scales, time scales, and energy scales. For a more detailed review of the transition region and its problems, see Mariska (1992).

2. Geometrical Structure

The structure observed at the limb of the Sun in spectroheliograms in the radiation of any transition region ion shows a finely striated form like fur, at the limit of transverse resolutions of order one second of arc or less. While structures on the limb are no doubt dominated by spicule or spicule–like features, they clearly indicate a more–or–less vertical magnetic field (there is little evidence for horizontal fields, but these fields may be masked if they are lower in the atmosphere). Typical gas pressures found in the transition region are of order 0.1 dynes cm^{-2}, if we assume we can derive meaningful temperatures and densities from the emission line spectra (e.g. chapter four of Mariska, 1992; but, cf. Judge *et al.*, 1997). The surface average vertical magnetic field on the quiet Sun is about one Gauss, so the average magnetic pressure is of the same order as the gas pressure. This near equipartition of energy densities is probably not coincidental. Since the solar field in the chromosphere is highly organized (cf. Fig. 2 in the paper by Judge and Peter, 1998, in this volume), we might expect furry structures.

Furthermore, we may infer that the transition region is much more finely structured than we observe at one second of arc. Observed structures have volumes of order $V_{obs} = 300^3$ km^3. If we assume that every photon created in emission line λ within such a volume escapes in a random direction, the observed flux at Earth should be

$$F_\lambda = \frac{V_\lambda}{4\pi D^2} N_u h \nu_\lambda A_{ul}, \tag{1}$$

where V_λ is the actual emitting volume within the resolution–limited V_{obs}, D is the distance to the Earth, N_u is the number density of ions in the upper state of the transition, and A_{ul} is the radiative decay rate. The upper state density is a calculable function of the electron density and temperature, which can in turn be determined from line ratios and/or the region pressure. Typical filling factors V_λ/V_{obs} derived from such studies are of order 0.01. If we assume that the 'fur' is linearly resolved, the cross section length scales are then of order 30 km or less.

These numerical estimates are susceptible to error, as they rely on the diagnostic value of the line emission; however, both energetic considerations and the observation and modeling of the underlying magnetic field originating in the photosphere also suggest very fine geometrical structuring in the transition region.

3. Energy Balance

The line radiation and the inferred temperatures and densities are the most quantifiable parameters of the transition region. A general formula for the energy balance may be written

$$\rho \frac{D\mathcal{E}}{Dt} + P \nabla \cdot \mathbf{v} + \nabla \cdot \mathbf{F} = 0, \tag{2}$$

where $\mathcal{E}$ is an internal energy per gram and $\mathbf{F}$ is an energy flux passing through the material. Both $\rho \mathcal{E}$ and P have values of order 0.14ζ ergs cm^{-3}, where $\zeta = N_{11} T_4$ (numerical subscripts n refer to values in units 10^n; N_{11} is the number density of particles in units 10^{11} cm^{-3}).

We may decompose the flux divergence into

$$\nabla \cdot \mathbf{F} = N_e^2 \Lambda(T) + \nabla \cdot (\kappa \nabla T) - \Gamma(?). \tag{3}$$

The first term represents the radiative cooling (which I have written in the coronal approximation, but published coronal cooling functions $\Lambda(T)$ may not apply), the second term accounts for some form of thermal conductivity, and the third term accounts for any postulated heating mechanisms whether they be magnetic, viscous, or the like. The order of magnitude of the radiative cooling is

$$N_e^2 \Lambda(T) = \zeta^2 T_4^{-2} \Lambda_{-22}(T) \text{ ergs cm}^{-3} \text{ s}^{-1}, \tag{4}$$

which implies cooling time scales of

$$\tau = \frac{\rho \mathcal{E}}{N_e^2 \Lambda} = 0.14 \zeta^{-1} T_4^2 \Lambda_{-22}^{-1} \text{ s}, \tag{5}$$

and dynamic length scales of

$$a\tau = (\gamma k T/\mu)^{1/2} \tau = 2\zeta^{-1} T_4^{3/2} \Lambda_{-22}^{-1} \text{ km}. \tag{6}$$

Here a is the isothermal sound speed.

For many years the so-called *emission measure* has been used as a way of trying to characterize the structure and energy balance of the transition region. Its derivation makes assumptions about the local statistical equilibrium and the relevance of unique temperatures which may be derived from brightnesses and ionizations. Returning to equation (1), for an emission line from a simple two-level atom in statistical equilibrium we may write for the upper state number density

$$N_u = N_l N_e C_{lu}(T)/A_{ul}, \tag{7}$$

where N_l is the number density in the ground state and $N_e C_{lu}(T)$ is the upward transition rate induced by electron collisions, per ion in the ground state. The line flux at the emitting surface becomes

$$\mathcal{F}_\lambda = \frac{V_\lambda}{A_\lambda} N_e^2 \mathcal{N}_{l/e}(T) C_{lu}(T) h\nu_\lambda, \tag{8}$$

where A_λ is the surface area of the emitting volume and $\mathcal{N}_{l/e}(T)$ is the ratio of the ground state ion density to the electron density, a number which has appreciable value only over a small range in temperature. All of the factors to the right of N_e^2 in equation (8) are functions of temperature and atomic data only. The emission measure consists of the first factors in equation (8), $V_\lambda N_e^2 / A_\lambda$, which has units cm^{-5} and is a measure of the distribution of material as a function of temperature. For any given line with an observed flux, the locus of emission measure as a function of temperature reaches a relatively narrow minimum at the temperature where the ion abundance is highest. The 'true' emission measure is taken to be a minimum envelope formed by the ensemble of observed line loci, such that observed fluxes are reproduced when properly integrated over all temperatures. In general, observers consistently find that the 'true' emission measure is approximately proportional to $T_4^{3/2}$ for $T_4 > 20$, and approximately proportional to T_4^{-3} for $T_4 < 10$ (cf. Raymond and Doyle, 1981, and Doschek, 1997). Many workers conclude that very different processes are responsible for the energy balance in these two regimes.

Thermal conduction fluxes parallel and perpendicular to the magnetic field in a zeroth order Maxwellian plasma are given by (Spitzer, 1962)

$$\kappa_\| \frac{dT}{dz} = 10^8 T_4^{7/2} \frac{d\ln T}{dz}, \tag{9}$$

$$\kappa_\perp \frac{dT}{dx} = 2 \times 10^8 \zeta^2 B^{-2} T_4^{-3/2} \frac{d\ln T}{dx} \text{ ergs cm}^{-2} \text{ s}^{-1}. \tag{10}$$

At high temperatures, parallel conduction will dominate unless the perpendicular gradients are exceedingly large. If we assume that the thermal equilibrium results from a balance between parallel conduction divergence and radiative cooling, and both are relatively small, then the parallel conductive flux remains constant along field lines: $F_c = 10^8 T_4^{7/2} d\ln T/dz$. This flux can be related to the emission measure. In the emission measure, the volume divided by the area is a characteristic length over which the temperature changes: $V_\lambda A_\lambda^{-1} \approx dz/d\ln T$. Therefore we may write:

$$\frac{V_\lambda}{A_\lambda} N_e^2 = 10^8 F_c^{-1} \zeta^2 T_4^{3/2}. \tag{11}$$

This relationship agrees with the observed temperature dependence of the emission measure for $T_4 > 20$, and leads to a conduction flux F_c downward from the corona of order 10^6 erg cm^{-2} s^{-1}, and characteristic distances over which the temperature varies of order

$$\Delta z \approx 10^2 T_4^{7/2} \text{ cm!} \tag{12}$$

Such a strong temperature dependence suggests that the appearance of Solar plasmas dominated by thermal conductivity should also be strongly temperature (or ion) dependent. Indeed, low coronal plasmas are much more geometrically diffuse than mid and low transition region plasmas.

The total radiation flux from the upper transition region $T_4 > 10$ is about $0.1 F_c = 10^5$ ergs cm^{-2} s^{-1} (Timothy, 1977), which is consistent with the idea that the conductive flux is constant. Thus, most of the evidence suggests that the upper transition region is energetically connected to the corona via classical electron conduction, in agreement with the first of our two schools of thought.

The total radiation flux from the lower transition region $T_4 < 10$ (including Lyα) is about $0.5 F_c$ (Timothy, 1977), which suggests that it, too, derives from the downward conduction. However, the emission measure behavior is inconsistent with that proposition. Why does this inconsistency arise? This question is at the root of work on the lower transition region for the last twenty years or more, but no convincing answer has been found.

Athay (1990) proposes that if conduction carries the right flux, we should give it the benefit of the doubt. Following a suggestion by Rabin (1986), he allows perpendicular conduction to redistribute flux horizontally across fingers of cold material penetrating into near coronal plasma. The surface area of these fingers exceeds the surface area of the sun by many orders of magnitude. Values for $V_\lambda A_\lambda^{-1}$ derived from the emission measure are much larger than $ds/d\ln T$ because the surface is very convoluted and very thin ($ds/d\ln T$ is the thickness parallel to the temperature gradient). Intuitively this idea is attractive because even small fluctuations in the energy density of the corona will translate to very steep horizontal gradients arising from the meter–scale parallel temperature gradients at $T_4 \simeq 1$. If one assumes the energy balance is set by conduction divergence and radiative cooling, the energy equation becomes

$$\zeta^2 T_4^{-2} \Lambda_{-22} + 2\frac{d}{dx} 2 \times 10^8 \zeta^2 B^{-2} T_4^{-3/2} \frac{d\ln T}{dx}$$
$$+ \frac{d}{dz} 1 \times 10^8 T_4^{7/2} \frac{d\ln T}{dz} = 0. \tag{13}$$

Athay (1990) uses dimensional arguments to show that a simple picket fence pattern of fingers can reproduce observed emission measures. Ji, Song, and Hu (1996) provide a more quantitative model based on saw–tooth isothermal surfaces. However, both of these papers suffer from solving the energy equation *locally* in an assumed geometry, when it should be solved *globally* with reasonable boundary conditions. After all, conduction is conservative.

Cally (1990) suggests another 'conduction' approach. A variety of observational evidence points toward nonthermal turbulent motions in the low transition region (see below). Cally also summarizes several physical arguments for the presence of turbulence. In a mixing–length approximation, one may write turbulent conductivity as

$$\kappa_t = \phi \rho C_p u l, \tag{14}$$

where u is an rms turbulent velocity, l is the mixing length, C_p is the specific heat at constant pressure, and ϕ is a constant of order unity. Given observed line–broadening velocities, Cally finds that a thermal dependence $l \propto T^{-1.5}$, with $l \approx 100$ km at $T_4 = 2$, reproduces the low transition region emission measure.

The previously cited works based on conduction fall into our first school. They are not entirely satisfactory because they postulate unconfirmed geometry and/or more–or–less arbitrary mixing parameters. However, we *do* have evidence that the geometry *is* complex, and that there *are* nonthermal motions carrying the emitting ions.

Other work trying to explain the emission measure from low–temperature regions includes an arbitrary heating term $\Gamma(?)$ adjusted to fit the emission measure. This work falls into our second school of thought. Often these explanations rely on stochastic rather than steady heating processes (cf. Sturrock *et al.*, 1990). At the outset, one must remember that stochastic heating is potentially incompatible with the basic premise of the emission measure as a valid diagnostic of the emission volume; such heating implies nonequilibrium ionization and excitation if the time scales are shorter than the ionization/recombination time scales. However, stochastic heating with short time scales is not in itself required; only heating mechanisms unrelated to conduction need be postulated. Much of this work derives from a proposal by Feldman (1983) that radiation from the low transition region comes from structures magnetically isolated from the corona, which he calls *unresolved fine structures*. Antiochos and Noci (1986) suggest a variety of closed magnetic loops with various heights and related maximum temperatures. While the physics of such loops is reasonable, the number of loops at each height is not determined *a priori*, but must be chosen to reproduce the emission measure. Rabin and Moore (1984) and Roumeliotis (1991) suggest heating by electric currents $\Gamma = J^2/\sigma$, $\mathbf{J} = \nabla \times \mathbf{B}/\mu_o$. However, once again there is no *a priori* determination of the distribution of electric currents. These and other writers point out that the coronal cooling function $\Lambda(T)$ reaches a maximum at $T_4 \approx 20$, which prevents the stability of magnetic loops when $T_4 > 20$. Associating the break in the emission measure with the behavior of $\Lambda(T)$ is attractive. However, with the inclusion of more and more accurate atomic data in the radiative cooling function, the maximum at $T_4 \approx 20$ becomes less and less obvious. In addition, $\Lambda(T)$ is not by itself responsible for the energetics or the stability. We require better knowledge of $\Lambda(T)$, $\Gamma(\rho, T, B)$, ζ, and $\mathbf{B}$ to address this problem.

4. Dynamics

Clearly there are energetic events having durations of order 10's of seconds which occur within the transition region over the quiet sun (cf. Moses *et al.*, 1994). These events are likely to be microflares due to magnetic reconnection. They appear to be concentrated near the chromospheric magnetic network, and avoid network cell centers. However, estimates by Cook *et al.* (1987) suggest they do not provide enough energy to heat the transition region *in situ*.

Emission lines formed within the transition region on average present a net redshift (cf. Mariska, 1992). These redshifts reach a maximum value of about 10 km s^{-1} for ions characteristic of temperatures of about 10^5 K, and decrease at both higher and lower ionization. The emission lines also exhibit nonthermal broadening (cf. Mariska *et al.*, 1978). The broadening is about 5 to 10 km s^{-1} for the lowest ionizations, increases up to about 35 km s^{-1} for ions characteristic of 3×10^5 K, and remains constant or declines slightly for higher ionization (Warren *et al.*, 1997). Such characterization of the broadening as 'nonthermal' implies that the ionization *is* (or nearly is) thermal; it may of course be that ionizations and/or particle velocity distributions are themselves nonthermal.

The net redshifts imply downflows which would be enough to empty the corona in a matter of days. Where is the upward moving material? Various explanations in transition region models generally require further *ad hoc* conditions. The most successful (from an *a priori* physics point of view) explanations involve brightness correlations with downward velocity. For example, Brynildsen et al. (1996) show that there is a connection between the C IV redshift and the chromospheric network; they favor the redshift originating in downward propagating Alfvén pulses (Hansteen, 1993; Hansteen and Maltby, 1992).

5. Statistical Equilibrium

In the presence of steep temperature gradients and mass flows, one might expect departures from standard statistical equilibrium. these departures may be either diffusive (fast–moving particles penetrating into cooler regions), advective (nonequilibrium ionization–recombination and/or velocity distributions for particles rapidly pushed through the transition region), or simply a result of rapidly time dependent phenomena.

Fontenla et al. (1991) construct plane–parallel semi–empirical models for the low transition region which contain only conduction, radiative transfer, and diffusion. They find that for temperatures above 25,000 K, conduction dominates the heating, while for lower temperatures the diffusive transport of ionization energy dominates the heating. The latter process smoothes out the ionization gradient and increases the Lyα formation temperature from 20,000 K to 40,000 K. As a result, Lyα significantly increases the cooling function $\Lambda(T)$ at temperatures approaching 10^5 K. While these models were constructed in an attempt to understand Lyα and other low transition region emissions, one should expect to find similar physics in the middle transition region.

Woods and Holzer (1991) explore rapid advection through a thermal boundary. They find that in downflows, there are two effects which may contribute to both emission measures and a brightening of redshifts over blueshifts. First, the electrons cool substantially faster than ions because of their more frequent collisions with each other, and second, the thermal force imparts upward momentum to minority species. The latter effect acts like a selective dam, behind which minority abundances increase sharply. Woods and Holzer's modeling assumes an electron temperature profile derived from the very emission measure they are suggesting is invalid, but the physics is relevant if not the actual results. The thermal force is not intuitive, but is relatively easy to explain (Woods and Holzer 1991; Frankel 1940). In the rest frame of the bulk motion, consider a surface across which there is a temperature gradient. Particles crossing the surface from hot to cold are, on average, moving faster than those crossing from cold to hot. The net force exerted by hydrogen on any species s is roughly $\rho v^- \nu^-_{Hs} - \rho v^+ \nu^+_{Hs}$ in the direction of the temperature gradient. Here the ($^-$) superscripts indicate contributions from particles coming from the cool side, the ($^+$) superscripts indicate contributions from particles coming from the hot side, and ν_{Hs} is the momentum transfer collision frequency between hydrogen ions and s ions. Since we are in the rest frame of the bulk motion, $\rho v^- = \rho v^+$. However, the collision frequency has a v^{-4} dependence

FIP FRACTIONATION: THEORY

JEAN-CLAUDE HÉNOUX
Observatoire de Paris; DASOP/LPSH(URA2080)
92195 Meudon Principal Cedex, France

Abstract. In this review, the main models of ion-neutral frationation leading to an enhancement of the low FIP to high FIP abundance ratio in the corona or in the solar wind, are presented. Models based on diffusion parallel to the magnetic field are discussed; they are highly dependent on the boundary conditions. The magnetic field, that naturally separates ions from neutrals moving perpendicular to the field lines direction, when the ion-neutral frequency becomes lower than the ion gyrofrequency, is expected to play an active role in the ion-neutral separation. It is then suggested that ion-neutral fractionation is linked to the formation of the solar chromosphere, i.e. in magnetic flux-tubes at a temperature between 4000 and 6000 K.

Key words: Element Abundances, FIP effect

1. Introduction

General reviews on the element abundances in the solar atmosphere have been published (Feldman, 1992; Saba, 1995; Meyer, 1993a; Meyer, 1993b; Fludra *et al.*, 1998). Relative abundances were measured in the corona, transition region, and recently in the low chromosphere at the site(s) of solar flares, and compared to the mean photospheric abundances. Efforts have been made to determine variations of relative solar element abundances with specific solar features – quiet sun, active regions, sunspots, coronal holes, polar plumes and flares – showing variations between these structures.

On average, elemental abundances in the solar wind, and solar energetic particles abundances measured in-situ, are in agreement with spectroscopic measurements of abundances in the upper solar atmosphere (Meyer, 1996a, 1996b). With respect to the photospheric composition, the low FIP elements are enriched relatively to the high FIP elements by a factor of about four. The discontinuity in abundances, allows to separate two classes of low FIP and high FIP elements and takes place between 10 and 11 eV.

2. Main Characteristics of the FIP Fractionation Models

Theoretical models of ion-neutral fractionation in the solar atmosphere have been built in order to explain the observed FIP dependence of the variation, from photosphere to corona, of the relative abundance of the solar atmosphere elements. They either consider a steady-state situation or follow the time evolution of the fractionation.

Most models assume element fractionation to take place in the chromospheric temperature plateau. They use parameters selected empirically to describe an atmosphere uniformly horizontally stratified. However, the great diversity of solar atmospheric magnetic structures, open, closed, quiet or active, which seems to be associated with various amplitudes of the fractionation effect, implies, that there is no need for a unique FIP fractionation model.

Models of fractionation have been suggested where the magnetic field does not play any role, except guiding the ions once they have been formed. In these models, fractionation results from diffusion along field lines, under the effect of collisions with vertically moving neutral hydrogen atoms and protons, of minor species irradiated by UV photons. The precise models, take into account the consecutive diffusion of the generated ions and hydrogen ionisation. They differ mainly by the boundary conditions used.

The observed magnetic structuration of the solar atmosphere, both in the very low chromosphere and in the corona, suggests that the magnetic field may play a key role in the FIP fractionation. At densities low enough for collisions not to couple ionized and neutral atoms, neutral high FIP elements can escape from a magnetic structure perpendicularly to the lines of force, which is not the case for ionized low FIP elements. For magnetic fields strong enough for this condition to be satisfied, these densities are chromospheric. Therefore, definite progress in understanding FIP fractionation may be linked to the understanding of the conditions of formation of the chromosphere.

3. Basic Equations

All models use the same basic equations, i.e. the continuity and momentum balance equations:

3.1. Continuity Equations

In 3D, the continuity equations (Burgers, 1969; Schunk, 1975) can be expressed as

$$\frac{\partial}{\partial t} n_j + \nabla \cdot (n_j \mathbf{u}_j) = \frac{\delta}{\delta t} n_j \tag{1}$$

where the quantity on the right hand side of eqn. (1) is the rate of change of density as a result of collisions, and n_j and $\mathbf{u}_j$ are respectively the number density and velocity of the particle j. Charge exchange does not modify the densities of the neutral and ionized fractions of a species. Consequently, and limiting ourself to the 1D case, writing respectively j^o and j^+ the neutral and ionized species j (including in j^+ all the states of ionization), the continuity equations can be rewritten as:

$$\frac{\partial}{\partial t} n_{j^o} + \frac{\partial}{\partial s}(n_{j^o} u_{j^o}) = \gamma_{j^+ j^o}\, n_{j^+} - \gamma_{j^o j^+}\, n_{j^o} \tag{2}$$

$$\frac{\partial}{\partial t} n_{j^+} + \frac{\partial}{\partial s}(n_{j^+} u_{j^+}) = \gamma_{j^o\, j^+}\, n_{j^o} - \gamma_{j^+\, j^o}\, n_{j^+} \quad (3)$$

where the recombination and the ionization rates are respectively $\gamma_{j^+\, j^o}$ and $\gamma_{j^o\, j^+}$.

3.2. MOMENTUM BALANCE EQUATIONS

The momentum balance equations take respectively the forms:

for a neutral species j^o:

$$\left(\frac{\partial}{\partial t} + (\mathbf{u}_{j^o} \cdot \nabla)\right) \mathbf{u}_{j^o} = -\frac{1}{m_j n_{j^o}} \nabla p_{j^o} + \mathbf{g} - \sum_k \nu_{j^o\, k}(\mathbf{u}_{j^o} - \mathbf{u}_k) \quad (4)$$

for a singly ionized species j^+:

$$\left(\frac{\partial}{\partial t} + (\mathbf{u}_{j^+} \cdot \nabla)\right) \mathbf{u}_{j^+} = -\frac{1}{m_j n_{j^+}} \nabla p_{j^+} + \frac{e}{m_j}(\mathbf{E} + \mathbf{u}_{j^+} \times \mathbf{B}) + \mathbf{g}$$

$$- \sum_k \nu_{j^+ k}(\mathbf{u}_{j^+} - \mathbf{u}_k), \quad (5)$$

where $\mathbf{g}$ is the gravity vector, p_j the partial pressure of an element j, and $\mathbf{E}$ and $\mathbf{B}$ are the electric and magnetic field vectors. For most of the existing fractionation models, the component of the velocity field perpendicular to the magnetic field is supposed to be null. For models for which this assumption is not made, the $\mathbf{E} + \mathbf{u}_{j^+} \times \mathbf{B}$ term comes from boundary conditions.

4. Models Where the Magnetic Field Does not Play any Role

Models have been proposed (Marsch *et al.*, 1995; Peter, 1996, 1998a; Wang, 1996), based on diffusion of minor species along a vertical magnetic field.

Since matter is fully ionized at the top, the fractionation of an element j relative to an element k is

$$f_{jk} = \frac{(n_{j^+}/n_{k^+})^{\text{top}}}{((n_{j^o} + n_{j^+})/(n_{k^o} + n_{k^+}))^{\text{bottom}}}. \quad (6a)$$

Matter is supposed to be fully neutral at the bottom, leading to

$$f_{jk} = \frac{(n_{j^+}/n_{k^+})^{\text{top}}}{(n_{j^o}/n_{k^o})^{\text{bottom}}}. \quad (6b)$$

In 1D diffusion models, ionized species have the same velocities at the top i.e. $u_{j^+} = u_{k^+}$ and therefore

$$f_{jk} = \frac{(\phi_j/\phi_k)^{\text{top}}}{(n_j/n_k)^{\text{bottom}}} = \frac{(\phi_j/\phi_k)^{\text{bottom}}}{(n_j/n_k)^{\text{bottom}}} = \left(\frac{u_j}{u_k}\right)^{\text{bottom}} \quad (7)$$

i.e. the fractionation is just the ratio of the velocities of the minor species at the bottom.

In the first two models, minor species slow down, at chromospheric level, in collisions with the main element that is neutral hydrogen. All the ions formed are supposed to be pumped away on the top of the chromosphere by the solar wind. For minor species j of identical mass, two parameters play a dominant rôle, that are the ionization time τ_j and the collision frequency ν_{jH} between a neutral element j and a neutral hydrogen atom. The shorter the ionization time relative to the time interval $1/\nu_{jH}$ between two successive collisions, the higher the velocity of the ion generated is. Consequently, for comparable number densities at the base, a higher flux of minor elements can be extracted and injected into the solar wind for minor elements of higher $\tau_j \nu_{jH}$ ratio, leading to a relative enhancement of the abundances of these elements.

The first two models require minor species not to be accelerated but rather slowed down in collisions. Therefore, they imply that, at the base, either the minor species move up faster than the main hydrogen gas (Peter, 1996, 1998a), or the main gas is moving down (Marsch et al., 1995). Neither of these conditions seems to be appropriate. The third model (Wang, 1996) differs from the first two, since, as seen in 4.2, based on the diffusion of ionized minor species in a upward flow of protons.

4.1. Steady-state Models of Fractionation by Diffusion

The Marsch et al. and Peter's models assume time stationarity. The properties of the species vary along a single coordinate s and the speed of the flow is supposed to be subsonic. Therefore, the left hand sides of eqns. (4) and (5) are null. In the continuity equation, only the ionization processes are considered, and in the momentum balance equation only the collisions of minor species with protons and hydrogen atoms are considered. In both models, diffusion is taking place in the chromospheric temperature plateau at temperature T=10^4 K and density N_H=10^{16} m^{-3}. The two models differ mainly by their boundary conditions. However, in both models:
– at the base ($s = 0$), the density of the ionized species is null ($n_{j^+}(0) = 0$) and the density of the neutral species is given ($n_{j^o}(0) = n_o$);
– at the top ($s = S$), the plasma if fully ionized and all ions have the same velocity ($u_{j^+}(S) = u_S$).

4.1.1. *Steady Hydrogen Velocity Dependent Fractionation (Peter, 1996)*

With the hypothesis listed above, assuming that the minor species are interacting mainly with neutral hydrogen (H) and protons (p), and (Peter, 1996, 1998a) imposing at bottom the boundary condition:

$$\frac{\partial u_n}{\partial s} = 0, \tag{8}$$

the continuity and momentum balance equations can be rewritten as:

$$u_j \frac{\partial}{\partial s} n_j = -\frac{1}{\tau_j} n_j \tag{9}$$

and

$$\frac{\partial}{\partial s} n_j = \frac{\nu_{jH}^o}{v_j^2} n_j (U_H - u_j), \tag{10}$$

where the ionization time τ_j is equal to $1/\gamma_{j^o,j^+}$. The solution of eqns. (9) and (10) is

$$u_j = \left(U_H \mp \sqrt{(U^2 + 4w_j^2)} \right) / 2 \tag{11}$$

where w_j is the ionisation-diffusion speed (Marsch *et al.*, 1995; Peter, 1996) defined by:

$$w_j = \frac{v_j}{\sqrt{(\tau_j \nu_{jH})}}, \tag{12}$$

$v_j = \sqrt{(k_B T_j / m_j)}$ being the thermal velocity. For upwards motion of the main gas, the solution with the minus sign must be rejected, corresponding to opposite directions of motions of the minor species and of the main gas. For high hydrogen velocities ($U_H \gg w_j$) no fractionation takes place. In the limits of low hydrogen velocity ($U_H < w_j$), the solution with the positive sign leads to

$$\lim_{(U_H \to 0)} f_{j,k} = \lim_{(U_H \to 0)} \frac{u_j}{u_k} = \frac{w_j}{w_k} = \sqrt{\frac{T_k m_k \nu_{kH}}{T_j m_j \nu_{jH}}} \tag{13}$$

where the dominant contribution comes from the ionization times ratio, i.e. the lower τ_i, the higher the relative abundance will be at the top.

The solution derived appears rather unphysical, requiring the minor species to move up faster than the main hydrogen gas. This comes directly from the lower reasonable boundary condition $\partial u_n / \partial s = 0$. That condition leads to eqn. (9) that implies that for upwards motions ($u_{j^o} > 0$), $\partial n_n / \partial s$ is negative. In order to satisfy this condition, the momentum balance equation requires the friction force to be directed downwards, i.e. $U_H - u_j$ must be negative. In other words, as stated at the beginning of section 4, minor species must enter into the chromospheric plateau with upwards velocities higher than the upwards velocity U_H of the main gas.

The height h_1 is greater than about 2100 to 2250 km for the elements considered to be ionized. The lower the height h_0, the higher the abundance enhancement. This model leads to an enhancement of the low FIP elements qualitatively in agreement with observations. The limits of the model are in the orientation of the lines of forces, since in general the field lines are not horizontal and ions fall back to the photosphere along the lines of force, and in the difficulty to explain the ejection of plasma into the interplanetary medium with such a mechanism.

5.2. VERTICAL MAGNETIC FIELD – TIME DEPENDENT FRACTIONATION IN A HOMOGENEOUS MAGNETIC FIELD (VON STEIGER AND GEISS, 1989)

von Steiger and Geiss (1989) considered ion-neutral separation perpendicular to an uniform magnetic field in the solar chromosphere under the effect of gravity or/and of a slight pressure gradient.

At a time $t = 0$ an initial mixture of neutral gas j^o (H, He and minor species) is put, with a gaussian density profile $n_{j^o}(r)$, in a slab of lateral extension $2\,r_0$ parallel to a magnetic field **B**. The gas is then submitted to ionizing UV irradiation. The plasma density ($n_o = 10^{15}$ m^{-3}) has been selected low enough for the ionizing UV radiation (Lyα and Lyman continuum, and the most prominent lines of HeI and HeII) not to be attenuated over distances of the order of 100 km. Under the effect of gravity or/and of the small density gradient, the neutrals diffuse with time across the magnetic field lines, while the plasma density and magnetic field intensity were selected in such a way that the ion gyrofrequency exceeds the ion-neutral collision frequency making the new born ions to be held fixed around the lines of force.

With the conditions used by the authors, the gas is fully ionized on a time scale close to one minute. At a time t, the enrichment or depletion $f_t(j)$ of a species j relative to Si in a slab of thickness $2\,r_o$ is

$$f_t(j) = \frac{[j](t)/[\text{Si}](t)}{[j](0)/[\text{Si}](0)} \tag{19}$$

where

$$[j](t) = \int_0^{r_o} \sum_i n_{ji}(r,t), \tag{20}$$

is the the integrated density of all states i of species j inside a slab of thickness $2\,r_o$.

In order to compute the populations of the various species, the continuity and momentum balance equations were solved. Since the ions do not cross the field lines, charge exchange between neutral and ionized element leads, when significant, to an exchange of momentum. This effect was taken into account in the momentum balance equations for hydrogen, oxygen and helium. The continuity equations were taking into account photoionization, recombination, and considering not only the ground states of atoms and ions but also the excited atomic states that contribute significantly to ionization and recombination. The excited levels to consider for the atoms were taken from Geiss and Bochsler (1985). The diffusion of neutral hydrogen atoms and protons was treated first and the result obtained used to study the behaviour of the minor species.

Assuming that the gas inside the slab of thickness $2r_o$ stay a time t_{end} at chromospheric temperature (T =6000 K) and then, being taken in an upward stream, becomes fully ionized at coronal height, the enrichment in the corona relative to the chromosphere (where the abundances are supposed to be photospheric) is just $f_{t_{end}}(j)$. By selecting an appropriate value of the parameters t_{end} and r_o, and including a contribution of the gravity force by inclining by 15° the slab relative to the vertical, a FIP dependence close to the observations was derived (Figures 5b and 7 in von Steiger and Geiss (1989). The limits of the model are in the low density used (10^{15} m^{-3}), compared to the chromospheric density 10^{19} m^{-3} given by usual quiet sun temperature models for T =6000 K, and in the need of having only a very narrow nearly vertical slab, of 4 km width, extracted from the chromosphere towards the corona. A low density is required in the model in order for the photoionizing radiation to penetrate the medium. However, in the chromosphere the hydrogen ionization does result from both collisional excitation to level 2 and photoionization by the Balmer continuum which can reach the very low and dense chromosphere (Athay, 1981). Consequently, similar computations can presumably be done with higher densities and more realistic magnetic field and pressure distribution.

5.3. Vertical Magnetic Field – Ion-neutral Fractionation in Magnetic Flux Tubes (Hénoux and Somov, 1997)

Hénoux and Somov (1997) did not estimate quantitatively the variation of abundance from the photospheric level to the corona. Their flux-tube model just predicts qualitatively the formation of closed or open structures with higher temperature, ionization state and higher low FIP to high FIP elements abundance ratios than the surrounding. A quantitative estimation of these abundance ratios has still to be done.

Strong pressure gradients across magnetic field lines can be present in magnetic flux tubes where electric currents are circulating (Hénoux and Somov, 1991; Hénoux and Somov, 1997). Since they produce two of the ingredients that are required for ion-neutral fractionation by magnetic fields, i.e. small scales and strong pressure gradients perpendicular to the magnetic field lines (Hénoux and Somov, 1992; Hénoux, 1995), these currents can lead to ion-neutral fractionation. In the model of flux-tube proposed, azimuthal motions of the partially ionized photospheric plasma, at the boundary of flux tubes, generate a system of currents flowing in opposite directions, such that the azimuthal component of the field vanishes at infinity. This result can be easily derived in the case of a fully ionized atmosphere where the field lines are frozen in the plasma. However, the study of a partially ionized atmosphere gives insight into questions that cannot be tackled in the hypothesis of a fully ionized plasma, i.e. the possible difference in velocities perpendicular to the magnetic field lines of neutrals and ions.

The internal current system and the azimuthal component of the magnetic field create an inward radial force $B_\theta j_z$ that enhances, by pinch effect, the pressure inside the internal current system. This pinch effect is present from the photosphere to the chromosphere and its effects are different in these two regions. At photospheric level, collisions couple ions and neutrals; they do not cross the field lines. Then, due to the exponential decrease of the density and, as a result, of the ion-neutral friction force with height, the difference in radial outward velocities of neutrals and ions increases with height. The current densities and magnetic fields in the flux tube are such that, at hydrogen densities lower than 10^{19} m^{-3}, the collisional coupling is low enough to allow the neutrals to cross the lines of force and to escape from the internal current shell with high velocities. In usual atmospheric models, the fractionation starts in the temperature minimum region at a temperature of about 4000 K. So the population of ionized low FIP species begin to be enhanced inside the internal current sheet just at heights where plane parallel atmosphere models place the chromospheric temperature rise and where the separation between the hot and cool component of Ayres (1996) bifurcation model starts to take place.

Between the two opposite currents flowing vertically, upwards electromagnetic forces $j_r B_\theta$ are present. Since the change of the direction of the vertical currents goes with the change of direction, from the photosphere to chromosphere, of the transverse current j_r carried by ions, the $j_r B_\theta$ force always produces a net ascending force. The intensity of this force is compatible with ejection of matter up to heights of about 10 000 km, and therefore with the formation of spicules. This force acts in a shell, between the two neutralizing currents, where the gas pressure and collision friction forces are reduced; it acts on ions and may then lead to a FIP effect in spicules by rising up preferentially the ionized low FIP species. However, here also, a quantitative study of all these effects remains to be done.

6. Conclusion

The various existing FIP fractionation models differ mainly by the role, active or passive, of the magnetic field and by the boundary conditions used. So the progress in the understanding of FIP fractionation are linked to the understanding of the formation of these boundary conditions and of the associated solar structures. More generally, they are linked to the understanding of the generation of the solar chromosphere. Since the solar chromosphere appears to be rooted in magnetic structures, an active role of the magnetic field is highly plausible. Moreover, models where the magnetic field is passive require rather unrealistic or weakly justified boundary conditions to reproduce the observed fractionation.

The FIP fractionation is not limited to the 10^4 K temperature plateau; it presumably starts lower in the solar atmosphere at densities as high as 10^{19} m^{-3} and temperatures that may be as low as 4000 K, i.e. just in the temperature minimum region at the base of the solar chromosphere. As quoted in Meyer (1996a, 1996b) and references herein, a number of coronal spectroscopic observations show extra enhancement of very low FIP species suggesting that ion-neutral fractionation takes place at the temperature minimum of 4000 K. If FIP fractionation is associated with the formation of the solar chromosphere, enhancement of the abundance of low FIP elements must be present at chromospheric temperatures. It must be pointed out that γ ray line observations during solar flares have shown that the low FIP/high FIP element abundance ratio can reach coronal values in the low chromosphere where nuclear γ ray lines are formed (Share and Murphy, 1995; Mandzhavidze and Ramaty, 1998).

FIP fractionation varies with height – diffusion models predict even a relative increase of the relative abundances of high FIP elements at intermediate heights – and any determination, from observations, of the height variation of abundance, from photosphere to corona, would allow a better understanding of the FIP effect. However, the rôle played in the formation of line intensities by the atmospheric temperature and density, makes such derivation rather improbable.

The FIP fractionation effect is not uniform over the solar surface. Therefore, there is no need for a unique model of ion-neutral separation. Moreover, spatial observations show that the solar corona and high chromosphere is not steady, with open structures like jets and macrospicules lasting no more than a few minutes. The dynamical aspect of the open structures like polar plumes, in which high fractionation is observed, suggests acceleration by electromagnetic forces that will have to be taken into account by any theoretical model of fractionation in open structures.

References

Athay, R.G.: 1981, *Astrophys. J.* **250**, 709.
Ayres, T.R.: 1996, 'Thermal bifurcation of the solar chromosphere', *Stellar Surface Structure,* IAU Symposium **173**, Kluwer Academic Publishers, 371–384.
Burgers, J.M.: 1969, *Flow Equations for Composite Gases*, Academic Press, New York.
Feldman, U.: 1992, *Physica Scripta* **46**, 202.
Fludra, A., Saba, J.L.R., Hénoux, J-C., Murphy, R., Reames, D., Lemen, J.L., Strong, K.T., Sylwester, J., and Widing, K.: 1998, 'Coronal Abundances', *The Many Faces of the Sun*, Strong, K.T., Saba, J.L.R, Haisch, B.M., and Schmelz, J.T. (eds.), Springer-Verlag, in press.
Fontenla, J.M., Avrett, E.H. and Loeser, R.: 1993, *Astrophys. J.* **406**, 319.
Geiss, J. and Bochsler, P.: 1985, 'Ion composition in the solar wind in relation to solar abundances', *Rapports isotopiques dans le systeme solaire*, Paris: Cepadues-Editions, 213–228.
Hénoux, J.-C.: 1995, 'Models for explaining the observed spatial variation of element abundances – a review', *Adv. Space Res.* **15**, (7)23–(7)32.
Hénoux, J.-C. and Somov, B.V.: 1991, 'The photospheric dynamo: I. Magnetic flux-tube generation', *Astron. Astrophys.* **241**, 613–617.
Hénoux, J.-C. and Somov, B.V.: 1992, 'First ionization potential fractionation', *Proceedings of the First SOHO Workshop*, ESA SP-348, 325–330.

Hénoux, J.-C. and Somov, B.V.: 1997, 'The photospheric dynamo: II. Physics of thin magnetic fluxtubes', *Astron. Astrophys.* **318**, 947–956.

Mandzhavidze, N. and Ramaty, R.: 1997, 'Gamma Rays from Solar Flares', *23th IAU General Assembly*, Kyoto.

Marsch, E., von Steiger, R. and Bochsler, P.: 1995, 'Element fractionation by diffusion in the solar atmosphere', *Astron. Astrophys.* **301**, 261–276.

Meyer, J.P.: 1993a, 'Element Fractionation at Work in the Solar Atmosphere', *Cosmic Abundance of Matter*, N. Prantzos, E. Vangioni-Flam and M. Cassé (eds.), Cambrige Univ. Press, 26.

Meyer, J.P.: 1993b, *Adv. Space Res.* **13**, (9)377.

Meyer, J.P.: 1996a, *Cosmic Abundances*, S.S. Holt and G. Sonneborn (eds.), ASP Conf. Series **99**, 127–146.

Meyer, J.P.: 1996b, *The Sun and Beyond,* J. Trân Thanh Vân, L. Celnikier, H.C. Trung and Vauclair (eds.), Gif-sur-Yvette: Editions Frontières, 27–46.

Peter, H.: 1996, 'Velocity-dependent fractionation in the solar chromosphere', *Astron. Astrophys.* **312**, L37–L40.

Peter, H.: 1998a, 'Element separation in the chromosphere', *Space Sci. Rev.*, this volume.

Peter, H.: 1998b, private communication.

Reames, D: 1998, 'Solar Energetic Particles: Sampling Coronal Abundances', *Space Sci. Rev.*, this volume.

Saba, J.L.R.: 1995, *Adv. Space Res.* **15**, (7)13.

Schunk, R.W.: 1975, 'Transport Equations for Aeronomy', *Planet. Space Sci.* **23**, 437–485.

Share, G.H. and Murphy, R.J.: 1995, *Astrophys. J.* **464**, 933–943.

von Steiger, R. and Geiss, J.: 1989, 'Supply of fractionated gases to the corona', *Astron. Astrophys.* **225**, 222–238.

Vauclair, S.: 1996, 'Element segregation in the solar chromosphere and the FIP bias: The Skimmer model', *Astron. Astrophys.* **308**, 228–232.

Wang, Y.-M.: 1996, 'Element separation by upward proton drag in the chromosphere', *Astrophys. J.* **464**, L91–L94.

Address for correspondence: Observatoire de Paris; DASOP/LPSH(URA2080) 92195 Meudon Principal Cedex, France

FIP EFFECT IN THE SOLAR UPPER ATMOSPHERE: SPECTROSCOPIC RESULTS

U. FELDMAN

E.O.Hulburt Center for Space Research, Naval Research Laboratory, Washington DC 20375, USA

Abstract. Recent spectroscopic measurements from instruments on the Solar and Heliospheric Observatory (SOHO) find that the coronal composition above a polar coronal hole is nearly photospheric. However, similar SOHO observations show that in coronal plasmas above quiet equatorial regions low-FIP elements are enhanced by a factor of ≈ 4. In addition, the process of elemental settling in coronal plasmas high above the solar surface was shown to exist. Measurements by the Ulysses spacecraft, which are based on non-spectroscopic particle counting techniques, show that, with the exception of He, the elemental composition of the fast speed solar wind is similar to within a factor of 1.5 to the composition of the photosphere. In contrast, similar measurements in the slow speed wind show that elements with low first ionization potential (FIP< 10 eV) are enhanced, relative to the photosphere, by a factor of 4–5. By combining the SOHO and Ulysses results, ideas related to the origin of the slow speed solar wind are presented. Using spectroscopic measurements by the Solar Ultraviolet Measurement of Emitted Radiation (SUMER) instrument on SOHO the photospheric abundance of He was determined as $8.5 \pm 1.3\%$ ($Y = 0.248$).

Key words: Sun – Solar Wind

1. Introduction

The solar wind composition is quite variable. During solar minimum periods two distinct types of winds, the slow speed wind and the high speed wind are observed. Along the equatorial direction the wind – the slow speed solar wind – reaches velocities of $v_s \leq 500$ km s^{-1} and is enriched by a factor of 4–5 with ions whose first ionization potential (FIP) is less than 10 eV. In contrast, in the solar polar directions a much faster solar wind – the fast speed solar wind – is observed. The fast speed solar wind reaches velocities of $500 < v_f < 800$ km s^{-1} and, with the exception of He, its elemental composition is similar to within a factor of 1.5, to that of the solar photosphere. For a discussion on the composition of the solar wind and on the processes that may cause it see papers by Geiss (1998), Geiss and Bürgi (1986), and von Steiger, Geiss and Gloeckler (1997).

Recent results from observations on board the Solar and Heliospheric Observatory (SOHO) provided new insight into the composition of the solar upper atmosphere (SUA). Below I will discuss new observational facts regarding the composition of the SUA above quiet and coronal hole regions. Ideas regarding the sources of the solar wind will also be presented. The SUA is the only source from which the He, Ne and Ar solar abundances can be obtained by spectroscopic means. A discussion on photospheric abundances of these three elements will be presented in an appendix.

Models attempting to explain the structure of the SUA were first developed several decades ago, when space research was in its infancy and observational facts were limited. With insufficient knowledge, theoreticians were forced to make reasonable assumptions regarding the SUA structure. For a summary of such models see Athay (1976). Today we know that the SUA is much more complex than originally assumed and that many of the reasonable assumptions that were central in establishing the prevailing solar models have not withstood the test of time (Feldman, 1993). Since any general conclusion regarding the composition of the SUA requires a basic understanding of its structure I will begin with a brief review on SUA structures above quiet and coronal hole regions.

2. The Structure of the Upper Solar Atmosphere

Empirical solar models, originally developed shortly after space observations began, depicted the SUA above quiet and coronal hole regions as continuous, starting with the cold chromosphere ($T_e \leq 2 \times 10^4$ K) and continuing into the hot corona ($T_e \geq 9 \times 10^5$ K). The models assumed that the plasma in the SUA is static, with no elemental abundance gradients, no unusual atomic processes taking place (i.e., the electron velocity distribution is Maxwellian), the geometry is plane parallel, the pressure between the chromosphere and corona is constant and the thermal conductive energy flux in the $10^5 - 10^6$ K range is constant. The interface between the chromosphere and corona was described as a narrow layer of less than one hundred kilometers that was named the transition region. With advances in space research instrumentation it became apparent that the original depiction of the SUA needed modification. High spectral and spatial resolution images of the SUA, taken at a variety of temperatures, show that SUA structures are very complex and cannot be approximated by the simple formations just described. Measurements in the vicinity of the solar limb show that the dimensions of structures having temperatures in the 2×10^4 K $< T_e < 9 \times 10^5$ K are much larger than predicted by the prevailing models. Using the Harvard instrument on Skylab which had a $5'' \times 5''$ spatial resolution Huber et al. (1974) derived the full width of half-intensity (FWHI) of the solar limb at a temperature of $\approx 4 \times 10^5$ K. According to them, the FWHI above coronal hole regions is $29''.1 \pm 1''.3$ while above the quiet region adjacent to the coronal hole the FWHI is $21''.1 \pm 1''.2$. Using the NRL normal incidence spectrometer on Skylab which had a $2'' \times 60''$ spatial resolution Doschek et al. (1976), Feldman et al. (1976), and Kjeldseth Moe and Nicolas (1977) verified the Huber et al. (1974) measurements. However, because of the higher resolution of their instrument they were able to show that emissions from temperatures of 1×10^5 K and 2×10^5 K which were predicted to be separated by only 10 km are actually separated by about 10^3 km. In an attempt to remedy the large discrepancy between the models and observations Gabriel (1976) proposed a modification of the models. His new model assumed a magnetic field that at photospheric levels is concentrated by supergranular flow into the network elements and then expanded at higher levels until it finally fills the corona. Although calculations based on the Gabriel (1976) model resulted in a wider transition region the problem was not resolved. Electron densities, plasma pressure and non-thermal mass motions that are derived in the $2 \times 10^4 \text{K} < T_e < 9 \times 10^5$ K temperature domain are also not consistent with those at higher or lower temperature domains. From the above and from other similar measurements Feldman (1983, 1987) and Feldman and Laming (1994) concluded that the SUA is not the continuous plasma domain the model predicted. Instead, they claimed that the majority of the emission in the $2 \times 10^4 \text{K} < T_e < 9 \times 10^5$ K range, traditionally referred to as the transition region, comes from unresolved fine structures (UFS) that are not part of the chromosphere-corona structure. These structures are much smaller in size than typical coronal structures and their peak temperature is less than 9×10^5 K. When projected on the solar surface UFS appear as clumps of emission that seem to trace the chromospheric network. On the limb they appear as brightened well defined rings (see the Ne VII image in Figure 1). Any emission from the chromosphere-corona transition (interface) layer must be significantly fainter than the emission originating in the UFS. Thus far, the true interface layer plasmas have not been identified. Composition measurements that will be discussed below reaffirm the conclusion that the UFS and the corona are different structures.

Morphologically, above quiet and coronal hole regions, the SUA can be divided into two independent domains, the UFS and the corona. UFS are best seen in solar images taken in the 465 Å Ne VII line (Figure 1) by the Skylab SO82A spectroheliograph. On the disk UFS appear as bright unresolved clumps of emission, and on the limb they form the prominent limb brightening rings (Bohlin and Sheeley, 1978). Not being able to resolve the individual structures, it is difficult to determine their lifetimes. Since UFS seem to maintain their shapes for more than an hour and less than a day it is perhaps reasonable to assume that their lifetimes are between 10^4 and 10^5 s.

The corona above quiet regions which starts very near the solar surface and extends outward (see the solar image taken in the 368 Å Mg IX line, Figure 1) consists of long lasting closed loop-like formations of $T_e \geq 1 \times 10^6$ K. Images obtained by the Large Angle Coronagraph (LASCO) on SOHO (Brueckner *et al.*, 1995) and by its predecessors show that during solar minimum coronal structures above quiet regions (streamers) extend to several solar radii, and last for many days.

The coronal temperature above coronal hole regions does not extend above $T_e \approx 9.0 \times 10^5$ K (see the solar image taken in the 368 Å Mg IX line, Figure 1). Since its electron density is lower than the density above quiet regions, it is considerably fainter. Contours of the magnetic field structure above coronal hole regions are recognizable from the shapes of the polar plumes that are embedded in them.

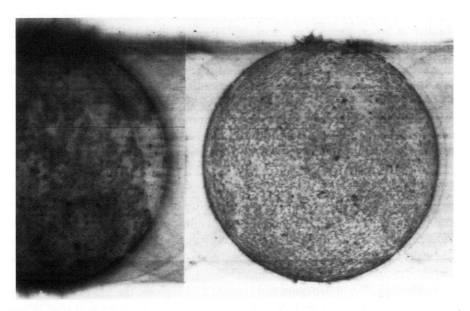

Figure 1. The solar image in the lines of Ne VII 465 Å (full image) $\approx 5 \times 10^5$ K and Mg IX 368 Å (half image) recorded by the SO82a spectroheliograph on Skylab. North is to the right.

3. Low- to High-FIP Ratios in the Vicinity of the Solar Surface

The observational methods and some of the results regarding the composition of the SUA in the vicinity of the solar surface were discussed in a number of review articles (e.g., Feldman, 1992; Meyer, 1985, 1993, 1995). Due to space limitations I will not discuss in detail observational methods employed in deriving abundances by spectroscopic means. Keeping with the convention used in the field, I will refer to changes in the low- to high-FIP abundance ratios relative to photospheric values as the FIP bias. In regions where the SUA and photospheric abundances are the same, the FIP bias is defined as 1 and in the slow speed solar wind the FIP bias is 4–5.

3.1. THE COMPOSITION OF THE UFS ABOVE QUIET AND CORONAL HOLE REGIONS

Feldman and Widing (1993) used Mg VI and Ne VI lines near 400 Å that are typical of $T_e \approx 4 \times 10^5$ K plasmas, to determine the abundance ratio between the low-FIP Mg and the high-FIP Ne. They have found that above quiet regions the FIP bias particles, and to measure the relative abundances of the heavier noble gases in ancient solar wind flows. Since the moon intercepts the solar wind at low heliospheric latitude, we may assume that the solar wind particles in lunar soils mainly represent the low speed wind. The observed, relatively small secular changes in the solar wind abundances of Kr and Xe are discussed by Wieler et al. (1993), Wieler and Baur (1995) and Wieler (1998).

From mid-1992 to the spring of 1993, Ulysses went into and out of the Southern high speed stream. Using a superposed epoch method, Geiss et al. (1995a,b) and Wimmer-Schweingruber et al. (1997) showed that the freeze-in temperatures of O and C and the Mg/O and Fe/O abundance ratios changed simultaneously at the boundaries between the high speed stream and the slow wind (Figures 2 and 3).

Since freeze-in temperatures reflect the electron temperature profile in the low corona (Geiss et al., 1995b; Ko et al., 1997), this shows that the dividing line between high and low FIP effect regions was identical (in the limits of spatial resolution) with the boundary of the low electron temperature regime of the southern coronal hole.

So far, the FIP effect in the solar wind has mainly been investigated in the two major types of flow, the high speed streams and the average slow wind. The slow wind abundances presented in Figure 1 are mainly derived from SWC-Apollo, ICI-ISEE3 and SWICS-Ulysses data. These experiments have all a relatively low time resolution and therefore, the discussions and conclusions in this paper relate primarily to relatively quiet or average solar wind conditions. For short-time variations or local differences in the FIP abundance pattern, we refer to the presentation of coronal abundances obtained from UV spectroscopy by Feldman (1998), or to forthcoming results from CELIAS-SOHO and SWICS/SWIMS-ACE.

3. Ionisation at the Solar Surface

Since atomic behaviour is often directly or indirectly controlled by the first ionisation potential (FIP), the loose relationship between elemental abundances and FIP as shown in Figure 1 does not reveal much about the mechanism that causes the FIP effect, and indeed, quite different mechanisms have been proposed for explaining this effect (cf. Hénoux, 1998). However, a strict relationship between abundances and FIP does not exist. This is not surprising, since such a relationship would imply thermodynamic equilibrium, which is unlikely to exist in those solar surface layers (cf. Vernazza et al., 1981), where the FIP mechanism could operate.

Geiss and Bochsler (1985), von Steiger (1988) and von Steiger and Geiss (1989) proposed that the FIP effect is caused in an atom-ion separation process characterized by a competition between ionisation and ion-atom separation. These authors noted that ionisation rates are much more element specific than separation rates.

Assuming the ionisation to proceed in an optically thin medium at the solar surface, and adopting the "quiet-sun" spectral intensity averaged over the solar sur-

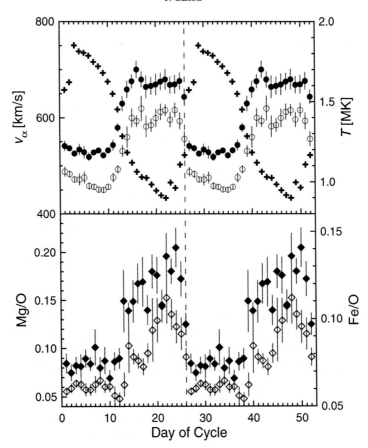

Figure 2. Superposed epoch plot of Ulysses-SWICS data from the 1992/1993 epoch when Ulysses went regularly into and out of the Southern high-speed stream once every solar rotation (after Geiss *et al.*, 1995b). This method has allowed to superpose the data of 20 crossings of the boundary between the high-speed stream and the slow solar wind (Geiss *et al.*, 1995a). The data are repeated after the dotted line to facilitate the recognition of the entire pattern. The crosses represent the speed of the He^{2+} ions. The freeze-in temperatures given for oxygen (full circles) and carbon (open circles) represent electron temperatures at different levels in the corona. In the lower pannel the Mg/O (full diamonds) and Fe/O (open diamonds) ratios are given. The coinciding changes in freeze-in temperatures and abundance ratios demonstrate that the drop in the electron temperature and in the strength of the FIP effect occur at or near the same place.

face, they calculated the ionisation as a function of time for a number of elements (Figure 4), taking into account ionisation and excitation by photons and electrons, charge exchange reactions and radiative recombination. Such ionisation curves of individual elements can then be used to calculate $\tau_{\text{St-Ion}}$, the standard ionisation time for an element which we define as

$$\tau_{\text{St-Ion}} = \tau_{1/2} / \ln 2, \tag{1}$$

where $\tau_{1/2}$ is the time after which a degree of ionisation of 50 % is reached for the element under consideration. Standard ionisation times for atomic species are

CONSTRAINTS ON THE FIP MECHANISMS FROM SOLAR WIND ABUNDANCE DATA   245

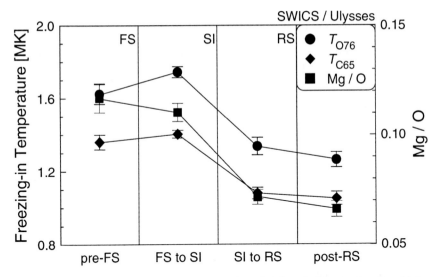

Figure 3. The traverse of Ulysses from the slow solar wind domain into the Southern high speed stream is shown here with better time resolution than in Figure 2 (after Wimmer-Schweingruber *et al.*, 1997). The superposed epoch technique is applied to measure average abundances and freeze-in temperatures in four regions: pre-FS (pre forward shock), FS to SI (forward shock to stream interface), SI to RS (stream interface to reverse shock), and post RS (post reverse shock). T_{O76} and T_{C65} are the freeze-in temperatures of oxygen (derived from O^{7+}/O^{6+}) and carbon, (derived from C^{6+}/C^{5+}), respectively.

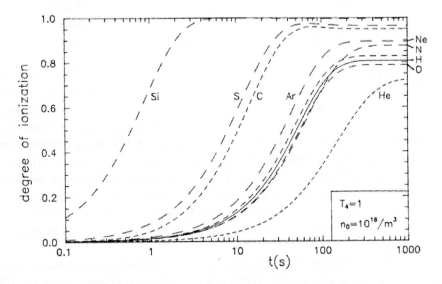

Figure 4. Ionisation at the solar surface for zero optical depth, quiet-sun conditions, a total hydrogen density of 10^{16} m^{-3}, an electron temperature (T_e) of 10000 K and vanishing initial ion densities (after von Steiger and Geiss, 1989). The ionisation time τ_{St-Ion} is defined in Equation (1). The τ_{St-Ion} values given in Table I were derived from such ionisation curves, using an electron density of 10^{16} m^{-3} and an electron temperature of 6000 K (cf. Bochsler and Geiss, 1985; von Steiger, 1988; von Steiger and Geiss, 1989; Marsch *et al.*, 1995).

Table I

Standard Ionisation Times τ_{St-Ion} at the solar surface for zero optical depth, and quiet sun conditions, as defined in Equation (1). A total hydrogen density of 10^{16} m^3, and an electron temperature (T_e) of 6000 K were assumed. For H, He, C, N, O, Ne, Si, S, Ar, Kr, and Xe the τ_{St-Ion} values were adapted from the ionisation curves calculated by Geiss and Bochsler (1985), von Steiger (1988), von Steiger and Geiss (1989), Geiss et al. (1994), Marsch et al. (1995). The Standard Ionisation Times for Na and Al were calculated in the present work using the photoionisation cross sections given by Verner et al. (1996) and the solar spectrum given by von Steiger (1988). τ_{St-Ion} values with asterisk (*) are estimates (cf. Swider, 1969).

Element	τ_{St-Ion} [s]	Element	τ_{St-Ion} [s]
He	260	S	11.6
Ne	81	Xe	10.1
O	74	Fe	2*
H	70	Na	1.5
N	68	Si	0.64
Ar	50	Mg	0.3*
Kr	20.3	Al	0.02
C	20		

given in Table I. τ_{St-Ion}, as defined here, is close to the ionisation times used by most authors. It corresponds to the inverse of the rate of ionisation, but only if ionisation occurs solely from the ground state (e.g. for neon), and only if the effect of recombination is minor.

Under the conditions described above, ionisation of the noble gases is essentially due to solar EUV absorption at the ground state, but for many other elements, excitation by photons and electrons affect τ_{St-Ion} significantly (Geiss and Bochsler, 1985; von Steiger, 1988; Joos, 1989). Thus, the temperature and density of the electrons has to be prescribed for the definition of τ_{St-Ion}. We choose a constant electron temperature (T_e) of 6000 K and a constant total hydrogen density (H-atoms plus protons) of 10^{10} cm^{-3}. Since the degree of ionisation of hydrogen increases with time (cf. Figure 4), the electron density is rising during the ionisation process. This has been taken into account in calculating the ionisation curves in Figure 4 and in the standard ionisation times τ_{St-Ion} given in Table I (cf. von Steiger and Geiss, 1989).

When abundances are plotted against the ionisation time τ_{St-Ion} instead of the ionisation energy, a monotonous relationship is obtained (Figure 5). If the FIP effect is caused by a competition between ionisation and ion-atom separation and if the diffusion approximation is valid, elemental abundances should be a function of $D\tau_{St-Ion}$, where D is an appropriate diffusion constant (Geiss and Bochsler, 1986; Marsch et al., 1995; Peter, 1996). Since, however, the ionisation time is more element specific than the ion-atom separation time, the simple, standardized ionisa-

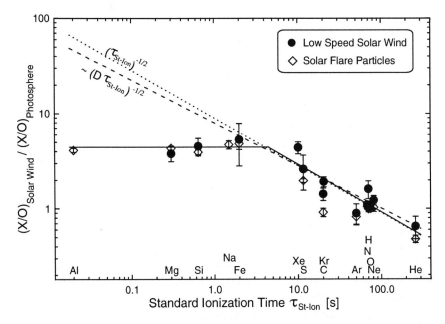

Figure 5. Element abundances in the average slow solar wind and of flare particles as a function of the standard ionisation time $\tau_{\text{St-Ion}}$ (cf. Table I). Sources of data as in Figure 1. Abundances are given relative to oxygen and normalised to the photospheric abundances given by Anders and Grevesse (1989).

tion time $\tau_{\text{St-Ion}}$ is a useful parameter to order the observational data, and it has become an often used basis for constructing FIP models. The pattern of elemental abundances indicates that the FIP effect operates at an electron temperature of the order of 10^4 K (Geiss, 1982). Above 10^5 K ionisation by electrons dominates (cf. Joos, 1989), and, since a considerable fraction of the electrons has then a kinetic energy of more than 10 eV, the difference in ionisation potential of high-FIP and low-FIP elements would loose its significance. In fact, the abundance systematics seen in Figure 5 is being destroyed if $T_e > 3 \times 10^4$ K is chosen for the calculation of $\tau_{\text{St-Ion}}$.

4. General Constraints on the FIP Separation Mechanism

In this chapter we summarize constraints on the FIP mechanism that follow from general considerations and from observations, primarily those derived from solar wind data. For more comprehensive discussions of the flare particle data and UV observations in the corona we refer to the articles in this volume by Williams *et al.* (1998), Reames (1998) and Feldman (1998). We consider the following list of constraints and observations to be important for finding the right FIP mechanism.

1) A one-dimensional steady-state theory cannot produce the FIP effect (see also McKenzie *et al.*, 1998). Since the sum of the ionic and atomic flows of an

element through a flux tube is conserved, any element enrichment achieved at the outflow boundary must also exist at the inflow boundary of the flux tube. Thus, whatever produces the separation is not covered by such a theory. There are of course one-dimensional, quasi-steady-state mechanisms that achieve a high degree of separation, an example being the thermal escape from a planetary atmosphere. In such cases, however, the processes at and below the lower boundary are crucial and should be analysed and included in the theory.

2) Two- or three-dimensional steady-state theories are kinematically valid approximations. Although the chromosphere and the transition region are characterized by fine and rapidly shifting structures (cf. Judge and Peter, 1998), steady state models should provide useful constraints on the mechanisms that produces the FIP effect. Figure 4 may illustrate this: The degrees of ionisation achieved at any given time is inversely related to the ionisation times τ_{St-Ion} of the elements. Therefore, an ion-atom separation process operating in a highly fluctuating environment would still favour elements with lower τ_{St-Ion} over those with higher τ_{St-Ion}.

3) Many elements are strongly ionized at all solar depths. If species are assumed to be initially neutral, it should be tested whether such an initial condition or boundary condition significantly influences the result.

4) The pattern of elemental abundances in the slow wind and in flare particle populations (Figures 1 and 5) indicates that the FIP mechanism operates at electron temperatures of $\sim 10^4$ K. At the corresponding altitude in the solar atmosphere, temperature gradients may be very high. Thus, the thermal diffusion force, strong for heavy ions in partly ionised hydrogen (cf. Geiss and Bürgi, 1986, 1987), could be a significant factor in the separation process.

5) The charge state distributions of low- and high-FIP elements in the solar wind are consistent with a flow through the low corona at an electron temperature of 1 to 2 million K, and consequently, elements heavier than helium are observed to be highly charged under all solar wind conditions. This implies that evaporation from grains of non-solar origin falling into the corona (e.g. from cometary or asteroidal matter) does not significantly contribute to the overabundance of low-FIP elements in the solar wind, because the material released into the corona would attain a charge distribution that reaches down to low charges (Geiss et al., 1992).

6) The FIP effect is much stronger in the slow wind than in the high speed streams. When flare particle abundances are corrected for mass discrimination (cf. Meyer, 1981, 1985; Breneman and Stone, 1985; Stone, 1989), their abundance is the same as in the low speed wind (Figures 1 and 5). This is consistent with the observation that flares normally occur at low or medium solar latitude, where the low speed solar wind is thought to originate.

7) In the slow wind and in flare particle populations, elemental abundances are approximately proportional to $(\tau_{St-Ion})^{-1/2}$ or $(D\tau_{St-Ion})^{-1/2}$ for $\tau_{St-Ion} > 10$ s, but they are constant for $\tau_{St-Ion} < 10$ s (Figure 5).

8) The positions of Kr and Xe on the FIP and τ_{St-Ion} scales are very different relative to the positions of the other elements. Since the Kr and Xe abundances fit

much better into the $\tau_{\text{St}-\text{Ion}}$ than into the FIP systematics (compare Figures 5 and 1), these elements provide specific evidence that the ionisation time plays a direct role in the mechanism that produces the FIP effect.

9) The average He/H ratios in the slow solar wind and in high speed streams are 0.035 and 0.05, respectively (Neugebauer, 1981). These values are much lower than the ratio of 0.085 in the Outer Convective Zone of the sun (Pérez Hernández and Christensen-Dalsgaard, 1994). Helium is also found to be depleted, relative to the CNONe group, in large flares (Reames, 1991). The helium depletion in the high speed streams and probably a part of the depletion in the slow wind and in flare particle populations is not produced in the corona, but below (Geiss, 1982). Since $\tau_{\text{St}-\text{Ion}}$ for helium is much higher than for all other elements in Figure 5, a FIP mechanism based on ionisation time produces not only an overabundance of the low FIP elements (relative to H, O, N, Ne and Ar group), but also an underabundance of helium. Whether these enrichments are achieved simultaneously in one process or in successive processes needs further study.

10) The aluminum, sodium and magnesium abundance can be used to test a FIP mechanism for its predictions at very short ionisation times. Figure 5 shows that particularly the aluminum data are incompatible with mechanisms that produce abundances proportional to $(D\tau_{\text{St}-\text{Ion}})^{-1/2}$.

11) Whereas excitation by photons significantly reduces $\tau_{\text{St}-\text{Ion}}$ of hydrogen, excitation by electrons is important for elements such as C or N that have metastable states with excitation energies below a few eV (Geiss and Bochsler, 1985). For an electron temperature of $\sim 10^4$ K, a large fraction of the ionisation of C goes via the ^1D state (at 1.26 eV). Thus, careful modelling and abundance measurements of carbon constrain the electron temperature at which the FIP mechanism operates.

12) Within the limits of the available time resolution the transitions from high to low electron temperature and from strong to weak FIP effect occur simultaneously when traversing the boundary of a polar coronal hole.

The list given here is certainly not complete, but it shows how important it is to have a comparison between model predictions and observations for a sizeable number of elements covering a wide range of atomic properties. The available abundance observations in the low speed wind and in flare particle populations are quite adequate for testing the validity of a FIP mechanism. We note however, that possible short-time variations (e.g. Coronal Mass Ejections) or spatial variations in coronal holes are not readily recognizable in the solar wind data published so far. Such variations in the FIP pattern or FIP amplitude might well be revealed by future studies. In the meantime we suggest that the existing data base is sufficient for identifying the mechanism producing the FIP effect, as it is normally observed in the slow solar wind. Such an identification would help us to understand the processes by which the sun looses matter to the galaxy via the corona and the solar wind.

Acknowledgements

I am grateful to Daniel Rucinski for his advice and help with photoionisation cross sections. I thank George Gloeckler and Rudolf von Steiger for discussions and Ursula Pfander and Silvia Wenger for preparing the manuscript.

References

Anders, E., and Grevesse, N.: 1989, 'Abundances of the Elements: Meteoritic and Solar', *Geochim. Cosmochim. Acta* **53**, 197–214.
Bame, S.J., Asbridge, J.R., Feldman, W.C., Montgomery, M.D., and Kearney, P.D.: 1975, 'Solar Wind Heavy Ion Abundances', *Sol. Phys.* **43**, 463–473.
Bochsler, P., Geiss, J., and Kunz, S.: 1986, 'Abundances of Carbon, Oxygen and Neon in the Solar Wind During the Period from August 1978 to June 1982', *Sol. Phys.* **102**, 177.
Bochsler, P.: 1987, 'Solar Wind Ion Composition', *Physica Scripta* **T18**, 55.
Bochsler, P., and Geiss, J.: 1989, 'Solar System Plasma Physics', in *Solar System Plasma Physics, Geophysical Monograph* **54**, (eds. Waite, J.H.Jr., Burch, J.L., and Moore, R.L.), 133–141.
Breneman, H.H., and Stone, E.C.: 1985, 'Solar Coronal and Photospheric Abundances from SEP Measurements', *ApJ* **299**, L57–L61.
Cerutti, H.: 1974, *Die Bestimmung des Argons im Sonnenwind aus Messungen an den Apollo-SWC-Folien*, Ph.D. Thesis, University of Bern.
Coplan, M.A., Ogilvie, K.W., Bochsler, P., and Geiss, J.: 1984, 'Interpretation of ^{3}He Abundance Variations in the Solar Wind', *Sol. Phys.* **93**, 415–434.
Feldman, U.: 1998, 'FIP Effect In The Solar Upper Atmosphere: Spectroscopic Results', *Space Sci. Rev.*, this volume.
Galvin, A.B., Gloeckler, G., Ipavich, F.M., Shafer, C.M., Geiss, J., and Ogilvie, K.W.: 1993, 'Solar Wind Composition Measurements by the Ulysses SWICS Experiment During Transient Solar Wind Flows', *Adv. Space Res.* **13**, (6)75–(6)78.
Garrard, T.L., and Stone, E.C.: 1993, 'New SEP-Based Solar Abundances', in *Proc. 23rd Int. Cosmic Ray Conf.* **3**, 384.
Geiss, J., Eberhardt, P., Bühler, F., Meister J., and Signer P.: 1970, 'Apollo 11 and 12 Solar Wind Composition Experiments: Fluxes of He and Ne Isotopes', *Geophys. Res.* **75**, 5972–5979.
Geiss, J., Bühler, F., Cerutti, H., Eberhardt, P., Filleux, Ch.: 1972, 'Solar Wind Composition Experiment', *Apollo 16 Prel. Sci. Report*, Sect. 14, NASA SP-**315**, 3375–3398.
Geiss, J.: 1982, 'Processes Affecting Abundances in the Solar Wind', *Space Sci. Rev.* **33**, 201.
Geiss, J., and Bochsler, P.: 1985, 'Ion Composition in the Solar Wind in Relation to Solar Abundances', in *Proc. Rapports Isotopiques dans le Système Solaire*, Paris: Cepadues-Editions, 213–228.
Geiss, J., and Bochsler, P.: 1986, 'Solar Wind Composition and What We Expect to Learn From Out-of-Ecliptic Measurements', in *The Sun and the Heliosphere in Three Dimensions* (ed. Marsden, R.G.), Dordrecht: Reidel, 173–186.
Geiss, J., and Bürgi, A.: 1986, 'Diffusion and Thermal Diffusion in Partially Ionized Gases in the Atmospheres of the Sun and Planets', *Astron. Astrophys.* **159**, 1–15.
Geiss, J., and Bürgi, A.: 1987, 'Diffusion and Thermal Diffusion in Partially Ionized Gases: The Case of Unequal Temperatures', *Astron. Astrophys.* **178**, 286–291.
Geiss, J., Ogilvie, K.W., von Steiger, R., et al.: 1992, 'Ions with Low Charges in the Solar Wind as Measured by SWICS on Board Ulysses', in *Solar Wind Seven* (eds. Marsch, E., and Schwenn, R.) COSPAR Coll. Ser. **Vol. 3**, 341–348.
Geiss, J., Gloeckler, G., and von Steiger, R.: 1994, 'Solar and Heliospheric Processes from Solar Wind Composition Measurements', *Phil. Trans. R. Soc. Lond. A* **349**, 213–226.
Geiss, J., Gloeckler G., and von Steiger R.: 1995a, 'Origin of the Solar Wind From Composition Data', *Space Sci. Rev.* **72**, 49.

Geiss, J., Gloeckler, G., von Steiger, R., et al.: 1995b, 'The Southern High Speed Stream: Results from the SWICS Instrument on Ulysses', *Science* **268**, 1033–1036.

Gloeckler, G., Ipavich, F.M., Hamilton, D.C., Wilken, B., and Kremser, G.: 1989, 'Heavy Ion Abundances in Coronal Hole Solar Wind Flows', *EOS Trans. AGU* **70**, 424.

Gloeckler, G. et al.: 1992, 'The Solar Wind Ion Composition Spectrometer', *Astron. Astrophys. Suppl. Ser.* **92**, 267–289.

Hénoux, J.C.: 1998, 'FIP Fractionation: Theory', *Space Sci. Rev.*, this volume.

Ipavich, F.M., Galvin, A.B., Geiss, J., Ogilvie, K.W., and Gliem, F.: 1992, 'Solar Wind Iron and Oxygen Charge States and Relative Abundances Measured by SWICS on Ulysses', in *Solar Wind Seven* (eds. Marsch, E., and Schwenn, R.), COSPAR Coll. Ser. **Vol. 3** Goslar, Germany, Pergamon Press, 369–374.

Joos, R.: 1989, *Zusammensetzung des Sonnenwindplasmas; Eichung des Sonnenwindmassenspektrometers SWICS*, Ph.D. Thesis, University of Bern.

Judge, Ph., and Peter, H.: 1998, 'The Structure of the Chromosphere', *Space Sci. Rev.*, this volume.

Ko, Y.-K., Fisk, L.A., Geiss, J., Gloeckler, G., and Guhathakurta, M.: 1997, 'An Empirical Study of the Electron Temperature and Heavy Ion Velocities in the South Polar Coronal Hole', *Solar Physics* **171**, 345–365.

Marsch, E., von Steiger, R., and Bochsler, P.: 1995, 'Element Fractionation by Diffusion in the Solar Chromosphere', *Astron. Astrophys.* **301**, 261.

McKenzie, J.F., Sukhorukova, G.V., and Axford, W.I.: 1998, 'Structure of a Photoionization Layer in the Solar Chromosphere', *Astron. Astrophys* **332**, 367–373.

Meyer, J.-P.: 1981, 'A Tentative Ordering of all Available Solar Energetic Particles Abundance Observations', in *17th Intern. Cosmic Ray Conf., Paris* **3**, 145.

Meyer, J.-P.: 1985, 'Solar-Stellar Outer Atmospheres and Energetic Particles, and Galactic Cosmic Rays', *ApJ Suppl.* **57**, 151.

Meyer, J.-P.: 1993, 'Element Fractionation at Work in the Solar Atmosphere', *In Origin and Evolution of the Elements*, (eds. Prantzos, N., Vangioni-Flam, E., and Cassé, M.), Cambridge Univ. Press, 26–62.

Neugebauer, M.: 1981, 'Observations of Solar Wind Helium', *Fundamentals of Cosmic Physics* **7**, 131–199.

Ogilvie, K.W., Coplan, M.A., and Geiss, J.: 1992, 'Solar Wind Composition From Sector Boundary Crossings and Coronal Mass Ejections', in *Solar Wind Seven* (eds. Marsch, E., and Schwenn, R.), 399–403.

Pérez Hernández, F., and Christensen-Dalsgaard, J.: 1994, 'The Phase Function for Stellar Acoustic Oscillations - Part Three - The Solar Case', *MNRAS* **269**, 475.

Peter, H.: 1996, 'Velocity-dependent Fractionation in the Solar Chromosphere', *Astron. Astrophys.* **312**, L37–L40.

Reames, D.V., Richardson, I.G., and Barbier, L.M.: 1991, 'On the Differences in Element Abundances of Energetic Ions from Corotating Events and from Large Solar Events', *ApJ* **382**, L43–L46.

Reames, D.V.: 1992, 'Energetic Particle Observations and the Abundances of Elements in the Solar Corona', in *Proc. 1st SOHO Workshop*, Annapolis, Maryland, USA, ESA SP-**348**, 315–323.

Reames, D.V.: 1995, 'Coronal Abundances Determined from Energetic Particles', *Adv. Space Res.* **15**, No. 7, 45, 41–51.

Reames, D.V.: 1998, 'Solar Energetic Particles: Sampling Coronal Abundances', *Space Sci. Rev.*, this volume.

Schmid, J., Bochsler, P., and Geiss, J.: 1988, 'Abundance of Iron Ions in the Solar Wind', *ApJ* **329**, 956–966.

Shafer, C.M., et al.: 1993, 'Sulfur Abundances in the Solar Wind Measured by SWICS on Ulysses', *Adv. Space Res.* **13**, (6)79–(6)82.

Stone, E.C.: 1989, *In Cosmic Abundances of Matter* (ed. Waddington, C.J.), AIP Conf. Proc. **183**, 72.

Swider W., Jr.: 1969, 'Processes for Meteoritic Elements in the E-Region', *Planet Space Sci.* **Vol. 17**, 1233–1246.

Vernazza, J.E., Avrett, E.H., and Loeser, R.: 1981, 'Structure of the Solar Chromosphere III', *The Astrophysical Journal Supplement Series* **45**, 635–725.

Verner, D.A., Ferland, G.J., and Korista, K.T.: 1996, 'Atomic Data for Astrophysics. I. Radiative Recombination Rates for H-like, He-like, Li-like, and Na-like Ions over a Broad Range of Temperature', *ApJ* **465**, 487–498.

von Steiger, R.: 1988, *Modelle zur Fraktionierung der Häufigkeiten von Elementen und Isotopen in der solaren Chromosphäre*, Ph.D. Thesis, University of Bern.

von Steiger, R., and Geiss J.: 1989, 'Supply of Fractionated Gases to the Corona', *Astron. Astrophys.* **225**, 222–238.

von Steiger, R., Christon, S.P., Gloeckler, G., and Ipavich, F.M.: 1992, 'Variable Carbon and Oxygen Abundances in the Solar Wind as Observed in Earth's Magnetosheath by AMPTE/CCE', *ApJ* **389**, 791–799.

von Steiger, R., Geiss, J., and Gloeckler, G.: 1997, 'Composition of the Solar Wind', in *Cosmic Winds and the Heliosphere* (eds. Jokipii, J.R., Sonett, C.P., and Giampapa, M.S.), Tucson: University of Arizona Press, 581–616.

Wieler, R., Baur, H., and Signer, P.: 1993, 'A Long-term Change of the Ar/Kr/Xe Fractionation in the Solar Corpuscular Radiation', in *Proc. Lunar Planet Sci.* **24**, Houston: Lunar and Planetary Institute, 1519–1520.

Wieler, R., and Baur, H.: 1995, 'Fractionation of Xe, Kr, and Ar in the Solar Corpuscular Radiation Deduced by Closed System Etching of Lunar Soils', *Astrophys. J.* **453**, 987–997.

Wieler, R.: 1998, 'The Solar Noble Gas Record in Lunar Samples and Meteorites', *Space Sci. Rev.*, this volume.

Williams, D.L., Leske, R.A., Mewaldt, R.A., and Stone, E.C.: 1998, 'Solar Energetic Particle Isotopic Composition', *Space Sci. Rev.*, this volume.

Wimmer-Schweingruber, R.F., von Steiger, R., and Paerli; R.: 1997, 'Solar Wind Stream Interfaces in Corotating Interaction Regions: SWICS/Ulysses Results', *JGR* **102**, No. A8, 17407–17417.

ELEMENT SEPARATION IN THE CHROMOSPHERE
Ionization-Diffusion Models for the FIP-Effect

HARDI PETER
High Altitude Observatory / NCAR,* Boulder, USA   and
Max-Planck-Institut für Aeronomie, Katlenburg-Lindau, Germany

Abstract. Ionization-diffusion mechanisms to understand the first ionization potential (FIP) fractionation as observed in the solar corona and the solar wind are reviewed. The enrichment of the low-FIP elements (<10 eV) compared to the high-FIP elements, seen in e.g. slow and fast wind or polar plumes, is explained. The behaviour of the heavy noble gases becomes understandable. The absolute fractionation, i.e. in relation to hydrogen, can be calculated and fits well to the measurements. The theoretical velocity-dependence of the fractionation will with used to determine the velocities of the solar wind in the chromosphere.

Key words: Sun: abundances, chromosphere, solar wind

1. Introduction

The relative elemental abundances change significantly from the photosphere to the corona and the solar wind depending on their first ionization potential (FIP): low-FIP elements ($<$ 10 eV) are enriched compared to high-FIPs by a factor of 4 (2) in the slow (fast) solar wind. The present paper is restricted to the Sun. See e.g. Drake *et al.* (1997) and references therein for a discussion in the stellar context.

These abundance variations are crucial for e.g the diagnosis of the coronal spectra. The densities of the minor species are also important for the thermodynamics of the corona, because they dominate the radiative losses above $\approx 5 \cdot 10^4$ K. Furthermore the abundances of the solar (stellar) wind are of importance to map back and hence identify the source region of the wind.

The present paper will focus on the ionization-diffusion models for a flow along the magnetic field as worked out by Marsch *et al.* (1995) and Peter (1996, 1998) and will discuss the underlying basic processes. These models provide an understanding of the two-plateau structure of the fractionation in the slow *and* fast wind as well as in polar plumes. They can also match the observed absolute fractionation (i.e. in relation to hydrogen) and the behaviour of the heavy noble gases. The fractionation of helium cannot be explained in the present models — for a discussion see Peter and Marsch (1998) and the solar wind model of Hansteen *et al.* (1997).

The fractionation $f_{j,k} = (N_j/N_k)_{\rm SW}/(N_j/N_k)_{\rm ph}$ of an element j in relation to k compares the respective relative abundances e.g. in the solar wind (SW) and in the photosphere (ph). For a discussion of measurements see Geiss (1998) and von Steiger (1998), for a review of models Hénoux (1998) and references therein, in this volume.

* The National Center for Atmospheric Research is sponsored by the National Science Foundation

2. Basic Fractionation Mechanisms and Their Location

2.1. LOCATION OF THE RELEVANT PROCESSES AND GEOMETRY

As pointed out by Geiss (1982), the ion-neutral separation causing the fractionation is located somewhere in the chromosphere, in a layer where the material gets first ionized. In the presented models this fractionation layer is very thin, namely about 50 km (Marsch *et al.*, 1995; Peter and Marsch, 1998),

Concerning the source region of the *fast wind* the material is assumed to flow out of the coronal funnels and fractionation takes place the the base of the funnel. One may assume that through the transition region and the corona the abundances remain unchanged, which is due to efficient Coulomb-coupling rendering equal velocities. However, it should be noted that Hansteen *et al.* (1997) found strong diffusion in the transition region due to thermal forces. Mapping back the particle flux at 1 AU to the base of a funnel results in a velocity of the order of 500 m/s there (Peter and Marsch, 1998).

Recent SOHO observations of Sheeley *et al.* (1997) have suggested that the *slow wind* is fed by exploding loops. One can assume a "siphon flow" through the loops, with a velocity of about 0.3 km/s at the footpoints (Klimchuk and Mariska, 1988). Because the fractionation is independent of the flow direction, Sect. 2.2 and (3), in *both* footpoints a fractionation will occur: the whole loop will be fractionated.

In the footpoint region of both a loop and a funnel, the magnetic field is more or less vertical. Thus a one-dimensional multi-fluid model for a chromospheric region at below 10^4 K will be considered to describe the fractionation.

2.2. BASIC FRACTIONATION MECHANISMS

In Fig. 1 (left) the basic fractionation mechanisms are illustrated. According to von Steiger and Geiss (1989), the (EUV-) radiation dominates the ionization at 10^4 K and originates mainly from above. As the upper chromosphere is more or less optically thin in the EUV (except e.g. in Lyα), the photoionization rates are assumed to be approximately constant with depth for the minor species.

The low-FIPs have short *ionization times* and are coupled quickly to the solar wind flow of the main gas, ionized hydrogen. The high-FIPs remain longer in the neutral phase. Thus the low-FIPs are transported by preference out of the chromosphere into the interplanetary space. This simple picture renders the enrichment of the low-FIPs understandable.

At the top of the ionization layer, where the material is ionized, the very effective Coulomb-coupling causes equal velocities of the different species. But at the bottom of the layer, in the neutral phase, the collisions are less effective (by a factor of 1000, see e.g. Peter, 1998). Thus a *diffusion* at the bottom of the layer is possible (see also Sect. 4.3). This diffusion regulates the fractionation simply by regulating the velocity differences of the trace gases.

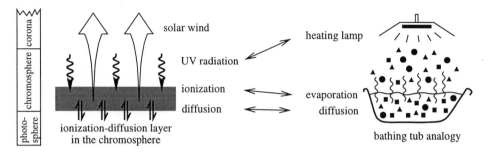

Figure 1. Basic fractionation mechanisms and the "bathing tub analogy".

The presented models are independent of the assumptions concerning e.g. the structure of the magnetic field topology and can be used for the explanation of the fractionation in a wide variety of solar structures, like coronal funnels, loops or polar plumes.

But there is yet another mechanism, leading to a *velocity-dependence* of the fractionation. If the *transit time* through the ionization layer is relatively short, the fractionation will be relatively weak, because the above described processes have less time to act. This mechanism, first pointed out by Peter (1996), leads to a *velocity-dependence* of the fractionation and is reflected in (3) below.

It is helpful to look at the analogy between the above fractionation and the evaporation of different kinds of perfumes dissolved in a bathing tub (Fig. 1, following an idea of R. Bodmer). If the water is heated, buoyancy will drive a flow of steam ($\equiv$ solar wind). The heating lamp causes the evaporation of the perfumes ($\equiv$ UV radiation). The volatile perfumes have a shorter "evaporation time" ($\equiv$ ionization time): they will be enriched in the steam. This enrichment of the volatile elements is regulated by their ability to diffuse through the water in the tub.

In this analogy it is immediately perceived that the abundances in the water ($\equiv$ photosphere) do not have to be the same as in the steam ($\equiv$ corona) and thus a fractionation is present. The ionization-diffusion layer in the solar chromosphere compares to the (thin) layer under the water surface.

This is only one possibility to generate a fractionation. See Hénoux (1998) in this volume for some other ideas.

3. Model Ingredients

As mentioned above one-dimensional multi-fluid equations along the magnetic field are used. See Peter and Marsch (1998) or Schunk (1975) for a detailed discussion. The equations of continuity and momentum for a species j read

$$(n_j u_j)' = \sum_{j'} \left(\gamma_{j'j} n_{j'} - \gamma_{jj'} n_j \right), \qquad (1)$$

$$u_j u_j' + \frac{(v_j^2 n_j)'}{n_j} + Z_j \frac{(v_j^2 n_e)'}{n_e} + g = \sum_k \nu_{jk}(u_k - u_j) + \sum_{j'} \frac{n_{j'}}{n_j} \gamma_{j'j}(u_{j'} - u_j). \quad (2)$$

All the symbols have their usual meanings: n_j, u_j – particle density, velocity ($\|\boldsymbol{B}$) of species j; $\gamma_{j'j}$ – ionization/recombination rates; $v_j = (k_B T/m_j)^{1/2}$ – sound speed, Z_j – charge number; ν_{jk} – elastic collisional rates.

Only collisions of the trace gas (neutral and ionized components) with the background (H and p), are considered, while the interaction between the different trace gases is neglected: every trace gas is treated as a collection of test particles.

Boundary conditions for the numerical models

In these kinematic models the velocities of the neutral and ionized component are assumed to be equal at the top as well as at the bottom of the layer. This is motivated by the fact that at the bottom the material is mostly neutral, at the top mostly ionized. Thus ionization and recombination cause the components to have the same speed.

At the bottom of the ionization-diffusion layer the abundance of an element and its degree of ionization (following the Saha equilibrium) are given.

It should be noted that no absolute value of the velocity or any diffusion velocity between two elements is presumed. The diffusion at the bottom of the ionization-diffusion layer results from the model (see Sect. 4.3).

Atomic data

A collection of atomic data, ionization/recombination and collisional rates as needed here can be found in Peter (1998). It should be noted that in the case of oxygen in the hydrogen background the efficient charge exchange has been considered.

4. Results

4.1. ANALYTIC MODEL: FRACTIONATION IN THE SLOW AND FAST WIND

By simplifying (1) and (2) Peter (1996) found a simple formula describing the velocity-dependent fractionation. To achieve this, gravity and recombination were neglected, a subsonic flow was assumed and only the interaction with the main gas was considered. At the bottom boundary all the material is assumed to be neutral and there is no acceleration for the neutrals. Note that this is different from the boundary condition for the following numerical models as discussed below, see also Sect. 4.3. The fractionation of an element j to k is then given by (U_H denotes the main gas, i.e. hydrogen, velocity)

$$f_{j,k} = \frac{1 + \sqrt{1 + 4(w_j/U_H)^2}}{1 + \sqrt{1 + 4(w_k/U_H)^2}}. \quad (3)$$

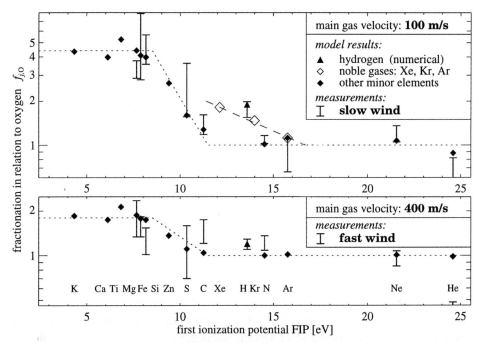

Figure 2. Calculated fractionation for two different main gas velocities in the chromosphere matching the measurements in the slow and fast wind. The values for the minors and for the noble gases are calculated from (3) for $N_H = 4 \cdot 10^{16}$ m^{-3} and $T = 10^4$ K. These conditions correspond to the model C of Vernazza *et al.* (1981). The fractionation of hydrogen is taken from the numerical study in Sect. 4.2. The bars show the measurements in the slow and fast wind (from von Steiger *et al.*, 1995; Wieler and Baur, 1995). The dotted lines indicate the low- and high-FIP plateaus, the dashed line the noble gas line. Xe and Kr are omitted in the 400 m/s case, because no measurements are available for them in the fast wind. The fractionation of O (relative to O) is by definition 1, and for this not shown.

The most important factor entering this formula is the *ionization-diffusion speed* w_j, defined as $w_j = v_j / \sqrt{\tau_j \tilde{\nu}_{jH}}$, where $\tilde{\nu}_{jH} = \nu_{jk} N_H / n_H$ is the (reduced) collision frequency between the neutrals of the species j and neutral hydrogen. This shows that photoionization (τ) and diffusion ($\tilde{\nu}$) together with the transit time ($\propto 1/U_H$) are determining the fractionation as pointed out in Sect. 2.2.

In the case of small main gas velocities the fractionation following (3) is simply given by the quotient of the respective ionization diffusion speeds, as already calculated by Marsch *et al.* (1995). Peter (1998) pointed out that the fractionation in the low velocity limit is given only by atomic properties, namely the cross-sections for photoionization and elastic collisions.

Two-plateau structure in the fast and slow wind. In Fig. 2 the resulting fractionation as following from (3) and the measurements in the solar wind are plotted versus the first ionization potential for a number of elements. For chromospheric velocities of 100 m/s and 400 m/s at a density of $N_H = 4 \cdot 10^{16}$ m^{-3} the models result in deed in a two-plateau structure with step heights matching the measurements in the slow and the fast wind quite well.

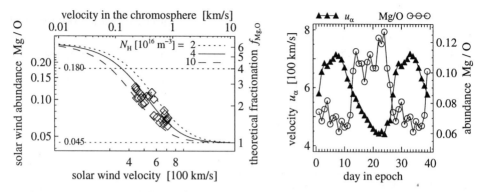

Figure 3. Velocity-dependence of the fractionation of Mg/O. The three curves on the left show the theoretical dependence as following from (3) for different densities (upper and right axis). The diamonds on the left show the measured data points from every day of one epoch (lower and left axis), as measured in the solar wind (right panel, from Geiss *et al.*, 1995).

Velocity-dependence. For the case of the fractionation Mg/O the velocity-dependence is shown in more detail in Fig. 3 for three different chromospheric densities. The comparison with the over-plotted solar wind data shows that the model compares well with the observed velocity-dependence.

Fractionation of the heavy noble gases. In Fig. 2 the theoretical results also for the noble gases as following from (3) are shown. The noble gases Xe, Kr and Ar are perfectly on a line, with Xe and Kr clearly enhanced to the high-FIP plateau. But the enhancement is much weaker than indicated by the observations of Wieler and Baur (1995). However the qualitative behaviour can be understood by the model.

4.2. NUMERICAL MODELS: ABSOLUTE FRACTIONATION AND POLAR PLUMES

To study the variation of the absolute abundances of the elements, e.g. the fractionation of hydrogen to oxygen, a numerical model has to be applied: the density of the trace gases has to be calculated on the background of the main gas. For this (1) and (2) are solved numerically in a given hydrogen-proton background (see Sect. 3 for the boundary conditions). Here the main gas models of Peter and Marsch (1998) are serving as a background. They calculated a variety of different hydrogen models which can be characterized by the typical mean velocity (taken at the maximum of the proton density) in the ionization-diffusion layer (their Sect. 5.3).

Fractionation of hydrogen. For the main gas models with (typical) velocities of $U_H \approx 100$ m/s and $U_H \approx 400$ m/s (the velocities for the slow and fast wind in the previous section) the fractionation of hydrogen to oxygen turns out to be $f_{H,O} \approx 2$ and $f_{H,O} \approx 1.2$ respectively at the top of the layer. This corresponds well with the in-situ measurements in the solar wind of von Steiger *et al.* (1995), see Fig. 2. One should note that spectroscopic measurements in the corona often give different results (Meyer, 1996), but owing to the non-uniqueness of the interpretation of spectral data these are not as reliable as the in-situ measurements.

Polar plumes. In the analytical study the fractionation of magnesium can only reach a maximum value of about 6 (see Fig. 3). In contrast to this a much stronger fractionation can be found in the numerical model: for very low mean background velocities of only ≈50 m/s the fractionation Mg/O can reach a factor of 10. This quasi-static solution compares well with the spectroscopic plume observations of e.g. by Widing and Feldman (1992).

4.3. VELOCITY PROFILE AND THE ROLE OF DIFFUSION

Diffusion, i.e. the fact that the elements have different velocities, plays a crucial role in these models. At this point only two basic results on the diffusion from the models should be highlighted. See Sect. 6 of Peter (1998) for a detailed discussion.

As a result of the numerical models a diffusion at the lower boundary occurs. The fractionation following from the diffusive equilibrium is *not* an artifact of the boundary conditions, because no assumption for the absolute or diffusive speed at the boundaries was made (see Sect. 3). The conclusion is that in deed a one-dimensional ionization-diffusion model can produce a fractionation.

The other point to mention is that in the numerical models the neutrals are entering the modelled layer at the bottom "smoothly", i.e. the gradient of the velocity of the neutrals at the bottom is very small. This proves the most crucial assumption in the analytical model in Sect. 4.1.

5. Conclusions

As in earlier fractionation models, these models give a two-plateau structure of the fractionation as a function of the first ionization potential. But furthermore it is possible to understand the fractionation in a great variety of phenomena:

– The fractionation in the slow *and* the fast wind can be understood with the theoretically predicted velocity-dependence matching the observations.
– A comparison of the theoretical fractionation in the chromosphere with the measured one in the solar wind leads to chromospheric velocities of 100 m/s and 400 m/s for the regions connected to the slow and fast wind respectively.
– With the help of the velocity-dependence it can be explained why sometimes in high speed streams in the wind no fractionation is found.
– The fractionation of the heavy noble gases can be understood (qualitatively).
– The numerical studies enable a determination of the absolute fractionation, i.e. the fractionation in relation to hydrogen: the position of H in the fractionation pattern fits well with the in-situ measurements in the slow and fast wind.
– The very strong fractionation as found in polar plumes can be modelled by assuming the plumes to be quasi-static as suggested by some measurements.

The main achievement of these ionization-diffusion models is to explain the above phenomena within *one* class of models basing on the well known processes of photoionization and diffusion.

Acknowledgements

Special thanks are due to Eckart Marsch for the many ingenious discussions we had. I would like to thank also Roland Bodmer for his basic idea for the bathing tub analogy.

References

Drake, J. J., Laming, J. M. and Widing, K. G.: 1997, 'Stellar Coronal Abundances V: Evidence for the FIP Effect in α Centauri', *Astrophysical Journal* **478**, 403–416.
Geiss, J.: 1982, 'Processes Affecting Abundances in the Solar Wind', *Space Science Reviews* **33**, 201–217.
Geiss, J., Gloeckler, G. and von Steiger, R.: 1995, 'Origin of the Solar Wind from Composition Data', *Space Science Reviews* **72**, 49–60.
Geiss, J.: 1998, *Space Sci. Rev.*, this volume.
Hansteen, V. H., Leer, E. and Holzer, T. E.: 1997, 'The Role of Helium in the Outer Solar Atmosphere', *Astrophysical Journal* **482**, 498–509.
Hénoux, J.-C.: 1998, *Space Sci. Rev.*, this volume.
Klimchuk, J. A. and Mariska, J. T.: 1988, 'Heating-Related Flows in Cool Solar Loops', *Astrophysical Journal* **328**, 334–343.
Marsch, E., von Steiger, R. and Bochsler, P.: 1995, 'Element Fractionation by Diffusion in the Solar Chromosphere', *Astronomy and Astrophysics* **301**, 261–276.
Meyer, J.-P.: 1996, 'Abundance Anomalies in the Solar Outer Atmosphere', *in* J. Trân Thanh Vân, L. Celnikier, H. Trung and S. Vauclair (eds), *The Sun and Beyond*, Editions Frontières, Gif-sur-Yvette, p. 27.
Peter, H.: 1996, 'Velocity-Dependent Fractionation in the Solar Chromosphere', *Astronomy and Astrophysics* **312**, L37–L40.
Peter, H.: 1998, 'Element Fractionation in the Solar Chromosphere Driven by Ionization-Diffusion Processes', *Astronomy and Astrophysics*, in press.
Peter, H. and Marsch, E.: 1998, 'Hydrogen and Helium in the Solar Chromosphere: A Background Model for Fractionation', *Astronomy and Astrophysics* **333**, 1069–1081.
Schunk, R. W.: 1975, 'Transport Equations for Aeronomy', *Planet. Space Sci.* **23**, 437–485.
Sheeley, N. R. Jr., Wang, Y. M., Hawley, S. H. et al.: 1997, 'Measurements of Flow Speeds in the Corona Between 2 and 30 $R_\odot$', *Astrophysical Journal* **484**, 472–478.
Vernazza, J. E., Avrett, E. H. and Loeser, R.: 1981, 'Structure of the Solar Chromosphere III: Models of the EUV Brightness Components of the Quiet Sun', *Astrophysical Journal, Supplement Series* **45**, 635.
von Steiger, R.: 1998, *Space Sci. Rev.*, this volume.
von Steiger, R. and Geiss, J.: 1989, 'Supply of Fractionated Gases to the Corona', *Astronomy and Astrophysics* **225**, 222–238.
von Steiger, R., Wimmer Schweingruber, R. F., Geiss, J. and Gloeckler, G.: 1995, 'Abundance Variations in the Solar Wind', *Adv. Space Res.* **15**, (7)3–12.
Widing, K. G. and Feldman, U.: 1992, 'Element Abundances and Plasma Properties in a Coronal Polar Plume', *Astrophysical Journal* **392**, 715–721.
Wieler, R. and Baur, H.: 1995, 'Fractionation of Xe, Kr and Ar in the Solar Corpuscular Radiation Deduced by Closed System Etching of Lunar Soils', *Astrophysical Journal* **453**, 987–997.

Address for correspondence: Hardi Peter, High Altitude Observatory/NCAR, PO-Box 3000, Boulder, CO 80307-3000, USA, E-mail: hpeter@hao.ucar.edu

TEMPORAL EVOLUTION OF ARTIFICIAL SOLAR GRANULES

S. R. O. PLONER and S. K. SOLANKI
Institute of Astronomie, ETH Zentrum, 8092 Zürich, Switzerland

A. S. GADUN
Main Astronomical Observatory of Ukrainian, Goloseevo, 252650 Kiev-22, Ukraine

A. HANSLMEIER
Institut für Astronomie, Universitätsplatz 5, A-8010 Graz, Austria

Abstract. We study the evolution of artificial granulation on the basis of 2-D hydrodynamical simulations. These clearly show that granules die in two different ways. One route to death is the well known bifurcation or fragmentation of a large granule into 2 smaller ones (exploding granules). The other pathway to death is characterized by merging intergranular lanes and the accompanying dissolution of the granule located between them. It is found that the lifetime and maximum brightness is independent of the way in which granules evolve and die. They clearly differ in size, however, with exploding granules being in general significantly larger.

Key words: Solar Convection, Solar Granulation, Radiation-Hydrodynamic Simulation

1. Introduction

Most solar elemental abundances refer to the solar photosphere. This lowest atmospheric layer is structured by overshooting convection visible as granulation (see Spruit *et al.*, 1990 for a review). The elemental abundances determined from photospheric spectral lines may be inaccurate, if the properties of the granules are not taken into account. We present here a theoretical investigation aimed at better understanding the evolution of granules and hence of the inhomogeneities responsible for introducing inaccuracies into abundance measurements.

Direct numerical integration of the radiation-hydrodynamic equations has provided new insight into the physical processes involved in solar granular convection (e.g. Nordlund, 1985; Stein *et al.*, 1989; Rast, 1993; Freytag *et al.*, 1996). A 2-D fully compressible radiation hydrodynamic simulation of solar granulation (free upper and lower and cyclic lateral boundary conditions) and of the corresponding synthetic spectra underlies the present investigation. Details of the simulation have been given by Gadun *et al.* (1995, 1996, 1998). In order to be able to follow the evolution of a sufficiently large number of granules we decided to reduce the number of spatial dimensions to 2. The computational domain covers horizontally ≈18 Mm real solar distance (512 grid points with 35 km spacing) and vertically ≈2 Mm (58 grid points with the same spacing). The evolution of the gas in the computational domain has been simulated for 5 hours real solar time in 0.3 s steps.

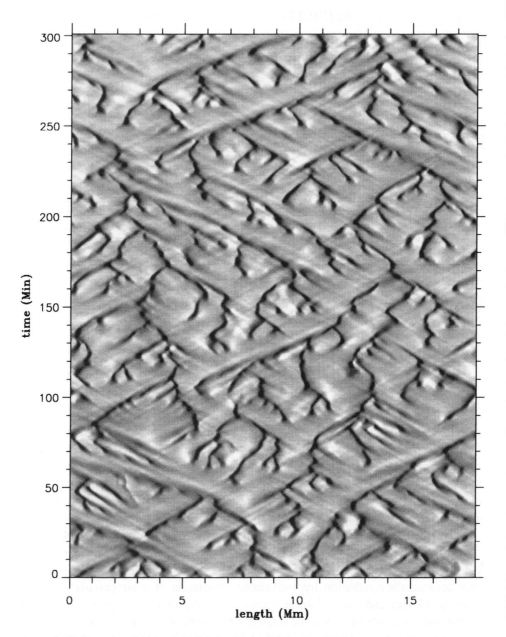

Figure 1. **Temporal evolution of the longitudinal velocity of artificial granulation.** Regions with dark shading indicate downflows, regions with light shading upflows. The intergranular lanes may start and merge but they never end. Merging intergranular lanes enclose a dying granule (which we call a *dissolving granule*). A starting intergranular lane marks both the death of a *fragmenting* or *exploding granule* and the birth of two new granules.

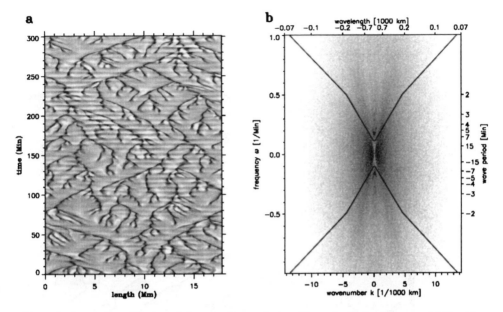

Figure 2. **Space–time filtering of the granular evolution.** Figure 2a displays the same as Fig. 1 but without space–time filtering. Figure b is the 2-D Fourier transformation of Fig. 2a. The neglect of the k-ω contribution in the inner part of the solid lines removes the p-modes. Figure 1 is the inverse Fourier transform of Fig. 2b, carried out after the filtering.

2. Analysis and Results

2.1. REMOVAL OF P-MODE OSCILLATION

Figure 1 displays the temporal evolution of the granules seen in the longitudinal velocity component taken at a constant height just slightly below the average formation height of the visible continuum. The thin dark structures correspond to downflows and the bright regions to upflows. The emergent continuum intensity has also been calculated (not shown). The pattern formed by the continuum intensity shows no significant differences relative to the longitudinal velocity structure. Figure 1 has been "cleaned" from the influence of p-modes. As seen in Fig. 2a, p-modes are prominent in the simulation and may locally falsify the granular structure by, e.g., the introduction of an effective upflow in an intergranular lane. The vertical velocity and also the pressure are parameters which are strongly influenced by the p-modes. Note that the influence of the p-modes on the atmosphere relative to that of granulation increases rapidly with height. In order to be able to study granulation unaffected by oscillations we applied a 2-D Fourier transformation to Fig. 2a. The transformed image is shown in Fig. 2b. Upon close inspection power ridges due to the 5 and 3 minutes oscillations can be identified in the k-ω diagram. The p-modes are then filtered by setting the transformation to zero in the part of the k-ω diagram lying in the inner part of the triangles (after apodization). Figure 1

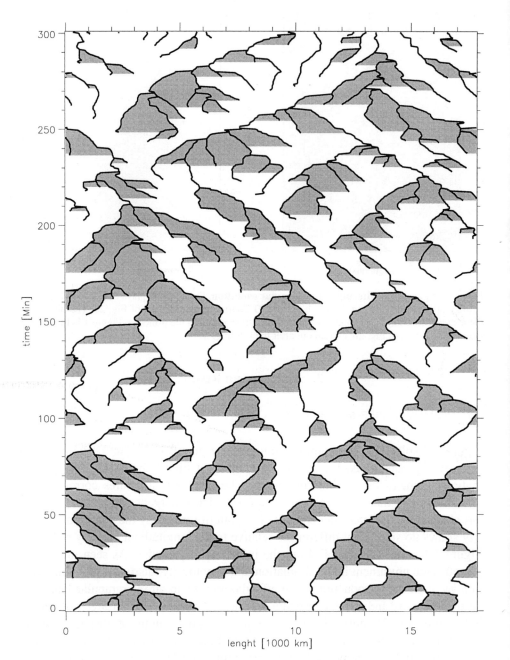

Figure 3. **Skeleton of the intergranular lanes** extracted from Figure 1 (solid curves). The shaded areas mark granules that end by dissolution while the remaining area is covered by fragmenting granules.

is then obtained by inverse Fourier transformation. This technique was pioneered for the analysis of high resolution observations (Title et al., 1986). Note that not only is the granulation now more clearly visible, but higher frequency propagating waves, which had earlier been swamped by the p-modes are now uncovered.

2.2. FRAGMENTING AND DISSOLVING GRANULES

In this presentation we focus on a statistical analysis of the temporal evolution of granulation. To this end we need to define the boundaries, as well as the birth and death of a granule. We have used the longitudinal velocity component taken at a fixed height to define these quantities. A granule is the upflowing structure between two adjacent downflow lanes (intergranular lanes). Hence for the purpose of the present analysis we can reduce the 3 dimensional data set (2 space and 1 time dimensions) to a more tractable 2 dimensional data set (1 horizontal and 1 time dimension).

Obviously, the granules and intergranular lanes exhibit a complementary temporal behaviour. Intergranular down-flowing lanes on the one hand often merge but never end although new lanes start. Granules, on the other hand, often end, but are only formed from the parts of fragmenting granules. Each time at the start of a new lane a granule splits into 2 (or more) smaller granules (fragmentation). This is a well studied route to granule death (e.g. Muller, 1989; Rast, 1993). Figure 1 also shows that most inter-granular lanes merge together. This corresponds to the death of the granule which is enclosed by the merging lanes. We call them dissolving granules (cf. Hirzberger et al., 1997, for an observational study of small and large granules).

Figure 1 naturally leads to the characterization of granules by their mode of death. It is also obvious that the time evolution of the size of fragmenters and dissolvers is opposite: fragmenting granules rapidly expand whereas dissolving granules shrink.

2.3. COMPARISON BETWEEN THE 2 GRANULE SPECIES

We consider now the relative significance and properties of dissolving and fragmenting granules. Figure 3 (solid curves) shows the skeleton of the intergranular lanes determined from Fig. 1. For each time step we searched for regions of downflows. The skeleton of the intergranular lane is then assigned to the location of maximal downflow. To keep the image from becoming cluttered only lanes that last more than 3 time-steps (= 1.5 Min) have been accepted. This sets a lower limit on the lifetime of dissolving granules that we consider, but we estimate that no such extremely short-lived dissolving granules were present in the simulation. In order to free the comparison between the two types of granules from bias fragmenting granules have also only been counted if their lifetime lies above 1.5 Min.

Table I

Granule Statistics. Granules are only considered if their lifetime is larger than 1.5 Min. The total number of fragmenters before this correction is displayed in brackets. The sum of the percentages of the area coverage is less than 100% due to the exclusion of short lived granules.

granule type	dissolving granules	fragmenting granules
total number	218	199 (218)
mean lifetime	8.6 Min	7.7 Min
mean size	865 km	2236 km
'area coverage'	35%	63%

Table I displays the basic statistics of the two granule types. The number and lifetime of fragmenting and dissolving granules are the same (before removing too short-lived fragmenting granules). Note that the number of granules is the same at the end and the beginning of the simulation. Remembering that fragmentation creates and dissolution removes 1 granule, it is obvious that on average there must be the same number of fragmenting and dissolving granules.

The mean lifetime is lower for fragmenting granules. This is mostly caused by rapid successive fragmentations, which give rise to very short lived granules (between 2 such splittings). Both the area coverage and the average size of the granules states that the dissolvers are significantly smaller than the fragmenters.

We then determined the average size, lifetime and maximum brightness of each granule. The comparison of granular lifetimes with mean sizes (Fig. 4a) confirms that dissolving granules ("•" in Fig. 4a) are smaller than fragmenting granules ("o" in Fig. 4a), with very little overlap in size between the two families of granules. The exploders exhibit a much larger range of sizes than dissolving granules. Dissolving granules show a clear relationship between size and lifetime, while for exploders such a relation is less clear. The larger the dissolving granule the longer it lives on the average (Fig. 4b). In contrast, the lifetime of the largest fragmenting granules is short. The two granule species basically agree in maximum brightness and lifetime. Figure 4b suggests that the brightness range of dissolving granules is slightly larger, certainly in the low intensity regime. The main effects visible from Fig. 4b is that the maximum brightness of granules increases on average with their lifetime and that the scatter of the maximum brightness values decreases towards larger lifetimes. Both Fig.s 4a and b also show that the two species cover the same lifetime range. The correlation of the average and maximum size of the granules seen in Fig. 4c gives an impression of the change of size during its evolution. A significant fraction of the granules lies below but near the slanted line. Granules lie exactly on the line when their size remains unchanged throughout their evolution. This indicates that over most of their lifetime their size changes only slightly.

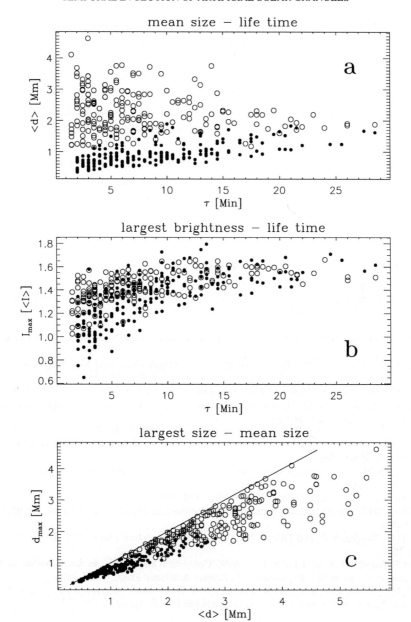

Figure 4. **Scatter plots** between the granule's average size $\langle d \rangle$ and lifetime τ (**a**), maximum brightness I_{max} and τ (**b**) and maximum size reached by the granule over its lifetime d_{max} versus $\langle d \rangle$ (**c**). Dissolving granules are marked by '•' (filled circle) and fragmenting granules by 'o' (open circle).

3. Summary

We have investigated the temporal evolution of artificial granulation. p-modes are prominent in the simulation and were filtered out in order to study granulation unaffected by oscillations. In addition to the well studied death of a granule by splitting (fragmenting or exploding granules) we have also studied granules that end by dissolution. This type of granule death is seen in Fig. 1 when two intergranular lanes merge. A comparison of the sizes, lifetimes and brightnesses shows that the dissolving granules differ mainly in size from the fragmenting granules, both types of granules having roughly the same lifetimes and maximum brightness.

Note that our study is limited to 2 spatial dimensions and that the properties of granulation in 3-D is different. Nevertheless, granules are known to split into different parts and some counteracting process must be at work in order to keep the number of granules finite. We therefore intend to continue studying the 2-D evolution of granules, in particular that of merging granules, with the aim of obtaining a better physical understanding of the underlying mechanisms.

References

Freytag, B., Ludwig, H.-G. and Steffen, M.: 1996, 'Hydrodynamical models of stellar convection. The role of overshoot in DA white dwarfs, A-type stars, and the Sun', *Astron. Astrophys.* **313**, 497.

Gadun, A.S. and Vorobyov, Yu.Yu.: 1995, 'Artificial Granules in 2-D Solar Models', *Sol. Phys.* **159**, 45.

Gadun, A.S. and Pikalov, K.N.: 1996, 'Two-Dimensional Hydrodynamical Modeling of Solar Granules: The Power Spectrum of Simulated Granules', *Astron. Rep.* **40**, 578.

Gadun, A.S., Solanki, S.K., Ploner, S.R.O., Hanslmeier, A., Pikalov, K.N., and Puschman, K.: 1998, 'Scale Dependent Properties of Artificial Solar Granulation Derived from 2–D Simulations', *Astron. Astrophys.*, in preparation.

Hirzberger, J., Vázquez, M., Bonet, J.A., Hanslmeier, A. and Sobotka, M.: 1997, 'Time Series of Solar Granulation Images. I. Differences between Small and Large Granules in Quiet Regions', *Astrophys. J.* **480**, 406.

Muller, R.: 1989, 'Solar Granulation: Overview', in *Solar and Stellar Granulation*, eds. Rutten, R.J. and Severino, G., Kluwer Academic Press, **C 263**, 101.

Nordlund, Å.: 1985, 'Solar Convection', *Sol. Phys.* **100**, 209.

Rast, M.P.: 1993, 'On the nature of "exploding" granules and granule fragmentation', *Astrophys. J.* **443**, 863.

Spruit, H.C., Nordlund, Å. and Title, A.M.: 1990, 'Solar Convection', *Annu. Rev. Astron. Astrophys.* **28**, 263.

Stein, R.F., Nordlund, Å. and Kuhn, J.R.: 1989, 'Convection and Waves', in *Solar and Stellar Granulation*, eds. Rutten, R.J. & Severino, G., Kluwer Academic Press, **C 263**, 381.

Title, A.M., Tarbell, T.D., Simon, G.W. and the SOUP Team: 1986, 'White-light movies of the solar photosphere from the soup instrument on space lab 2', *Adv. Space Res.* **6**, 253.

Address for correspondence: S. R. O. Ploner, Institute of Astronomie, ETH Zentrum, 8092 Zürich, Switzerland

THE LOWER SOLAR ATMOSPHERE
Rapporteur Paper II

ROBERT J. RUTTEN
Sterrekundig Instituut, Postbus 80 000, NL–3508 TA, Utrecht, The Netherlands

Abstract. This "rapporteur" report discusses the solar photosphere and low chromosphere in the context of chemical composition studies. The highly dynamical nature of the photosphere does not seem to jeopardize precise determination of solar abundances in classical fashion. It is still an open question how the highly dynamical nature of the low chromosphere contributes to first ionization potential (FIP) fractionation.

Key words: Solar composition, solar atmosphere

1. Lower Atmosphere: Context

This is a "consumer" report on the lower solar atmosphere (photosphere and low chromosphere, the latter defined as the regime where hydrogen is not yet fully ionized) in its role as a provider of diagnostics for solar composition studies. The topic is reviewed by Solanki (1998), Judge and Peter (1998) and Grevesse and Sauval (1998) elsewhere in this volume.

Let me begin by placing the lower atmosphere into its context. This meeting is unusual by combining research and researchers from the innermost to the outermost solar reaches — a wide variety sketched broadly in Figure 1.

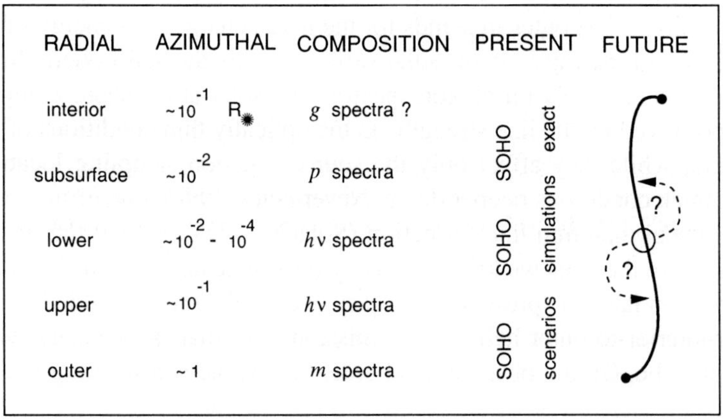

RADIAL	AZIMUTHAL	COMPOSITION	PRESENT	FUTURE
interior	$\sim 10^{-1}\ R_\star$	g spectra ?	SOHO	exact
subsurface	$\sim 10^{-2}$	p spectra	SOHO	simulations
lower	$\sim 10^{-2} - 10^{-4}$	hv spectra	SOHO	simulations
upper	$\sim 10^{-1}$	hv spectra	SOHO	scenarios
outer	~ 1	m spectra	SOHO	scenarios

Figure 1. Very broad overview of solar composition studies.

The first column labeled "radial" describes the overall structure of the workshop. It splits the sun and heliosphere into different radial domains that also characterize the splitting of solar physicists into separate estates. The lower atmosphere can be seen as the pivot. Inwardly, it provides the surface diagnostics (Doppler shifts, irradiance variations, magnetic field patterns) for the subsurface studies on how our star is made. Outwardly, it provides the boundary conditions (mechanical energy production, field structure, field pattern evolution) that make the solar atmosphere such an exciting laboratory to MHD and plasma physicists. Of course, it also provides the light that we see and that maintains our ecosphere, but the production of photospheric radiation is basic course material* rather than food for ISSI workshops these days.

The second column gives rough estimates of the lateral fine-structure scale that is typically discussed for each regime. The lower atmosphere stands out by being most detailed, down to the $\Delta r = 10^{-4} R_\odot = 0.1''$ holy grail of optical solar telescopes, sometimes nearly glimpsed by the SVST** and DOT*** telescopes on La Palma when the seeing is superb, but generally awaiting the development of image restoration techniques, in particular adaptive optics, and/or larger telescope apertures in space than that of SOHO's MDI/SOI[‡] and TRACE[‡‡]. The plethora of detail on the solar surface tends to make non-solar astrophysicists unhappy, but it does harbor rich diagnostics of astrophysical structures and processes at the intrinsic scales at which the actual physics operates – unresolvable outside the solar system. The problem is that the dynamical nature and juxtapositon of the atmospheric structures and processes, especially those having to do with the complexities of magnetism, require comprehensive diagnostics combining high angular, temporal, spectral and polarimetric resolution over large fields, long times and many wavelengths (heights) simultaneously — Judge and Peter (1998) typecast the chromosphere as the most difficult solar regime.

The third column indicates the diagnostics used to study solar composition. The sub-surface p stands for the pressure eigenmodes, with gravity modes (g) yet a SOHO tantalizer. The outer m stands for the mass, charge, and energy of particles, collected rather far away from the solar surface, within the heliosphere. In between, composition studies rely on photon spectra. The solar-atmosphere composition is directly portrayed by the line strengths in the optically thin conditions of the upper atmosphere, while they affect only the source function sampling location in the optically thick conditions deeper down. Nevertheless, thick line formation is much easier to model than thin line formation (Solanki, 1998), so that the photospheric abundances are the ones we know precisely (Grevesse and Sauval, 1998).

The column labeled "present" is an easy one! SOHO was heralded as an inside-out solar-interior-to-outer-heliosphere mission, and that is precisely what it has turned out to be. On the observational side, this whole-sun meeting is dominated

---

* lecture notes at http://www.astro.uu.nl/~rutten
** http://www.astro.su.se/groups/solar/solar.html
*** http://www.astro.uu.nl/~rutten/dot
‡ http://soi.stanford.edu
‡‡ http://www.space.lockheed.com/TRACE

by SOHO research, an obvious tribute to SOHO's success. On the interpretational side, there is an interesting gradient in which mathematical exactness in p-mode fitting gives way to the occasional realism reached by numerical hydrodynamical and MHD simulations of surface-layer structuring, and to the yet exploratory nature of the scenarios in vogue further out. For the lower atmosphere, numerical simulations contribute the largest advances at present. While the optical telescopes are (too) slowly approaching their $0.1''$ grail, spectacular progress is made numerically, in particular in the convection simulations of Nordlund and Stein (1990), the acoustic shock simulations of Carlsson and Stein (1997) and the fluxtube simulations of Steiner et al. (1998).

The final column labeled "future" is less easy. It seems likely that the domains must and will come together, in holism that exceeds just multi-diagnostic observing by integrating interior, atmospheric, and outer solar structure to a larger extent than the congregation assembled here is wont to (want). Such integral synthesis seems yet far off, but as a starter I suggest a low-chromosphere connection to both subsurface and outer-atmosphere phenomena below (dotted arrows).

2. Lower Atmosphere: Radial Scene

Figure 2 is a cartoon summarizing the radial stratification of the lower atmosphere. It also summarizes points discussed in the reviews of Grevesse and Sauval (1998), Solanki (1998) and Judge and Peter (1998) in this volume. It sketches "standard models of the solar atmosphere" where the term "model" implies just a temperature-depth relation because generally, hydrostatic equilibrium is assumed to derive the corresponding density stratification. The difference with stellar model atmospheres is that the latter tend to be based on the assumption of flux-constancy (radiative equilibrium above the convective regime), whereas solar models tend to be empirical in some respect.

This diagram portrays the sun as seen by spectroscopists who determine solar abundances in classical fashion (and then, rather boldly, call them "cosmic", *e.g.*, Allen, 1976). The most classic model portrayed here is the one by Holweger (1967), which is usually quoted as its slightly-modified HOLMUL reincarnation by Holweger and Müller (1974). Holweger inverted the observed line-core brightness temperatures of optical lines at disk center, especially Fe I lines, into a best-fitting $T(\tau)$ relation assuming LTE line formation (i.e., line formation under local thermodynamic equilibrium conditions). In simple Eddington-Barbier terms with $I(\mu) \approx B(T[\tau=\mu])$, each disk-center line core displays the temperature where it reaches total optical depth unity; by plotting these temperatures against known gf-value ratios for members of multiplets, Holweger obtained segments of a $T(\tau)$ model that he shifted together and combined with various continuous opacity sources and hydrostatic equilibrium to obtain monotonic $T(\tau)$ and $\tau(h)$ stratifications. Holweger's model has no chromospheric temperature rise because the optical Fe I lines

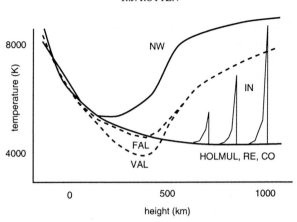

Figure 2. Cartoon of the radial stratification of the lower solar atmosphere outside active regions. See also Fig. 1 of Solanki's (1998) review and Fig. 1 of Judge and Peter's (1998) review. The sketched models VAL, FAL, HOLMUL and RE (radiative equilibrium) are all "classical" in describing static plane-parallel stratifications that obey hydrostatic equilibrium. Convective energy transport affects the deepest layers and produces slightly different stratifications in and between granules. The middle photosphere (i. e. around 200 km height) is closest to the classical plane-layer-plus-turbulence paradigm. The temperature minimum region around $h = 400$ km does not exist anymore; the apparent VAL–FAL chromospheric temperature rise in the non-magnetic internetwork regime (IN) is now thought to represent intermittent shocks in an otherwise cool atmosphere (Carlsson and Stein, 1995). The shocks do not heat or lift the time-averaged atmosphere which remains cool out to $h > 1000$ km and explains the deep cores of the infrared CO lines. In magnetic network structures (NW) the temperature rise occurs much deeper (see Fig. 6 of Solanki's 1998 review), but we don't know where or how or why. VAL stands for the continua-fitting model by Vernazza *et al.* (1981), FAL for its update by Fontenla *et al.* (1993), HOLMUL for the update by Holweger and Müller (1974) of the line-fitting model of Holweger (1967), RE for radiative-equilibrium models such as the one by Bell *et al.* (1976). CO designates yet underived time-averaged models based on the infrared CO line cores (cf. Avrett, 1995).

do not possess self-reversed emission cores (the only optical lines that do so are Ca II H & K, a separate story).

Chromospheric temperature rises are a fixture of the major other class of empirical models, those inverting the observed disk-center continua Eddington-Barbierwise into temperature stratifications. The technique is essentially the same as Holweger's; the difference is that the required variation in height sampling now comes from the variation of continuous opacity over the whole spectrum, with the chromosphere sampled by the ultraviolet shortward of 160 nm and the infrared longward of 160 μm. In the infrared LTE is a good assumption, but it fails in the ultraviolet where the bound-free ionization edges behave much like resonant scattering lines so that their source functions follow the angle-averaged intensity J_ν more closely than the Planck function B_ν. The inversion therefore has to take $J_\nu \neq B_\nu$ departures into account. This was done with impressive sophistication by Avrett and coworkers in Vernazza, Avrett and Loeser (1973, 1976, 1981 = VAL), Avrett (1985), Maltby *et al.* (1986) and Fontenla, Avrett and Loeser (1993 = FAL).

The main differences between VAL and HOLMUL are the VAL chromospheric temperature rise above $h = 400$ km and the lower VAL temperature just below this height. The latter produces appreciable $J_\nu > B_\nu$ excess in the violet and ultraviolet continua. It causes large NLTE (i. e. non-LTE) ionization departures for minority ionization stages such as Fe I (Lites, 1972). I have shown that one may actually derive the HOLMUL model from the VAL model by admitting Fe I NLTE departures in Holweger's LTE inversion procedure (Rutten and Kostik, 1982; cf. Rutten 1988, 1990). At the time, I felt that this disproved Holweger's sarcasm that "NLTE departures tend to arise in the computer when the modeling is incomplete" — I thought that his model arose erroneously from his own computer because he ignored departures from LTE. However, the deep dip of VAL went away when Avrett included more and more lines from Kurucz's gigantic atomic tabulations in his computer. The ultraviolet line haze diminishes J_ν and so brought the FAL photosphere close to HOLMUL while doing away with Fe I NLTE effects, adhering to Holweger's claim.

More recently, a similar change from FAL back to HOLMUL has affected the chromosphere. The successful numerical reproduction by Carlsson and Stein (1997) of the so-called Ca II H_{2V} grains observed by Lites *et al.* (1993) has led Carlsson and Stein (1995) to propose that the non-magnetic parts of the low chromosphere are actually cool, but that acoustic shocks that pass through intermittently are sufficiently bright in the ultraviolet that the time-averaged ultraviolet spectrum requires an apparent temperature rise when inverted in Avrett's continua-fitting procedure. The shocks are weak and seem not to lift or heat the chromosphere persistently, but produce enough temperature variation along the line of sight that the nonlinear $\partial B_\nu/\partial T$ response in the ultraviolet makes the average chromosphere appear hot even though it remains cool (cf. review by Judge and Peter, 1998). This change of view, from an ubiquitous temperature rise above $h = 400$ km to an extended cool, yet shock-ridden atmosphere must also explain the long-standing CO line-formation problem raised by Ayres (*e.g.,* Ayres, 1981; Ayres *et al.*, 1986; Ayres and Rabin, 1996). The final CO paper hasn't yet been written, but it is quite likely that the infrared CO lines display the cool time-averaged structure without much sensitivity to the temporary disruptions that are observed in Ca II H & K and that dominate the ultraviolet continuum emission. Similarly, the shocked-but-cool clapotispheric picture may explain various infrared, sub-mm and mm observations that indicate cool extent larger than VAL-predicted (*e.g.,* Labrum *et al.*, 1978; Horne *et al.*, 1981; Wannier *et al.*, 1983; Lindsey *et al.*, 1986; Roellig *et al.*, 1991; Deming *et al.*, 1992; Belkora *et al.*, 1992; Ewell *et al.*, 1993; White and Kundu, 1994; Solanki *et al.*, 1994).

The RE in Fig. 2 stands for stellar-like theoretical radiative-equilibrium models (cf. Gustafsson and Jorgensen, 1994). The fact that they follow HOLMUL closely suggests that the solar photosphere obeys radiative equilibrium closely. The differences are largest in the deepest layers where the theoretical model takes convective energy transport into account.

3. Lower Atmosphere: Azimuthal Scene

Figure 3 again portrays the lower atmosphere cartoon-wise, but now viewed from the side and emphasizing its inhomogeneities — just as in Fig. 2 of the review by Judge and Peter (1998). Even outside active regions (spots and plage, not illustrated here), the magnetic field is a major agent. New flux that emerges from below tends to quickly assume evacuated kilogauss fluxtube format and to congregate in the loosely defined network pattern (Hagenaar et al., 1997). The latter may be mapped with magnetographs (magnetic network), imaged in the Ca II H & K lines and in many ultraviolet lines (chromospheric network), is outlined by tiny "bright points" in the very best images taken in the Fraunhofer G-band around 430.5 nm, and outlines the cell borders of the supergranulation, although incompletely. The fast congregation results in the strong dichotomy between kilogauss-rich network and kilogauss-poor internetwork (the non-network areas).

The network fluxtubes are modelled in most detail in the Freiburg time-dependent 2D simulations of Steiner et al. (1998). They expand with height, magnetostatically in the "Zürich wine glass model" by Solanki and coworkers (e.g., Solanki and Steiner, 1990; Solanki et al., 1991; Bündte et al., 1993; Bruls and Solanki, 1995; Briand and Solanki, 1995) and dynamically in the Freiburg simulations. Depending on polarity and tube separations, the resulting canopies above which magnetic field pervades the entire medium should be located lower or higher in the chromosphere, or possess an open-out structure as sketched in Fig. 2 of Judge and Peter (1998).

Internetwork fields are even more elusive than overlying canopies (e.g., Zwaan, 1987). Initially described by Livingston and Harvey (1971) as ubiquitous bipolar patterns of random nature, in which the single-polarity patches extend over 2–10 arcsec, their properties and nature remain controversial. Keller et al. (1994) derived a sub-kiloGauss upper limit to the intrinsic field strength B while $B \approx$ 500 Gauss was reported by Lin (1995; cf. also Solanki et al., 1996). Internetwork features with this strength make up a relatively rare class, however (Wang et al., 1995; Wang et al., 1996; Lee et al., 1997; Meunier et al., 1998). The more general, weaker "salt-and-pepper" background is perhaps best displayed in the lowest-flux parts of Fig. 4 of Lites et al. (1996), who found an additional class of short-lived, compact internetwork features with predominantly horizontal field orientation (HIF's). Finally, a truly weak background field may exist as well (Faurobert-Scholl, 1993; Bianda et al., 1998 — but see Landi Degl'Innocenti, 1998).

The dynamical phenomena are strongly affected by the presence of magnetic fields. Even the granulation gets "abnormal", i.e. inhibited when there are too many fluxtubes as in a plage (Title et al., 1992; Berger et al., 1998). The chromospheric dynamics higher up is markedly different in the network and internetwork domains (Lites et al., 1993). The so-called three-minute oscillation dominates in the internetwork areas. It consists of a wide-band acoustical wave spectrum of which the higher frequencies have more upward propagation; shock steepening, shock overtaking and shock colaescence occur in the layers around $h = 800 - 1000$ km, where

ELEMENTAL ABUNDANCES IN CORONAL STRUCTURES

JOHN C. RAYMOND, RAID SULEIMAN and JOHN L. KOHL
Harvard-Smithsonian Center for Astrophysics, 60 Garden St., Cambridge, MA

GIANCARLO NOCI
Universitá di Firenze, I-50125 Firenze, Italy

Abstract. A great deal of evidence for elemental abundance variations among different structures in the solar corona has accumulated over the years. Many of the observations show changes in the relative abundances of high- and low-First Ionization Potential elements, but relatively few show the absolute elemental abundances. Recent observations from the SOHO satellite give absolute abundances in coronal streamers. Along the streamer edges, and at low heights in the streamer, they show roughly photospheric abundances for the low-FIP elements, and a factor of 3 depletion of high-FIP elements. In the streamer core at 1.5 $R_\odot$, both high- and low-FIP elements are depleted by an additional factor of 3, which appears to result from gravitational settling.

Key words: Abundances, Coronal Streamers

1. Introduction

Though there is still some controversy about the reality of abundance variations in the solar corona (Jordan *et al.*, 1997), fairly strong evidence has accumulated in recent years. Peter Young and Uri Feldman discuss UV observations (Young and Mason, 1998; Feldman, 1998). This paper summarizes X-ray evidence for abundance variations, before turning to recent SOHO measurements. The emphasis will be on spectra of streamers obtained with the Ultraviolet Coronagraph Spectrometer (UVCS). At 1.5 $R_\odot$, the gas appears to be surprisingly close to isothermal. Images of quiescent equatorial streamers show a striking difference between Lyα and O VI morphologies. The Lyα image is strongly center-brightened, while the O VI lines are brightest near the edges (Noci *et al.*, 1997; Kohl *et al.*, 1997). This turns out to be an abundance effect, most easily attributed to gravitational settling (Raymond *et al.*, 1997a). This paper discusses UVCS observations of streamers at smaller and larger heights.

2. Abundances from X-ray Observations

Several methods used to obtain abundances from X-ray observations have been reviewed by Saba (1995). They have various advantages and disadvantages in sensitivity to observational and theoretical uncertainties.

2.1. LINE-TO-CONTINUUM RATIOS

Some of the earliest indications of abundance variations came from changes in the line-to-continuum ratios in solar flares (Sylwester *et al.*, 1984). Provided that one observes a helium-like ion (which accounts for nearly all the concentration of the element in question over a broad temperature range) this method is relatively insensitive to both atomic rate uncertainties and instrument calibration errors. Given than more than 90% of an element is in the He-like stage over 0.5 dex in $\log T$, uncertainties in ionization and recombination rates are not relevant. Also, the He-like ions are simple and well-studied, so the excitation rates are known more accurately than those of most ions. Instrumental calibration is not a problem, in that the line is compared with the continuum level at the same wavelength. The continuum (mostly H bremsstrahlung) and the line are formed by electrons of essentially the same energy, so there is relatively little uncertainty from temperature sensitivity, and the result is an absolute abundance (the abundance relative to hydrogen). The potential problems are that the continuum may be very faint, that it may be contaminated by non-thermal continuum or by continuum from far hotter gas, and that fluorescence in a Bragg crystal spectrometer might give a false continuum.

The overall conclusion of these studies was an indication that the abundance of Ca varies from one flare to another, possibly depending on the fraction of flare material evaporated from the chromosphere. Another flare study done by Fludra and Schmelz (1995) found that calcium is enhanced by factors 1.5–2.0, the iron abundance can be photospheric or enhanced by factors up to 1.5, other low-FIP elements (silicon and magnesium) have photospheric abundances, while oxygen is depleted by a factor of 4.

2.2. CORONAL VS. PHOTOSPHERIC IRON

It is also possible to compare coronal and photospheric abundances by comparing the fluorescent iron K emission with the Fe XXV line formed in the flare. The fluorescent emission is produced when X-rays above the K edge of nearly neutral Fe cause inner shell ionization, followed by emission of a Kα and Kβ photons. As for the method above, the result is insensitive to ionization and recombination rates, and the excitation rates and fluorescence yields are accurately known. Again, the lines are so close in wavelength that instrumental calibration is not a problem. The potential dangers are that a higher energy continuum might contribute to the innershell ionization of photospheric iron, or that energetic particles may reach the chromosphere and cause inner shell ionization.

Phillips *et al.* (1994) applied this method to several flares, and they found that the coronal abundances were the same as the photospheric abundances to within a factor of 2.

2.3. RELATIVE ABUNDANCES

Many studies have obtained relative abundances of various pairs of elements by comparing the intensities of strong lines. It is possible either to derive an emission measure curve and use theoretical ionization balances and excitation rates, or to try to choose a pair of lines which have almost the same temperature dependence, so that the temperature of the flare of active region need not be known. Examples of this method include the study of Ca/Fe ratios in flares which indicated factor of 2 variations among flares (Antonucci and Martin, 1995) and studies of the Ne/Fe ratio in active regions based on the ratio of Ne IX to Fe XVII lines (Schmelz *et al.*, 1996; Phillips *et al.*, 1997). The latter results have been controversial, as recent ionization balance calculations predict different Fe XVII concentrations (Arnaud and Raymond, 1992, *vs.* Arnaud and Rothenflug, 1985) and there is disagreement about the importance of resonance scattering of the strongest Fe XVII line.

3. Streamer Abundances from UVCS

The UVCS instrument is described by Kohl *et al.* (1996), and early results are summarized in Kohl *et al.* (1997) and Noci *et al.* (1997). Raymond *et al.* (1997a) analyzed the absolute elemental abundances in streamers at $1.5\,R_\odot$. Particular attention was devoted to the quiescent equatorial streamer observed on the W limb on July 25, 1996. Spectra at 1.75 and $2.0\,R_\odot$ were obtained immediately after the $1.5\,R_\odot$ observation. Here, we review the results at $1.5\,R_\odot$, discuss the observations of the higher heights, and summarize a subsequent observation at $1.3\,R_\odot$ from August 21, 1997.

The July 25, 1996 observations at 1.75 and $2.0\,R_\odot$ were made with the same position angle ($-90°$) and grating positions used at $1.5\,R_\odot$. Exposure times were the same, except that the observations with the grating position covering the shortest wavelengths at $2.0\,R_\odot$ had to be cut short. As for the $1.5\,R_\odot$ data, we extracted bright and faint O VI sections along the spectrograph slit corresponding to the streamer legs and core. The faint O VI portion becomes narrower with height, and the intensity drops quickly. Therefore, the fainter lines are difficult to measure at the higher positions. Here we consider only the brightest lines; Lyα, Lyβ and the O VI doublet.

UVCS can measure the intensities of about 40 spectral lines in the ranges 940–1120 Å, 1170–1350 Å, 470–560 Å, and 580–635 Å. Most lines are produced by electron collisional excitation, but the H I Lyman lines and the O VI doublet contain large contributions from scattered disk photons (e.g. Noci, Kohl and Withbroe, 1987). We separate the radiative and collisional contributions as in Raymond *et al.* (1997). For small outflow speeds, the collisional ratios I(Lyα)/I(Lyβ) and I(O VI 1032)/I(O VI 1037) are 7.57 and 2.0, respectively, while the scattered radiation ratios are 910 and 4.0. Simple algebra gives the radiative and collisional fractions.

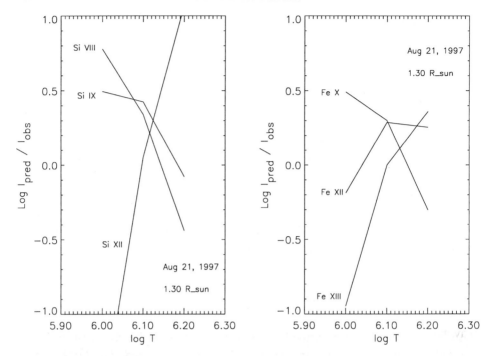

Figure 1. Predicted to Observed ratios of intensities of Si and Fe lines to the collisional component of Lyβ. These ratios assume photospheric abundances. The temperature is determined by the temperature where the lines cross, and the abundance from the vertical offset.

We then determine the electron temperature of the streamer in two ways. The proton kinetic temperature is easily found from the Lyα line width (using the widths of lines of more massive ions to correct for the small non-thermal contribution to the widths), and the result for the July 25 data is $\log T = 6.2$. In coronal holes, T_p can be much different that T_e, but in the higher density, nearly static plasma of the streamers there is ample time for electron-ion equilibration. Another way to obtain electron temperatures is to use the relative intensities of lines of different ions, together with a theoretical calculation of the ionization state (Arnaud and Rothenflug, 1985; Arnaud and Raymond, 1992). Figure 1 shows the comparison for Si and Fe for the August 21 data. These plots indicate a temperature $\log T = 6.13$ for this streamer. Comparison of lines formed over a broad range of temperatures (O VI to Si XII and Fe XIII]) indicates that the streamer is remarkably isothermal. UVCS has also measured T_e from the width of the electron-scattered Lyα line, but not for these streamers.

To find abundances for the higher positions in the streamer, we must consider projection effects. As described in Raymond *et al.* (1997b), a simple model which matches the overall geometry of the O VI and Lyα emission implies that a significant fraction of the O VI observed in the center of the streamer actually arises from the streamer legs seen in projection. More complete modelling efforts are under-

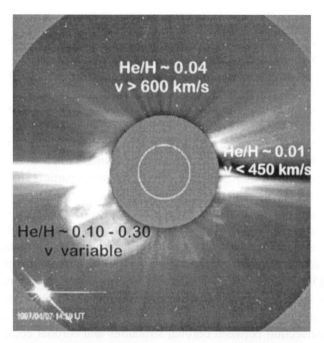

Figure 1. The structured solar corona as seen on April 7, 1997 with the LASCO C2 coronagraph on SOHO combined with typical He/H abundance ratios in associated solar wind flows. High speed solar wind emanating from the open field line structures in the polar coronal holes has a typical helium content of 5% by number (Bame *et al.*, 1977), whereas solar wind flowing near the equatorial current sheet is frequently depleted in He (Borrini *et al.*, 1981). On the other hand, coronal mass ejecta (lower left part) contain plasma of variable composition with occasional enrichment of helium relative to average solar wind: He/H ratios > 0.10 are sometimes observed (Borrini *et al.*, 1982).

wind and they expand slowly (Wang and Sheeley, 1991). On the other hand, in the vicinity of the equatorial current-sheet the large scale magnetic fields exhibit more complex structures. The innermost loops are closed and no plasma can escape into the interplanetary space. The slow speed solar wind is thought to originate on the fringes of these closed structures. Flux tubes along the fringes of these structures expand rapidly within the first few solar radii and, consequently, the interaction of protons and alphas with each other and with minor species can be drastically reduced. Bürgi (1992) has investigated this effect in detail and shown that strong depletions of helium and also of minor ions can occur in superradially expanding flow regimes due to inefficient Coulomb drag. A new and alternative view on streamer-belt associated, low-speed solar wind has been forwarded by Noci *et al.* (1997). The authors note that the magnetic configuration of the streamer belt as observed with SOHO/LASCO is usually more complex than the structure formed from a simple current sheet. In a multiple current sheet, however, also converging but open flux tubes must exist, and such structures may cause slow and hot solar wind flows. The authors also suggest that such a converging configuration could lead to the depletion of heavy elements due to the limitation of the driving proton flux. The idea that converging structures prevailing at rather large distances from the solar surface can lead to a significant depletion of heavy elements due to insufficient Coulomb drag from protons should be checked in detailed models. Possibly the depletion of heavy species still has to be ascribed to the flux tube divergence before the zone of convergence.

Consider now the other case, i.e. coronal hole-type wind. In contrast to the case of rapidly diverging flows in the vicinity of current-sheets discussed above, Coulomb drag could be more efficient in the associated weakly diverging flux tubes. Hence, helium and minor ions are tightly coupled to the main fluid and carried forward without further depletion. Moreover, wave particle interaction seems to be so vigorous in coronal hole-associated high speed streams (Kohl *et al.*, 1997) that helium and minor ions reach temperatures of the order of 10^7 K or even 10^8 K, sufficient to escape from the solar gravity field without help of the proton drag.

Unfortunately, it is rather difficult to combine optical off-limb observations as the ones of (Noci *et al.*, 1997) with in situ particle measurements carried out at the location of SOHO. Possibly, one has to await the combination of particle measurements from a solar orbiter which will frequently cruise above the solar limb and optical observations made from a near Earth spacecraft, such as SOHO, in order to discuss the connection between coronal structures and coronal composition more conclusively.

The other important ingredient of elemental and isotopic fractionation is the FIP/FIT effect with which neutrals are converted into ions and then fed into the corona. According to the model of Marsch *et al.* (1995) this conversion can be visualized as a sequence of two processes (see Figure 2) which together determine the overall efficiency in a steady state scenario. The first important process operating in the upper chromosphere is ionization of neutral species by exposure to the ambient UV photon bath and to UV from the overlaying transition region and corona. Further, it is assumed in this model that ions are immediately transferred into the corona by some undefined mechanism which collects all charged particles efficiently without introducing additional bias. The second rate determining step is the supply of neutrals through the ambient neutral hydrogen of the photosphere. The simple approach of Marsch *et al.* (1995) has been successful in reproducing the observed abundance pattern in low speed solar wind and in Solar Energetic Particles (SEPs). An explanation for the fact that high speed solar wind seems to be considerably less enriched in low FIP elements (von Steiger *et al.*, 1997) was later supplied by Peter (1996) who generalized the model of Marsch *et al.* (1995) and pointed out that the solar wind ion flux is largely determined by the prevailing UV photon flux. Evidently and as indicated before, varying UV intensities will have an impact on the ionization rate of minor species and, hence, also on the bulk abundances in the solar wind. Since UV intensities are largely correlated with the appearance of certain solar surface structures it is not surprising to find a strong interdependence between solar wind composition and coronal structures.

THE SOLAR NOBLE GAS RECORD IN LUNAR SAMPLES AND METEORITES

R. WIELER

ETH Zürich, Isotope Geology, NO C61, CH-8092 Zürich, Switzerland, wieler@erdw.ethz.ch

Abstract. Lunar soil and certain meteorites contain noble gases trapped from the solar wind at various times in the past. The progress in the last decade to decipher these precious archives of solar history is reviewed. The samples appear to contain two solar noble gas components with different isotopic composition. The solar wind component resides very close to grain surfaces and its isotopic composition is identical to that of present-day solar wind. Experimental evidence seems by now overwhelming that somewhat deeper inside the grains there exists a second, isotopically heavier component. To explain the origin of this component remains a challenge, because it is much too abundant to be readily reconciled with the known present day flux of solar particles with energies above those of the solar wind. The isotopic composition of solar wind noble gases may have changed slightly over the past few Ga, but such a change is not firmly established. The upper limit of $\sim$5% per Ga for a secular increase of the ^{3}He/^{4}He ratio sets stringent limits on the amount of He that may have been brought from the solar interior to the surface (cf. Bochsler, 1992). Relative abundances of He, Ne, and Ar in present-day solar wind are the same as the long term average recorded in metallic Fe grains in meteorites within error limits of some 15–20%. Xe, and to a lesser extent Kr, are enriched in the solar wind similar to elements with a first ionisation potential < 10 eV, although Kr and Xe have higher FIPs. This can be explained if the ionisation time governs the FIP effect (Geiss and Bochsler, 1986).

1. Introduction

During the past four billion years the dust layer on the moon has been stirred by meteorite bombardment, such that almost every grain trapped ions from the solar wind on the immediate lunar surface (e. g. Pepin *et al.*, 1970a; Reynolds *et al.*, 1970; Eberhardt *et al.*, 1970). The moon therefore conserves an unequaled record of solar history. A similar record is available in meteorites which represent compacted dust from asteroids (Wänke 1965; Eberhardt *et al.*, 1965). The solar contribution can only be detected for those few elements which are otherwise extremely rare in the samples, mainly the noble gases. On the other hand, the high number of solar noble gas atoms accumulated in extraterrestrial samples results in a relatively high precision compared to measurements of present-day solar corpuscular radiation. Lunar samples provide even the only data on solar Kr and Xe available so far. The extraterrestrial record is disturbed, however, and needs to be carefully read, as should become clear in this review. I focus on data obtained by the gas extraction technique of in-vacuo etching, which largely avoids noble gas fractionation in the laboratory.

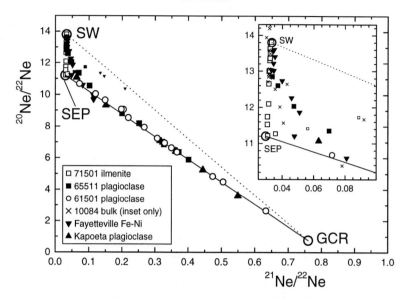

Figure 1. Ne data from 4 lunar and 2 meteorite samples with solar noble gases. Gases were released by in-vacuo etching, except for samples 10084 (stepwise heating) and 61501 (total extraction of aliquots etched off-line). In all samples, besides solar wind Ne (SW) a second solar component appears to be present, labelled SEP. Note that the ^{20}Ne/^{22}Ne ratios above the SW point in the first steps of 10084 indicate isotopic fractionation during gas release, which is not observed for gas extraction by acid etching. GCR indicates the composition of Ne produced by galactic cosmic rays in plagioclase. Data sources: 71501: Benkert et al. (1993); 65511: Wieler et al. (1986); 61501: Etique et al. (1981); 10084: Hohenberg et al. (1970); Fayetteville: Murer et al. (1997); Kapoeta: Pedroni (1989)

2. Two Solar Noble Gas Components in Extraterrestrial Samples

Fig. 1 shows several typical Ne data sets. A mixture of two components is represented by a data point on the straight line connecting the two "endmembers", and an n-component mixture plots within the respective polygon. We may therefore recognize different noble gas components by multiple analyses, e. g. by measuring grain-size fractions of a sample or by degassing a sample in several steps. The latter is usually done by increasing the temperature until the sample melts, but this may cause isotopic fractionation.

For several data sets in Fig. 1, the points of the first extraction steps fall close to the composition labelled "SW", representing solar wind Ne collected in foils exposed during the Apollo missions (Geiss et al., 1972; Table I). Later steps define a path that first points towards the composition labelled SEP (Solar Energetic Particles, a terminology discussed later). The data for an ilmenite separate of lunar soil 71501 actually reach the SEP point, whereas for all other samples the datapoints of the last steps fall onto the straight line connecting points SEP and GCR (the latter representing Ne produced by spallation of target atoms by the Galactic Cosmic Radiation). For two data sets in Fig. 1, and many others not shown, the large number of points along this line is very remarkable. The figure indicates that lunar samples and meteorites apparently contain a second trapped Ne component besides SW-Ne. This SEP-Ne resides at larger depth than the few hundred Ångstrom typical for SW-Ne. The very first lunar sample analyses were interpreted as providing evidence for the existence of SEP-Ne with ^{20}Ne/^{22}Ne ~ 11.2 (Pepin et al., 1970a; Reynolds et al., 1970). However, these workers later cautioned that the data (e. g. "10084" in Fig. 1) might also indicate isotopic fractionation of a single SW component by noble gas diffusion during stepwise heating (Pepin et al., 1970b; Hohenberg et al., 1970). This ambiguity led to the development of in-vacuo etching, where grains are etched layer by layer in a vacuum device connected to a mass spectrometer (Wieler et al., 1986). This should minimize isotopic or elemental fractionation in the laboratory, since gases are released essentially at room temperature by destroying the carrier by HF or HNO$_3$. The results verified this expectation and most data discussed here were obtained by stepwise etching. We can therefore rule out now that the SEP component is an artifact of diffusion in the laboratory.

We will discuss below that the SEP component is much too abundant to be easily reconciled with the known flux of particles above SW energies. We therefore have to ask whether SEP may be a fake component caused by some other artifact, such as SW gas that migrated towards the grain interior or because the heavier SW isotopes are implanted slightly more deeply than the lighter ones. This is very unlikely, because such effects would not lead to a uniform composition over a large depth range and in many different minerals. Noble gas diffusion is highly mineral-dependent, as elemental ratio patterns show. For example, solar He and Ne do not diffuse in metallic iron-nickel (section 4), and yet, SEP-Ne is found in all Fe-Ni samples (Murer et al., 1997; Fig. 1). Furthermore, diffusion should enrich rather than deplete the light isotope at depths where SEP-Ne resides. Ion implantation at a single solar wind energy simulated by the TRIM code (Ziegler, 1985) shows that the light isotope will be underabundant at larger depths, but no constant isotopic composition over a large depth range will result (R. Wieler, unpublished data). Moreover, the SEP component is found in depths of up to several μm, orders of magnitude deeper than the solar wind range. In summary, while the search for possible artifacts should continue, the extensive data set available today strongly favours the view that SEP-Ne is a genuine component, isotopically heavier and implanted with higher energy than SW-Ne. A component with the same characteristics has been observed for the other noble gases also (Benkert et al., 1993; Wieler and Baur, 1994; Pedroni and Begemann, 1994).

What is this SEP component then? Once it was thought to represent particles from solar flares (Black, 1972; Etique et al., 1981). This inference was supported by the first direct measurements of Ne isotopes emitted during solar flares, which yielded ^{20}Ne/^{22}Ne ratios ~35% below the SW value (e. g. Mewaldt et al., 1984). Recent satellite measurements suggest that this ratio is variable but may on average well be lower than the SW composition (Selesnick et al., 1993, Mewaldt, 1998). Part of the SEP-Ne in lunar grains probably indeed resides in depths of 10 μm or so, implying energies on the order of MeV/amu. However, the SEP component is much too abundant to be due mainly to the rare energetic particles measured in space. SEP-Xe nominally represents roughly 20% of the total solar Xe in many samples studied, usually mineral separates (Wieler et al., 1986; Benkert et al., 1993; Murer et al., 1997). This can be compared to the contribution of $10^{-4} - 10^{-5}$ estimated for > 0.1 MeV/amu particles based on heavy-ion tracks (crystal lattice damages) in lunar rocks (Wieler et al., 1986). The mineral grains will lose the surface-sited SW component more readily than the more deeply sited SEP, e. g. by grain surface sputtering (we consider Xe here since this gas is not lost by diffusion, see below). This may enhance the SEP/SW ratio by ∼2 orders of magnitude, if we assume for SEP a mean implantation depth of ∼ $3\,\mu m$, i. e. 100 times the SW depth. The $3\,\mu m$ correspond to ion energies of 0.1 MeV/amu, according to TRIM. With these assumptions the SEP flux should be about 0.2% of the total solar particle flux, or 20 to 200 times higher than the flux > 0.1 MeV/amu. An even much higher SEP/SW flux ratio might be derived by considering bulk samples, for which SEP-Xe often also accounts – nominally – for 10% or more of the total solar gas. It appears that the lunar regolith as a whole retains all solar Ar, Kr, and Xe that ever hit the lunar surface, although not necessarily in the original trapping sites (Geiss, 1973, Wieler, 1997). If true, preferential removal of the SW component cannot be invoked for bulk samples and one would have to conclude that the flux of SEP particles accounts for ten percent or so of the total solar flux. Such estimates are obviously plagued by many uncertainties, but it seems in any case that the SEP component stems largely from particles < 0.1 MeV/amu. Whether the known flux of suprathermal particles is sufficient to explain the high concentrations of SEP noble gases is highly questionable, and we must ask whether the sun periodically emits a higher flux of particles above solar wind energies than it does today. This problem remains a challenge to students of both the present-day sun and the record of the ancient sun left in extraterrestrial samples. Note that the label SEP for the second noble gas component in lunar samples and meteorites is a misnomer in the sense that it should not imply that this component is identical to the Solar Energetic Particles measured by space science experiments.

For all five noble gases the ratios of the isotopic ratios of SEP to SW are related by the same law, being about equal to the square of the corresponding mass ratios (Wieler and Baur, 1994). In contrast, relative elemental abundances in SEP and SW are indistinguishable from each other within experimental uncertainties of some 20–30% (Pedroni and Begemann, 1994; Wieler and Baur, 1995; Murer et al., 1997). On the one hand, these observations leave no doubt that the SEP component is solar in origin. On the other hand, if the $(m_1/m_2)^2$ law would be valid for elemental ratios also, this would lead to a fractionation between SW and SEP for e. g. Ar/Kr and Kr/Xe of 80% and 60%, respectively. It is thus a further challenge to conceive a fractionation process leading to large isotope effects unaccompanied by corresponding elemental effects.

led early workers to conclude that lunar samples do not conserve quantitative information on heavy noble gas abundances in the solar wind (e. g. Bogard et al., 1973). A more optimistic view was expressed by Kerridge (1980), who noted that Kr/Xe ratios do not scatter randomly but correlate with the ^{40}Ar/^{36}Ar ratio. He deduced from this that the samples testify to a secular increase of the Kr/Xe ratio in the solar wind. I now discuss recent experiments confirming this conclusion. Lunar samples allow us to deduce a secularly variable enhancement of Xe – and partly of Kr – in the solar wind similar to that observed for elements with a low first ionisation potential (FIP).

These conclusions are based on in-vacuo etch studies on the one hand and measurements of single lunar grains on the other. Wieler and Baur (1995) showed that He/Ar and Ne/Ar ratios in the first steps of etch runs are 1–2 orders of magnitude below the solar wind values but approach them towards the end of the runs. Therefore, essentially only the shallowly sited SW component has lost He and Ne, whereas the SEP component conserves nearly unfractionated He, Ne and Ar abundances. In striking contrast, Ar/Kr and Kr/Xe ratios are almost constant throughout etch runs. These flat ratio patterns and the fact that not even the mobile He and Ne are strongly fractionated in the SEP regime are evidence that the samples conserve the true Ar/Kr and Kr/Xe ratios in the solar wind (Wieler and Baur, 1995). The single grain data corroborate this conclusion (Wieler et al., 1996), because all grains from a given soil have the same Ar/Kr, and Kr/Xe values, in sharp contrast to He/Ar or Ne/Ar ratios which vary by orders of magnitude in the same grains.

The fractionation factors for Kr and Xe in the solar wind deduced from four samples are shown in Fig. 3. The solar wind data are normalized to the solar abundances of Anders and Grevesse (1989), and Kr and Xe are anchored to the Ar point given by these authors. The errors for Kr and Xe are a rather conservative estimate of the analytical accuracy but do not include the uncertainties of the solar Ar, Kr, and Xe abundances, which are ~20–25% each (Anders and Grevesse, 1989). Xe is enriched in all four samples by about the same factor as elements with a FIP < 10eV in the slow solar wind (Meyer 1985), whereas Kr is partly enriched. This is remarkable, since Kr and Xe have a FIP above the threshold for which no enhancement is observed for other elements. Geiss and Bochsler (1986) proposed that the ionisation time in the chromosphere is the parameter that actually governs the FIP effect. In such models, Xe and Kr enrichments similar to those shown in Fig. 3 are indeed predicted (Geiss et al., 1994; Marsch et al., 1995; Geiss, 1998; Peter, 1998).

The two groups of "young" and "old" samples in Fig. 3 display different fractionation patterns, with the Kr/Xe ratio differing by about a factor of two between the groups. This is the trend already observed by Kerridge (1980) from bulk sample analyses. The two "young" samples both received their solar wind within the last 50–100 Ma. The two "old" samples contain on average solar wind that is ~1 Ga and ~4 Ga old, respectively, whereby the solar-wind-age of their individual grains is not well constrained but possibly varies by hundreds of millions of years. The

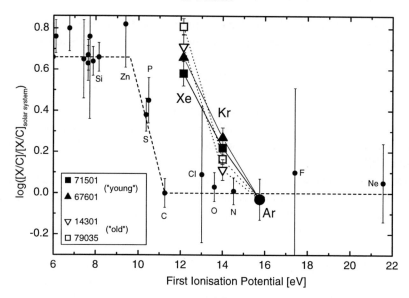

Figure 3. Small symbols represent elemental abundances in solar energetic particles relative to photospheric or "solar system" values (Anders and Grevesse, 1989). The abscissa shows the first ionisation potential. Large symbols show Xe and Kr abundances in the solar wind deduced from lunar samples. Kr and Xe values are normalized to the solar abundances of these elements inferred by Anders and Grevesse and are anchored to the Ar point given by these authors. Solid symbols represent solar wind trapped in the last ∼100 Ma, open symbols ancient solar wind of ∼1 Ga (79035) and ∼4 Ga (14301), respectively. Xe in the solar wind is enriched similar to low-FIP elements, and Kr is partly enriched. The enrichment factors show a secular trend.

Kr/Ar fractionation seems to show a slight monotonic increase over the lifetime of the sun, whereas the Xe/Ar fractionation may have been largest about 1 Ga ago (Wieler *et al.*, 1996). In that paper we speculate that a higher proportion of (less fractionated) high-speed solar wind in the past might explain the Kr/Ar trend, but the opposite temporal trends of Xe and Kr enhancements require additional processes. Note that the FIP effect and its temporal variations affect SW and SEP components to the same extent, because elemental ratios of both components are always indistinguishable.

5. Summary

The precious record in lunar soils and meteorites of solar noble gases from the recent and remote past needs and deserves carefully designed experiments to become decipherable. Evidence for the presence of two isotopically different solar noble gas components in all these samples is now overwhelming, but the so-called "SEP component" is still enigmatic because it is much more prominent than expected from known solar particle flux spectra. The mean isotopic and elemental composition of He, Ne, and Ar have not changed discernably in the last ∼100 Ma and the isotopic composition at best very slightly in the past several Ga. Xe (and partly Kr) are enriched in the solar wind similar to low-FIP elements and these enrichments show a secular trend.

Acknowledgements

I thank Peter Signer and all his crew for many years of fruitful cooperation and Johannes Geiss, Peter Bochsler and their coworkers for many discussions. Comments by R. H. Becker and an anonymous referee helped to improve the manuscript. The work at ETH Zurich summarized here has been supported by several grants from the Swiss National Science Foundation.

References

Anders, E. and Grevesse, N.: 1989, 'Abundances of the elements: meteoritic and solar', *Geochim. Cosmochim. Acta* **53**, 197.
Becker, R. H., Schlutter, D. J., Rider, P. E., and Pepin, R. O.: 1998, 'An acid-etch study of the Kapoeta achondrite: Implications for the argon-36/argon-38 ratio in the solar wind', *Meteoritics Planet. Sci.* **33**, 109.
Benkert, J.-P., Baur, H., Signer, P., and Wieler, R.: 1993, 'He, Ne, and Ar from the solar wind and solar energetic particles in lunar ilmenites and pyroxenes', *J. Geophys. Res. (Planets)* **98**, 13147.
Black, D. C.: 1972, 'On the origins of trapped helium, neon and argon isotopic variations in meteorites-I. Gas-rich meteorites, lunar soil and breccia', *Geochim. Cosmochim. Acta* **36**, 347.
Bochsler, P.: 1984, 'Helium and oxygen in the solar wind: dynamic properties and abundances of elements and helium isotopes as observed with the ISEE-3 plasma composition experiment', Habilitation Thesis, Univ. of Bern.
Bochsler, P.: 1992, 'Minor ions – tracers for physical processes in the heliosphere', *Solar Wind Seven*, eds. E. Marsch and R. Schwenn, Pergamon, 323.
Bogard, D. D., Nyquist, L. E., Hirsch, W. C., and Moore, D. R.: 1973, 'Trapped solar and cosmogenic noble gas abundances in Apollo 15 and 16 deep drill samples', *Earth. Planet. Sci. Lett.* **21**, 52.
Cerutti, H.: 1974, 'Die Bestimmung des Argons im Sonnenwind aus Messungen an den Apollo-SWC-Folien', PhD Thesis, Univ. of Bern.
Eberhardt, P., Geiss, J., Graf, H., Grögler, N., Krähenbühl, U., Schwaller, H., Schwarzmüller, J., and Stettler, A.: 1970, 'Trapped solar wind noble gases, exposure age and K/Ar-age in Apollo 11 lunar fine material', *Proc. Apollo 11 Lunar Sci. Conf.*, 1037.
Eberhardt, P., Geiss, J., and Grögler, N.: 1965, 'Über die Verteilung der Uredelgase im Meteoriten Khor Temiki', *Tschermaks Min. u. Petr. Mitt.* **10**, 535.
Etique, P., Signer, P., and Wieler, R.: 1981, 'An in-depth study of neon and argon in lunar soil plagioclases, revisited: implanted solar flare noble gases', *Lunar Planet. Sci.* **XII**, Lunar Planet. Institute, Houston, 265.
Geiss, J.: 1973, 'Solar wind composition and implications about the history of the solar system', *Int. Cosmic Ray Conf. 13th*, 3375.
Geiss, J.: 1998, 'Solar wind abundance measurements: constraints on the FIP mechanism', *Space Sci. Rev.*, this volume.
Geiss, J. and Bochsler, P.: 1986, 'Solar wind composition and what we expect to learn from out-of-ecliptic measurements', In *The sun and the heliosphere in three dimensions*, ed. R. G. Marsden, Reidel, 173.
Geiss, J., Bühler, F., Cerutti, H., Eberhardt, P., and Filleux, C.: 1972, 'Solar wind composition experiment', *Apollo 16 Prelim. Sci. Rep., NASA SP-315*, 14.1.

Geiss, J., Gloeckler, G., and von Steiger, R.: 1994, 'Solar and heliospheric processes from solar wind composition measurements', *Phil. Trans. R. Soc. London A* **349**, 213.

Goswami, J. N., Lal, D., and Wilkening, L. L.: 1984, 'Gas-rich meteorites: probes for particle environment and dynamical processes in the inner solar system', *Space Sci. Rev.* **37**, 111.

Hohenberg, C. M., Davis, P. K., Kaiser, W. A., Lewis, R. S., and Reynolds, J. H.: 1970, 'Trapped and cosmogenic rare gases from stepwise heating of Apollo 11 samples', *Proc. Apollo 11 Lunar Sci. Conf.*, 1283.

Kallenbach, R., et al.: 1998, 'Fractionation of Si, Ne, and Mg isotopes in the solar wind as measured by SOHO/CELIAS/MTOF', *Space Sci. Rev.*, this volume.

Kerridge, J. F.: 1980, 'Secular variations in composition of the solar wind: Evidence and causes', *Proc. Conf. Ancient Sun*, eds. R. O. Pepin, J. A. Eddy, and R. B. Merrill, 475.

Manka, R. H. and Michel, F. C.: 1971, 'Lunar atmosphere as a source of lunar surface elements', *Proc. Lunar Sci. Conf. 2nd*, 1717.

Marsch, E., von Steiger, R., and Bochsler, P.: 1995, 'Element Fractionation by Diffusion in the Solar Chromosphere', *Astron. Astrophys.* **301**, 261.

Mewaldt, R. A.: 1998, 'Solar energetic particle studies of coronal isotopic composition', *Space Sci. Rev.*, this volume.

Mewaldt, R. A., Spalding, J. D., and Stone, E. C.: 1984, 'A high-resolution study of the isotopes of solar flare nuclei', *Astrophys. J.* **280**, 892.

Meyer, J.-P.: 1985, 'The baseline composition of solar energetic particles', *Astrophys. J.* **57** Suppl., 151.

Murer, C. A., Baur, H., Signer, P., and Wieler, R.: 1997, 'Helium, neon, and argon abundances in the solar wind: in vacuo etching of meteoritic iron nickel', *Geochim. Cosmochim. Acta* **61**, 1303.

Pedroni, A.: 1989, 'Die korpuskulare Bestrahlung der Oberflächen von Asteroiden; eine Studie der Edelgase in den Meteoriten Kapoeta und Fayetteville', PhD Thesis, (Nr. 8880), ETH Zürich.

Pedroni, A. and Begemann, F.: 1994, 'On unfractionated solar noble gases in the H3-6 meteorite Acfer111', *Meteoritics* **29**, 632.

Pepin, R. O., Becker, R. H., and Rider, P. E.: 1995, 'Xenon and krypton isotopes in extraterrestrial regolith soils and in the solar wind', *Geochim. Cosmochim. Acta* **59**, 4997.

Pepin, R. O., Nyquist, L. E., Phinney, D., and Black, D. C.: 1970a, 'Isotopic composition of rare gases in lunar samples', *Science* **167**, 550.

Pepin, R. O., Nyquist, L. E., Phinney, D., and Black, D. C.: 1970b, 'Rare gases in Apollo 11 lunar material', *Proc. Apollo 11 Lunar Sci. Conf.*, 1435.

Peter H.: 1998, 'Element Separation in the Chromosphere', *Space Sci. Rev.*, this volume.

Reynolds, J. H., Hohenberg, C. M., Lewis, R. S., Davis, P. K., and Kaiser, W. A.: 1970, 'Isotopic analyses of rare gases from stepwise heating of lunar fines and rocks', *Science* **167**, 545.

Sarda, P., Staudacher, T. and Allègre, C.J.: 1988, 'Neon isotopes in submarine basalts', *Earth Planet. Sci. Lett.* **91**, 73.

Selesnick, R. S., Cummings, A. C., Cummings, J. R., Leske, R. A., Mewaldt, R. A., and Stone, E. C.: 1993, 'Coronal abundances of neon and magnesium isotopes from solar energetic particles', *Astrophys. J.* **418**, L45.

Wänke, H.: 1965, 'Der Sonnenwind als Quelle der Uredelgase in Steinmeteoriten', *Z. Naturforsch.* **20a**, 946.

Wieler, R.: 1997, 'Why are SEP noble gases so abundant in extraterrestrial samples?', *Lunar and Planeary. Science.* **XXVIII**, Lunar and Planetary Insitute, Houston, 1551.

Wieler, R. and Baur, H.: 1994, 'Krypton and xenon from the solar wind and solar energetic particles in two lunar ilmenites of different antiquity', *Meteoritics* **29**, 570.

Wieler, R. and Baur, H.: 1995, 'Fractionation of Xe, Kr, and Ar in the solar corpuscular radiation deduced by closed system etching of lunar soils', *Astrophys. J.* **453**, 987.

Wieler, R., Baur, H., and Signer, P.: 1986, 'Noble gases from solar energetic particles revealed by closed system stepwise etching of lunar soil minerals', *Geochim. Cosmochim. Acta* **50**, 1997.

Wieler, R., Kehm, K., Meshik, A. P., and Hohenberg, C. M.: 1996, 'Secular changes in the xenon and krypton abundances in the solar wind recorded in single lunar grains', *Nature* **384**, 46.

Ziegler, J. F.: 1985, 'The stopping power and ranges of ions in solids', in *The stopping and ranges of ions in matter*, Vol. 1, Pergamon Press, New York.

ATOMIC PHYSICS FOR ATMOSPHERIC COMPOSITION MEASUREMENTS

P.R. YOUNG and H.E. MASON
Department of Applied Mathematics and Theoretical Physics, Silver Street, Cambridge CB3 9EW, United Kingdom

Abstract. The atomic physics relevant to the interpretation of solar spectra produced by plasmas at temperatures $\gtrsim 10^5$ K are discussed. Methods for determining relative abundance ratios are presented and examples provided from the Coronal Diagnostic Spectrometer on board SOHO. In particular, the Fe/Si ratio in the corona is found to be close to photospheric; the Mg/Ne ratio in the transition region is found to vary by an order of magnitude in different solar features. The Mg/Ne ratios in supergranule cell centres and the network are separated for the first time, although no significant differences are found.

Key words: Corona, Transition Region, Abundances, CDS/SOHO

1. Atomic Physics Processes Relevant to the Solar Atmosphere

In the upper transition region and corona, the plasma has temperatures between 100 000 and 3 000 000 K, and electron number densities between 1×10^8 and 1×10^{12} cm^{-3}. Such a plasma gives rise to a predominantly emission line spectrum, with many of the lines lying in the extreme ultraviolet (EUV) part of the solar spectrum at wavelengths of 100–1000 Å. To infer physical parameters from such spectra, three key assumptions are usually made:

1. All lines are optically thin
2. The plasma is in steady state
3. Atomic processes affecting the ionisation state of an element can be separated from those affecting the level balance within an ion

To interpret the observed intensity of a line, an estimate of the expected power loss in the line is required. The *emissivity*, ϵ_λ, is defined as the energy lost in an emission line of wavelength λ by a unit volume of plasma each second, and is given by

$$\epsilon_\lambda = 0.83 \, Ab(X) \, \Delta E \, n_j A_{ji} \, N_e \, F(T) \tag{1}$$

where 0.83 is the ratio of protons to free electrons (a constant at high temperatures), $Ab(X)$ the abundance of element X relative to hydrogen, ΔE the energy of the emitted photon, n_j the fraction of ions that are in the emitting state j, A_{ji} the radiative decay rate for the transition, N_e the electron number density, and $F(T)$ the ionisation fraction. n_j is determined by solving the *level balance* equations for the ion, whereas $F(T)$ comes from solving the *ion balance* equations for the element. By virtue of assumption (3) above, these two quantities are independent of each other, while assumption (2) makes the solution of these equations relatively simple.

From assumption (1), the observed line intensity, I_λ, is directly proportional to the emissivity, with

$$4\pi I_\lambda = \int \epsilon_\lambda dh \qquad (2)$$

where h is a measure along the line-of-sight of the observing instrument.

Writing

$$\mathscr{G} = 0.83\, Ab(X)\, \Delta E\, \frac{n_j A_{ji}}{N_e}\, F(T) \qquad (3)$$

gives

$$4\pi I_\lambda = \int \mathscr{G} N_e^2 dh. \qquad (4)$$

At this stage one can make an approximation to the shape of the $\mathscr{G}$ function (e.g., Pottasch, 1963; Jordan and Wilson, 1971) to give

$$4\pi I_\lambda = \mathscr{G}_0 \int_R N_e^2 dh = \mathscr{G}_0\, EM(h) \qquad (5)$$

where $EM(h)$ is the column emission measure, and $\mathscr{G} = \mathscr{G}_0$ over a specified temperature range, with R the part of the line-of-sight that contains plasma at these temperatures.

An alternative method is to write

$$4\pi I_\lambda = \int \mathscr{G}\, \phi(T)\, dT, \qquad (6)$$

where $\phi(T)$ is the *differential emission measure*, where for a set of observed intensities, the aim is to invert the integral equation to derive the form of $\phi(T)$.

In emission measure analysis it is important to study lines for which the $\mathscr{G}$ functions are independent of density, so that the density does not have to be specified in the problem. Lines from the same ion that do have different dependencies on N_e are useful as density diagnostics.

1.1. Determining Element Abundances

There are three ways of determining abundance ratios
1. If, for two ions belonging to different elements, the $F(T)$ functions are very similar then one can assume that they are identical, and so one can directly compare emission measures to yield abundance ratios. I.e., from Eqn. 5, assume the two $EM(h)$ values are identical, and then the abundance ratio is directly proportional to the intensity ratio.
2. When there are consecutive ion stages of two elements that overlap significantly in temperature, emission measures can be derived for each individual element, and curves compared. This is used in Sect. 4.

3. In terms of the differential emission measure method, the abundances can be treated as a free parameter and those that provide the smoothest $\phi(T)$ curve should be adopted. A variation on this method is used in Sect. 3

1.2. LEVEL BALANCE—THE CHIANTI DATABASE

The fundamental atomic processes that dominate the level balance of an ion (i.e., the determination of the quantity n_j in Eqn. 1) are spontaneous radiative decay and electron excitation. The former require an accurate model for the structure of the emitting ion and are comparatively easy to calculate with accuracies of $\lesssim 10\,\%$. Electron excitation rates require models of both the emitting ion and how this interacts with the free electron, with accuracies of typically $\lesssim 30\,\%$. There is a huge volume of both radiative decay and electron excitation data in the literature and so it is convenient to have a database with this data in it. One such database is CHIANTI and is described in detail by Dere *et al.* (1997). The most important features of this database are:
- the database is updated when new data becomes available;
- the electron excitation rates for each transition are individually assessed using the techniques of Burgess and Tully (1992);
- CHIANTI is freely available over the internet.

The easy accessibility of the CHIANTI data makes possible the detailed analysis of solar spectra, and an example is provided in Young *et al.* (1998), who compare the theoretical CHIANTI line emissivities with the line intensities in the SERTS-89 active region spectrum of Thomas and Neupert (1994). The excellent agreement found for many ions gives confidence in the quality of the atomic data contained in CHIANTI.

1.3. ION BALANCE

The most widely-used ion balance calculations are those of Arnaud and Raymond (1992) for the iron ions, and Arnaud and Rothenflug (1985) for all other ions. There is a need in the astrophysics community for updates, particularly to the Arnaud and Rothenflug (1985) data. *The present ion balance calculations are likely the most significant source of errors in solar spectrum analysis.* There is a need for a freely-available, regularly updated database similar to CHIANTI but containing ionisation and recombination rates.

2. The Coronal Diagnostic Spectrometer

The Coronal Diagnostic Spectrometer (CDS) on board SOHO is described thoroughly in Harrison *et al.* (1997) and the main features are noted here. There are two distinct spectrometers: the Grazing Incidence (GIS) and Normal Incidence

(NIS). Only data from the NIS are presented here. The NIS obtains both spectral information and one dimension of spatial information simultaneously, with images obtained by rastering in the remaining spatial dimension (usually solar-X). For any one pointing of the instrument a 4×4 arc min^2 region can be viewed with around 2–3 arc sec resolution. Two spectral wavebands are observed simultaneously, 307–379 Å and 513–633 Å, and are referred to as simply NIS1 and NIS2. Spectral resolutions are approximately 0.32 Å and 0.54 Å, respectively. Due to telemetry constraints, the entire NIS1 and NIS2 wavebands are rarely taken; instead, specific spectral lines are selected prior to observations.

3. The Iron-to-Silicon Ratio

Si XII and Fe XIV are predicted to have considerable overlap in temperature and so are suitable for studying the Fe/Si abundance ratio in coronal plasma. Si XII emits a strong line at 520.7 Å in the NIS2 band, while Fe XIV has two lines in NIS1 at 334.2 Å and 353.8 Å. The $\mathscr{G}$ functions for the two Fe XIV lines show significant density sensitivity, but a sum of the two lines is insensitive, and so the two lines will be added together here. The Si XII function is insensitive to density. In Fig. 1, the $\mathscr{G}$ functions for Si XII and Fe XIV are plotted, assuming photospheric silicon and iron abundances (Anders and Grevesse, 1989; Grevesse et al., 1992). For the left-hand plot, the Arnaud and Rothenflug (1985, hereafter referred to as AR85) ion balance calculations are used for both elements, whereas in the right-hand plot, Arnaud and Raymond (1992, hereafter referred to as AR92) are used for iron.

Although the $\mathscr{G}$ functions for Si XII and Fe XIV are seen to overlap very well at lower temperatures, the Si XII function is found to have a significant high temperature tail and it is for this reason that a Fe XVI $\mathscr{G}$ function is plotted. Fe XVI emits a line in the NIS1 waveband at 360.8 Å and essentially gives a measure of how much high temperature material there is in the line-of-sight.

3.1. COMPARING ION BALANCE CALCULATIONS

For the Fe XVI line, the $\mathscr{G}$ function is considerably larger when the AR92 calculations are used and it is possible to use the CDS observations to find which of the two ion balance calculations are most appropriate. This is done by noting that the Si XII function extends to both lower *and* higher temperatures than Fe XVI (Fig. 1) and so the Fe XVI/Si XII ratio has a theoretical maximum: with AR85, this is 2.25 and occurs for an isothermal plasma at a temperature of $\log T = 6.5$, whereas for AR92 it is 9.17 at a temperature of $\log T = 6.6$. Fig 2 plots values of this ratio obtained from several active region observations. All but one of the observed ratios lie above the AR85 theoretical limit, whereas the AR92 limit is only exceeded in three observations. This clearly implies that the AR92 calculations are more appropriate, but that theory is still inconsistent with observation.

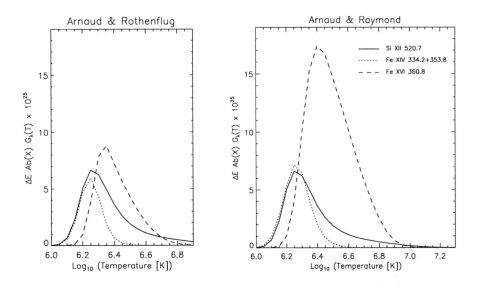

Figure 1. Comparison of the $\mathscr{G}$ functions for Si XII 520.7 Å, Fe XIV 334.2+353.8 Å and Fe XVI 360.8 Å. For Si XII, the Arnaud and Rothenflug (1985) ionisation balance has been used in each plot, but for the iron ions, Arnaud and Rothenflug (1985) has been used for the left-hand plot, while Arnaud and Raymond (1992) has been used for the right-hand plot.

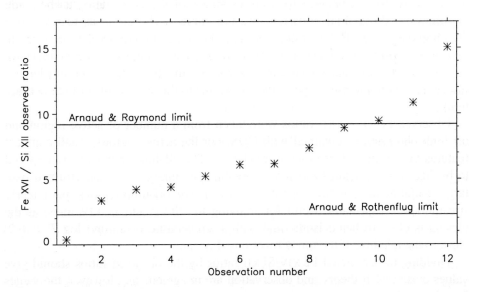

Figure 2. Fe XVI 360.8/Si XII 520.7 ratios derived from active region observations. The two limits shown are the theoretical maxima derived from the Arnaud and Raymond (1992) and Arnaud and Rothenflug (1985) ion balance calculations.

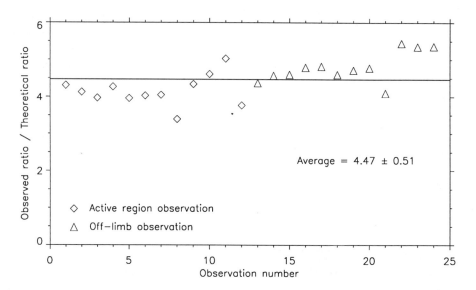

Figure 3. A plot of observed/theoretical Fe XIV/Si XII ratios derived from the active region and off-limb data-sets.

3.2. THE FE/SI RATIO

Using the AR92 ion balance for iron, estimates of the Fe/Si ratio can be made as follows: a differential emission measure curve is chosen so as to reproduce the observed Fe XVI/Fe XIV ratio, and then the same curve is used to *predict* the Fe XIV/Si XII ratio. There is considerable freedom in the choice of the d.e.m. curve but, if a physically reasonable curve is chosen, the predicted ratios are largely insensitive to the precise shape of the curve, due to the large widths of the $\mathcal{G}$ functions.

Observed Fe XIV/Si XII ratios were taken from a number of active region and off-limb observations obtained with CDS. For the active regions, small compact features visible in all three ions were chosen. The off-limb regions were required to be free from structures (such as loops) and summations were performed over large spatial areas to improve photon statistics. The off-limb regions, particularly streamers, are especially useful for studying the Si XII and Fe XIV lines, as the plasma is close to being isothermal with a temperature of around $\log T = 6.2$, close to the maxima of the ion $\mathcal{G}$ functions (Fig. 1).

Dividing the predicted Fe XIV/Si XII ratios by the observed ratios should give values close to 1 if theory and observation are in agreement. However, the values found here (Fig. 3) are substantially greater than 1, with an average of 4.5. In terms of abundances, this can be interpreted as iron being enhanced by a factor 4.5 over silicon. Such an enhancement is not expected, however, as both iron and silicon are low-FIP elements; could there be another explanation?

Fe XIV is a complex ion and Young *et al.* (1998) found several discrepancies between theory and the SERTS-89 observations. New electron excitation rates have recently been calculated for Fe XIV by Storey and Saraph (private communication), and including these rates in the Fe XIV level balance is found to increase the factor n_j in Eqn. 3 for the 334.2+353.8 blend by a factor of 1.9. This lowers the theory/observation discrepancy to 2.4.

Of additional interest is the finding of Landi *et al.* (1997, p. 564) that the current NIS1–NIS2 calibration may be in error by a factor of three, with the NIS2 lines found to be too weak compared to the NIS1 lines. This is what is found here with the Si XII and Fe XIV lines, and it would seem that this is giving rise to the remaining discrepancy.

3.3. CONCLUSIONS

The Fe XIV/Si XII ratio was found to be a factor of 4.5 stronger in the observations than predicted by theory. This would appear to be, not an abundance variation, but due to inaccurate atomic data for Fe XIV and an NIS1–NIS2 calibration problem. By assuming that the photospheric Fe/Si abundance ratio applies in the corona, the analysis suggests that NIS2 lines are currently too weak compared to the NIS1 lines by a factor of around 2.4.

The Arnaud and Raymond (1992) iron ion balance is found to be in better agreement with observations than the Arnaud and Rothenflug (1985) balance. Note that, with the revised NIS1–NIS2 calibration, the three discrepant points seen in Fig. 2 fall below the AR92 limit, giving increased confidence in the Arnaud and Raymond (1992) calculations.

4. Magnesium-to-Neon Ratio

A strength of the two NIS wavebands is the fact that consecutive ionisation stages of several elements are observed—see, e.g., Table I of Mason *et al.* (1997). In particular, lines from the ions Ne IV–VII and Mg V–VIII, which show considerable overlap in temperature, are seen. In terms of the first ionisation potential (FIP), neon has a FIP of 21.6 eV, whereas magnesium has a FIP of 7.6 eV, and so the elements are classed as high-FIP and low-FIP, respectively. It is to be noted that the magnesium ions all emit lines in the NIS1 waveband, whereas the neon ion lines are found in NIS2, and so the calibration problems noted in the previous section will affect the derived Mg/Ne abundance ratios—this is discussed in Sect. 4.5.

The Mg/Ne abundance ratio has been studied previously using a set of Mg VI and Ne VI lines near 400 Å (Sheeley, 1995, 1996). Although the NIS of CDS does not observe these lines, it does observe other Mg VI and Ne VI lines and abundance variations associated with an emerging flux region are reported in Young and Mason (1997). The results reported here are an extension of this work.

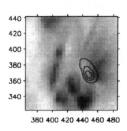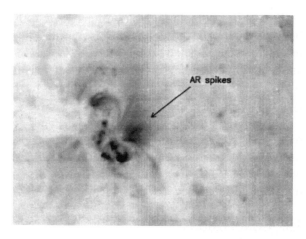

Figure 4. Images of AR 8038 from 13 May, 1997, showing active region spikes. The large image is from EIT, and was taken with the Fe IX/X filter at 19:00 UT. The smaller image is from CDS and was taken over the period 18:48–19:08 UT. CDS focussed on a small part of the active region (2 × 2 arc min^2). The displayed image is from the Ca X 557.8 Å line which is formed over a similar temperature range to Fe IX and features common to both the CDS and EIT images can be readily identified. Overplotted on the CDS image are contours showing the Mg VI 349.2 Å emission, which can be seen to come from the footpoints of the spikes. The EIT image is courtesy of the SOHO/EIT consortium.

4.1. METHOD OF ANALYSIS

The method (2) of those listed in Sect. 1.1 is used here. Line intensities for each of the ions are first measured from the spectra, and are converted into emission measures using the atomic data from the CHIANTI database and the Mg and Ne photospheric abundances of Anders and Grevesse (1989) and Grevesse *et al.* (1992), respectively. The emission measure points are plotted in Fig. 5. A $T^{3/2}$ dependence has been removed from the displayed curves to aid comparison—see, e.g., Sect. 5 of Jordan (1996). In addition to the neon and magnesium ions, the emission measure derived from the Ca X 557.8 Å line is displayed, and will be discussed later.

4.2. ACTIVE REGION SPIKES

The term 'spike' comes from the papers of Sheeley (1995, 1996) and refers to the thin, elongated features seen in Skylab S082A spectroheliograms in lines emitted by 200 000–1 000 000 K plasma. Such features would appear to be analogous to the *sunspot plumes* observed by the S055 instrument on Skylab (Foukal, 1975). Two key properties of plumes are their long lifetimes and their relationship to the surface magnetic field, which have not yet been investigated for the CDS observations and the so the term 'spike' is retained here. A crucial property of spikes is that they have Mg/Ne abundance enhancements of around a factor 10.

ATOMIC PHYSICS FOR ATMOSPHERIC COMPOSITION MEASUREMENTS 323

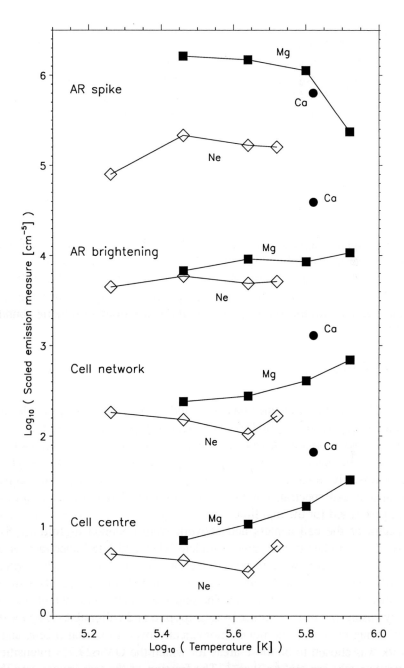

Figure 5. Emission measure curves for magnesium and neon derived from the ions Ne IV–VII and Mg V–VIII. Each set of curves has been scaled to aid display, and a $T^{3/2}$ dependency has been removed. For reference, the Ne IV log emission measures are 26.49, 27.54, 25.65, and 25.28 for the spike, brightening, network and cell centre. Also displayed are emission measures derived from Ca X.

Young and Mason (1997) presented an analysis of one spike, although the CDS study used did not observe all of the neon lines. For an observation of AR 8038 on 13 May 1997, however, the TREG22/v17 study was used which was specifically focussed towards the magnesium and neon ions. The particular CDS FITS file is s7901r00, and an image from this data is shown in Fig. 4, together with a larger image from EIT. The Mg VI emission can be seen to come from the footpoints of the spikes—it is for this region that the emission measure analysis was performed.

4.3. ACTIVE REGION BRIGHTENINGS

A particularly intense type of transition region brightening was identified by Young and Mason (1997), and was found to have a high density ($N_e \approx 10^{11}$ cm^{-3}) and show photospheric abundances. Another of such brightenings occurred in a CDS observation from 10 July, 1996, of a flaring active region and was captured in all of the neon and magnesium lines. The CDS study used was TREG22/v17 and the FITS file is s3485r00. The second set of curves in Fig. 5 shows the emission measures derived for this brightening, revealing close-to photospheric abundances.

4.4. SUPERGRANULE CELLS

The Mg/Ne relative abundance has been studied in a wide range of solar features with data from the Skylab S082A instrument (e.g., Feldman, 1992); however, the fact that images from this instrument overlapped meant that such studies were confined to discrete features for which spatial–spectral separation was straightforward. This, necessarily, ruled out the study of abundances in supergranule cells. With CDS, spatial and spectral dimensions are completely resolved and so this study can be performed for the first time.

Spectra of the cell centers and network were obtained as follows. Synoptic NISAT_S/v2 studies (which give complete NIS1 and NIS2 spectra over an area of 20×240 arc sec^2) were inspected and chosen with the following criteria: the raster had to be close to disk centre; there were no Si XII bright points in the field of view; there were no data drop-outs. The selected rasters were then cleaned of cosmic rays using the routine cds_clean_exp.pro and calibrated. The O V 629.7 Å line is strong enough to be fitted automatically over all spatial pixels, and so the network was chosen to be those areas for which the O V 629.7 Å intensities were greater than 600 erg cm^{-2} s^{-1} sr^{-1}. The fraction of the total raster area that was network varied from 40 % up to 65 %, depending on the particular data-set. In all, 27 rasters were chosen, and cell centre and network spectra were extracted from each and summed. The magnesium and neon lines were fitted and the derived emission measure curves are presented in Fig. 5.

4.5. Discussion

Fig. 5 shows the magnesium and neon emission measure curves in each of the four features, assuming photospheric abundances. Magnesium is clearly seen to be enhanced by a factor of around 10 in the active region spike, with the result clear in all of the magnesium and neon lines, except the Mg VIII line which falls off sharply on account of there being little plasma above $T = 10^{5.8}$ K in the observed region. For the active region brightening, a completely different situation is found, with the magnesium and neon curves showing excellent agreement, implying close-to photospheric abundances. Little difference is found between the supergranule cell centres and the network, with magnesium enhanced by a factor of around 2. Note that all of these abundance ratios are affected by the revision in the NIS1–NIS2 calibration factor suggested in the previous section. The new value would raise all of the neon curves by a factor of 2.4. In particular, this would imply that photospheric abundances apply at the cell centres and boundaries.

The Ca X 557.8 Å emission measure for the four features is also shown in Fig. 5 and it appears to suggest that calcium is enhanced by factors of up to 5 compared to magnesium. However, the Ca X ion is a member of the sodium isoelectronic sequence, for which the ionisation fractions are very weak compared to the neighbouring neon-like ions. Comparatively small changes in the ionisation or recombination rates could significantly affect the ion fraction of Ca X and so it is felt that the discrepancy seen here is due mainly to inaccurate atomic data.

5. Conclusions

Advances in the quality of both atomic physics and solar spectra allow abundance variations in the solar atmosphere to be studied in greater detail than ever before. The availability of a database such as CHIANTI makes the interpretation of the spectra much easier for solar physicists, while the ability of instruments like CDS to observe spectral lines in many different conditions means that the spectra can be used as checks on both the atomic data and the instrument calibration.

The silicon and iron data from CDS presented in Sect. 3 provide one such example where the Fe/Si ratio discrepancy is found to be due to both the atomic data and instrument calibration, rather than an abundance anomaly. Comparing CDS magnesium and neon spectra of different features, however, clearly reveals abundance anomalies.

Acknowledgements

The financial support of PPARC is acknowledged. SOHO is a project of international cooperation between ESA and NASA.

References

Anders, E., and Grevesse, N.: 1989, 'Abundances of the Elements: Meteoritic and Solar', *Geochim. Cosmochim.* **53**, 197–214.

Arnaud, M., and Raymond, J.C.: 1992, 'Iron Ionization and Recombination Rates and Ionization Equilibrium', *Astrophys. Jour.* **398**, 394–406.

Arnaud, M., and Rothenflug, R.: 1985, 'An Updated Evaluation of Recombination and Ionization Rates', *Astron. Astrophys. Supp. Ser.* **60**, 425–457.

Burgess, A., and Tully, J.A.: 1992, 'On the Analysis of Collision Strengths and Rate Coefficients', *Astron. Astrophys.* **254**, 436–453.

Dere, K.P., Landi, E., Mason, H.E., Monsignori-Fossi, B.M., and Young, P.R.: 1997, 'CHIANTI— an Atomic Database for Emission Lines. I. Wavelengths Greater than 50 Å', *Astron. Astrophys. Supp. Ser.* **125**, 149–173.

Feldman, U.: 1992, 'Elemental Abundances in the Upper Solar Atmosphere', *Phys. Scripta* **46**, 202–220.

Foukal, P.V.: 1975, 'The Pressure and Energy Balance of the Cool Corona over Sunspots', *Astrophys. Jour.* **210**, 575–581.

Grevesse, N., Noels, A., and Sauval, A.J.: 1992, 'Photospheric Abundances', in: Proc. of the First SOHO Workshop, ESA SP-348, 305–308.

Harrison, R.A., Sawyer, E.C., Carter, M. K., *et al*.: 1995, 'The Coronal Diagnostic Spectrometer for the Solar and Heliospheric Observatory', *Sol. Phys.* **162**, 233–290.

Jordan, C.: 1996, '*EUVE* Spectra of Coronae and Flares', in: 'Astrophysics in the Extreme Ultraviolet' (eds. S. Bowyer and R. F. Malina), 81–88, Kluwer, The Netherlands.

Jordan, C., and Wilson, R.: 1971, 'The Determination of Chromospheric–Coronal Structure from Solar XUV Observations', in: 'Physics of the Solar Corona' (ed. C. J. Macris), 219–236, Reidel, Dordrecht–Holland.

Mason, H.E., Young, P.R., Pike, C.D., *et al.*: 1997, 'Application of Spectroscopic Diagnostics to Early Observations with the SOHO Coronal Diagnostic Spectrometer', *Sol. Phys.* **170**, 143–161.

Landi, E., Landini, M., Pike, C.D., and Mason, H.E.: 1997, 'SOHO CDS–NIS In-flight Intensity Calibration with the Arcetri Diagnostic Method', *Sol. Phys.* **175**, 553–572.

Pottasch, S.R.: 1963, 'On the Interpretation of the Solar Ultraviolet Emission Line Spectrum', *Space Sci. Rev.* **3**, 816–855.

Sheeley, N.R.: 1995, 'A Volcanic Origin for High-FIP Material in the Solar Atmosphere', *Astrophys. Jour.* **440**, 884–887.

Sheeley, N.R.: 1995, 'Elemental Abundance Variations in the Solar Atmosphere', *Astrophys. Jour.* **469**, 423–428.

Young, P.R., and Mason, H.E.: 1997, 'The Mg/Ne Abundance Ratio in a Recently Emerged Flux Region Observed by CDS', *Sol. Phys.* **175**, 523–539.

Young, P.R., Landi, E., and Thomas, R.J.: 1998, 'CHIANTI: an Atomic Database for Emission Lines. II. Comparison with the SERTS-89 Active Region Spectrum', *Astron. Astrophys.* **329**, 291–314.

Address for correspondence: Department of Applied Mathematics and Theoretical Physics, Silver Street, Cambridge CB3 9EW, United Kingdom
P.R.Young@damtp.cam.ac.uk, H.E.Mason@damtp.cam.ac.uk

SOLAR ENERGETIC PARTICLES: SAMPLING CORONAL ABUNDANCES

DONALD V. REAMES

NASA/ Goddard Space Flight Center, Greenbelt, MD 20771, USA

Abstract. In the large solar energetic particle (SEP) events, coronal mass ejections (CMEs) drive shock waves out through the corona that accelerate elements of the ambient material to MeV energies in a fairly democratic, temperature-independent manner. These events provide the most complete source of information on element abundances in the corona. Relative abundances of 22 elements from H through Zn display the well-known dependence on the first ionization potential (FIP) that distinguishes coronal and photospheric material. For most elements, the main abundance variations depend upon the gyrofrequency, and hence on the charge-to-mass ratio, Q/A, of the ion. Abundance variations in the dominant species, H and He, are not Q/A dependent, presumably because of non-linear wave-particle interactions of H and He during acceleration. Impulsive flares provide a different sample of material that confirms the Ne:Mg:Si and He/C abundances in the corona.

1. Introduction

It is rather surprising that the most complete measurements of element abundances in the solar corona do not come from photons from the coronal plasma. They come from measurements of high-energy particles accelerated in the large solar energetic particle (SEP) events. The comparison of element abundances in SEP events with corresponding abundances in the photosphere led to the well-known dependence on the first ionization potential (FIP) of the elements (see *e.g.* Meyer 1985, 1993, 1996 and references therein). In fact, it is especially ironic that this "FIP effect" was sought for elements in the SEP events because a similar effect had been observed much earlier in the energetic particles of the galactic cosmic rays.

Initially it was thought that the energetic particles in the large SEP events were accelerated in solar flares. In recent years, however, it has become clear that in the large "gradual" events, particles are accelerated at shock waves that are driven out from the Sun by coronal mass ejections (CMEs) (Reames 1990, 1993, 1995b, 1997; Kahler 1992, 1994; Gosling 1993; Cliver 1996). We will see that these shocks accelerate the ions of the chemical elements in a fairly democratic manner. In contrast, the particles that are accelerated in impulsive solar flares have 1000-fold enhancements in the isotope ^{3}He relative to ^{4}He and enhancements of ~10 in heavy elements like Fe/O. These enhancements are believed to result from resonant wave-particle interactions during stochastic acceleration of the ions from the flare plasma.

In this paper we will consider three different populations of energetic particles: 1) gradual SEP events, 2) impulsive flare events, and 3) particles from co-rotating interaction regions (CIRs). Gradual SEP events provide the most extensive and complete abundance information. We will devote most of our attention to these events, to the physical processes involved, the abundances they provide, and the variations and uncertainties in interpreting them as "coronal." Impulsive events involve a different source and acceleration mechanism with well-defined coronal sites, and despite the enhancements, can give independent abundance information in specific cases, such as Ne/Mg/Si and He/C. In CIR events the acceleration occurs at heliospheric shocks from a high-speed solar-wind source. Knowing the abundances of both the solar wind and the energetic CIR ions allows us to investigate the shock acceleration mechanism and compare it to that involved in the gradual events. This paper intends to describe abundance observations and their variations and errors; it is not intended as an exhaustive review.

2. Gradual SEP Events

The earliest measurements of element abundances in SEP events were made on sounding rockets in the 1960s (Fichtel and Guss 1961; Bertsch, Fichtel, and Reames 1969). Over the next decades measurements gradually improved and it became clear that SEP abundances differed from those of the photosphere or meteorites in two primary ways (Meyer 1985; Brenneman and Stone 1985). First, there was a dependence on the charge-to-mass ratio, Q/A, of the ion that varied from event to event, and second, there was a persistent dependence of the abundances on FIP. The Q/A dependence was recognized to be a result of the acceleration of the ions that also varied with time and with the energy of observation (Mazur *et al.* 1992). The ions are highly ionized by a coronal electron temperature of ~2 MK (Luhn *et al.* 1985) at the time of acceleration. Thus, the FIP effect can only occur at a much earlier time and a much lower temperature when an ion-neutral separation is possible.

2.1 THE AVERAGED ABUNDANCES OF ELEMENTS

The effect of FIP on the element abundances of the SEPs is most clearly illustrated by simply accumulating the raw abundance measurements over many SEP events. Measurements of an ion entering a particle telescope consist of measuring the energy loss in a thin (dE) silicon detector together with the energy deposited in a second thicker (E) detector in which the particle stops. A typical plot of dE *vs.* E, with each particle plotted as a single point, is shown in Figure 1. Data for Figure 1 were accumulated over 49 gradual SEP events periods occurring during 14 years. Already in these "raw" data, it is apparent that the abundances of Ne, Mg, Si and Fe are all comparable, a characteristic feature of coronal abundances.

Element abundances have been derived from Figure 1 by using the instrument calibration to define the velocity or energy/nucleon along the track of each element and then simply counting the number of ions of each element within a fixed velocity interval (5 - 12 MeV/amu). The resultant SEP-coronal abundances are shown in Table 1. Combining these SEP abundances with the corresponding photospheric abundances of Grevesse, Noels, and Sauval (1996) also given in Table 1, we obtain the FIP plot shown in Figure 2.

The measured SEP abundances in this study (Reames 1995a) are in excellent agreement with those obtained 10 years earlier by Brenneman and Stone (1985) for a different sample of events observed by a different instrument on a different spacecraft. The SEP measurements are extremely stable and well determined within the stated errors. However, Brenneman and Stone used a wrong value for the photospheric abundance of Fe and were led to the erroneous conclusion that the SEP and coronal abundances were substantially different by ~50% in Fe/C (Garrard and Stone 1994).

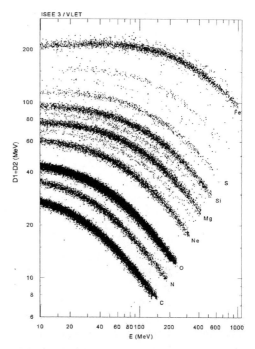

Fig. 1. Response of a detector to individual particles showing element resolution. Similar numbers of Ne, Mg, Si and Fe, typical of coronal abundances, are visible here in the raw data.

2.2 ACCELERATION

Shock acceleration is understood to occur when ions are scattered back and forth across the shock by scattering against magnetic turbulence (Alfvén waves) in the upstream and downstream regions. Because of the velocity discontinuity at the shock, the process can be viewed as scattering from "walls" that are approaching each other; ions receive an increment of velocity on each transit of the shock.

The interaction of a particle with electromagnetic fields is governed by the Lorentz equation, which may be written:

$$m_o \frac{d}{dt}(\gamma v) = \frac{Q}{A} e(\mathbf{E} + \mathbf{v} \times \mathbf{B}) \tag{1}$$

Table 1. Element abundances in gradual SEP events and the photosphere.

	Z	FIP (eV)	SEP Corona (O=1000)	(Gradual Events) (dex)	Photosphere (dex)
H	1	13.527	$(1.57\pm0.22)\times10^6$	12.07 ± 0.06	12.00 ± 0.0
He	2	24.46	57000 ± 3000	10.63 ± 0.023	10.99 ± 0.035
C	6	11.217	465 ± 9	8.54 ± 0.008	8.55 ± 0.05
N	7	14.48	124 ± 3	7.97 ± 0.010	7.97 ± 0.07
O	8	13.55	1000 ± 10	8.87 ± 0.004	8.87 ± 0.07
F	9	17.34	<0.1	<4.83	4.56 ± 0.3
Ne	10	21.47	152 ± 4	8.06 ± 0.011	8.08 ± 0.06
Na	11	5.12	10.4 ± 1.1	6.89 ± 0.046	6.33 ± 0.03
Mg	12	7.61	196 ± 4	8.17 ± 0.0089	7.58 ± 0.05
Al	13	5.96	15.7 ± 1.6	7.07 ± 0.044	6.47 ± 0.07
Si	14	8.12	152 ± 4	8.06 ± 0.011	7.55 ± 0.05
P	15	10.9	0.65 ± 0.17	5.69 ± 0.11	5.45 ± 0.04
S	16	10.3	31.8 ± 0.7	7.38 ± 0.0096	7.33 ± 0.11
Cl	17	12.952	0.24 ± 0.1	5.25 ± 0.18	5.5 ± 0.3
Ar	18	15.68	3.3 ± 0.2	6.39 ± 0.026	6.52 ± 0.10
K	19	4.318	0.55 ± 0.15	5.61 ± 0.12	5.12 ± 0.13
Ca	20	6.09	10.6 ± 0.4	6.90 ± 0.016	6.36 ± 0.02
Ti	22	6.81	0.34 ± 0.1	5.41 ± 0.13	5.02 ± 0.06
Cr	24	6.74	2.1 ± 0.3	6.20 ± 0.06	5.67 ± 0.03
Fe	26	7.83	134 ± 4	8.00 ± 0.013	7.50 ± 0.04
Ni	28	7.61	6.4 ± 0.6	6.68 ± 0.041	6.25 ± 0.01
Zn	30	9.36	0.11 ± 0.04	4.92 ± 0.16	4.60 ± 0.08

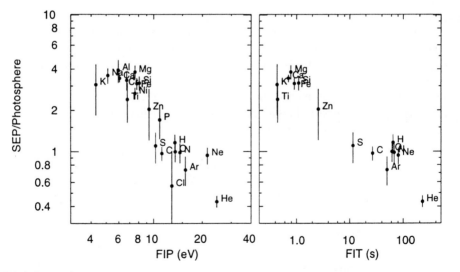

Fig. 2. Ratio of gradual SEP and photospheric abundances *vs.* first ionization potential (FIP) and *vs.* the first ionization time (FIT) of Marsch, von Steiger and Bochsler (1995).

where **E** and **B** are the electric and magnetic fields of the wave (or other accelerating fields), **v** is the ion velocity, Qe its charge, m_0A its mass, and γ is the Lorentz factor, $(1-v^2/c^2)^{-1/2}$. If the ions may be regarded as test particles that sample, but do not modify, the electromagnetic fields, then their behavior is completely described by two parameters, velocity and Q/A. According to equation (1), ions with the same value of Q/A are indistinguishable and will be accelerated identically. Since we always compare abundances in constant velocity intervals, Q/A is the only other variable that can distinguish different elements. Q/A determines the ion gyrofrequency and, hence, where it will interact with the wave-frequency spectrum.

However, before proceeding further, we should note that this simple picture is an approximation since the accelerated ions are *not* test particles. In fact, it is the H and He, streaming away from the shock that actually generate the waves that scatter subsequent particles (see *e.g.* Lee, 1983). Hence, it is unlikely that H and He abundances will depend solely on velocity and Q/A. Even the rarer elements can preferentially generate and absorb waves at their gyrofrequencies, modifying the electromagnetic environment slightly. However, we will see that variations in the abundances of the elements other than H and He are well described by Q/A.

Charge states of ions of the dominant elements from C through Fe were first measured in 12 large SEP events by Luhn *et al.* (1987). More recently, Boberg, Tylka and Adams (1996) summarized measurements of the ionization states of energetic Fe ions and compared them with measurements of Fe ions in the solar wind. They found that the ionization states of the accelerated Fe ions (Q_{Fe} ~ 15) are somewhat higher than those of the solar wind (Q_{Fe} ~ 11). The energetic Fe ion charges are similar to those of the solar wind plasma in the CME ejecta and in the sheath region that corresponds to coronal material swept up by the CME. This suggests that the "seed population" for the accelerated ions is coronal rather than solar wind material. Thus, coronal ions, initially accelerated to suprathermal speeds when the shock first encountered the base of each magnetic field line, continue to resupply the shock with "seed" particles as it moves outward along that flux tube.

2.3 ABUNDANCE VARIATIONS

We can use the measured ionization states of the elements to examine the Q/A-dependence of abundances in individual events. A plot of the relative enhancement of various elements *vs.* Q/A is shown for 3 events in Figure 3. The enhancements are relative to the average SEP corona given in the last section and Q/A is derived from measured values of Q of each element in each event (Luhn *et al.* 1985) at 0.3-2 MeV/amu. Two of the events shown in the figure, 1979 June 7 and 1978 September 23, represent the largest positive and negative slopes, respectively, of the 49 events studied (Reames 1995a). Note especially that C, N, O, Ne, and Mg usually have very similar values of Q/A, hence there is very little variation in the relative abundances of these elements. These elements are highly likely to exhibit the same abundances in SEP events that they had in the source plasma.

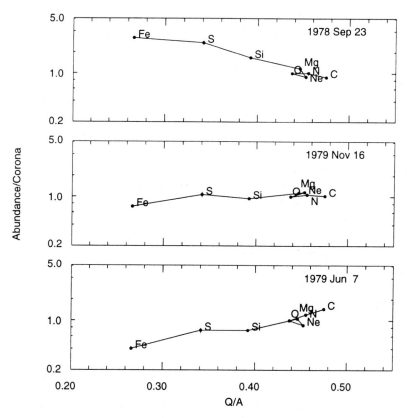

Fig. 3. *Q/A* dependence of the abundances in individual gradual SEP events relative to the corona. The upper and lower panels are the extreme values of 49 events that contributed to the 5-12 MeV/amu SEP average. The event in the middle panel is more typical. Note that the elements from C to Mg have similar values of *Q/A*.

It is a coincidence that the *Q/A* variations cancel when we sum over many events. That this is the case is shown by the close proximity of Fe to Mg and Si on the FIP plot in Figure 2; these three elements have similar values of FIP but much different values of *Q/A*. The average slope of the wave spectrum in the acceleration region affects the *Q/A* dependence. Ions of different *Q/A*, hence different gyrofrequencies, resonate with different parts of the wave frequency spectrum. The local slope of the wave frequency spectrum can thus determine the slope of the *Q/A* dependence. Of course, the ions can sample and average over a wide variety of conditions during the complete history of their acceleration.

There is some controversy as to whether a *Q/A* dependence like that shown in Figure 3 should be described as large or small. If we are allowed to fit smooth curves through the plots, changes of the charge state of Fe by a few units, for example, would change the observed abundance of Fe by ~20% in the worst case. For abundance measurements involving atomic spectral lines, changing the charge by a few units would make the line emission disappear completely. It is in this sense that we describe the SEP measurements as highly temperature insensitive and only weakly dependent on Q/A. However, Bochsler (1998, this volume) points out that the Q/A variations in the solar wind are much smaller than are those in SEP events. In fact, it is rather surprising that the more-modest acceleration of the solar wind produces a measurable Q/A dependence at all. A spectrum of Alfvén waves may be involved here as well.

When the enhancements are a smooth function of Q/A, and when each element's Q/A value varies little from event to event, another way to examine the dependence is by plotting ratios such as Si/O vs. Fe/O, for example. If Fe/O is enhanced or suppressed in a given event, Si/O is likely to follow proportionately. Such behavior is shown in the lower panels of Figure 4 where C/O and Si/O are highly correlated with Fe/O. In contrast, the upper 2 panels of the figure show little correlation between H/He or He/C and Fe/O. Clearly, the dominant elements, H and He, do not behave as test particles; they modify **E** and **B** in equation (1) in a nonlinear way and do not have the simple Q/A dependence of the heavier elements. The abundances of H and He depend upon the properties of the shock and on details of the particle and wave spectra.

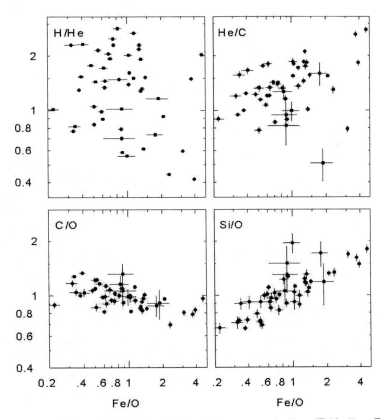

Fig. 4. Cross plots of H/He, He/C, C/O and Si/O relative to coronal values (Table 1) vs. Fe/O for 49 gradual SEP events at 1-4 MeV/amu for H/He and 5-12 MeV/amu otherwise (Reames 1995).

When we consider event-averaged data as shown in Figure 4, the scatter gives the impression of randomness in the variations that are independent of Q/A. In fact, early observers (Mason, Gloeckler, and Hovestadt 1984) thought that the time variations within an event were random, as the coronal shock crossed material with intrinsic abundance variations. However, more recent data, and especially new data from the Wind spacecraft, show that abundance variations with time are extremely systematic. Figure 5 shows data from the Wind spacecraft during the recent events associated with CMEs emitted on 1997 November 4 and 6 as marked by the vertical lines in the figure. The upper panel shows the intensity of 2.4–3.0 MeV/amu ^{4}He ions, while the lower panels show Fe/O and He/O in the same energy region. The event beginning on November 4 shows an Fe/O ratio that decreases smoothly with time while He/O remains constant. The classic Q/A varia-

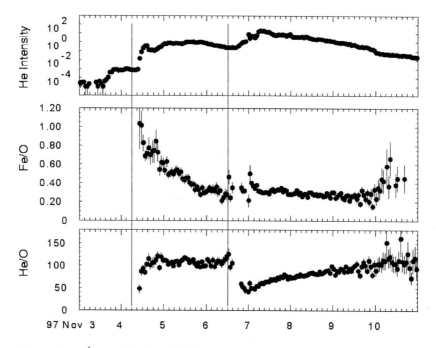

Fig. 5. Intensity of ^{4}He and Fe/O and He/O ratios during the 1997 November SEP events observed on the Wind spacecraft. During the event beginning on the 4th, Fe/O varies smoothly while He/O does not. For the event beginning on the 6th, He/O varies while Fe/O does not.

tions we have seen before occur smoothly here within a single event. The event that begins on November 6 has a constant value of Fe/O, but He/O increases smoothly by a factor of ~2 over a 3-day period. These new measurements make it clear that abundance variations in SEP events are systematic, depending upon the physics of the acceleration, and are *not* random as was once thought.

An example of even more spectacular systematic abundance variations during an event is provided by the 1998 April 20 event in Figure 6 (Tylka, Reames, and Ng 1998). Variations of this type can occur when wave generation at the shock

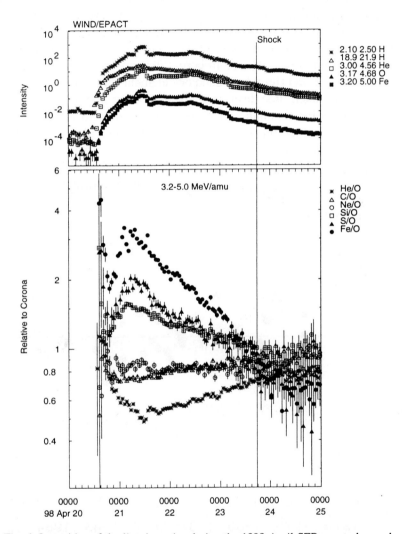

Fig. 6. Intensities of the listed species during the 1998 April SEP event observed on the Wind spacecraft are shown in the upper panel. The lower panel shows a spectacular systematic evolution in abundance ratios during the event.

leads to flattened spectra of the escaping ions at low energies. Ions with given velocity, v, but different Q/A resonate with different parts of the wave spectrum. The same variations in Q/A dependence that we saw in individual events (*e.g.* Figure 3) can now be seen to occur systematically with time during a single event as the point of magnetic connection moves along the face of the shock of varying strength and with differing wave intensities.

The energy dependence of abundances is a significant factor that we have not yet considered explicitly. Mazur *et al.* (1992) found that abundances were more likely to approach coronal values with smaller event-to-event fluctuations at the lowest energies, ~1 MeV/amu. At high energies, >10 MeV/amu, the effects of acceleration became more pronounced and event-to-event variations became much larger (see also Tylka, Dietrich, and Boberg 1997). With this in mind, we have focused on the lowest energies where adequate element resolution exists.

Having spent so much time discussing abundance variations, one might get the impression that the SEP abundances are not stable at all. However, for most elements with similar values of Q/A, SEP abundances are quite well determined. Event-averaged abundances of 5-12 MeV/amu Mg/Ne, normalized to the photospheric ratio, are shown in Figure 7 as a function of time spanning 7 years. The standard deviation of a single event from the mean is less than 20% for these events and the error in the mean itself is only ~3%. Mg/Ne is a principal determinant of the amplitude of the FIP effect.

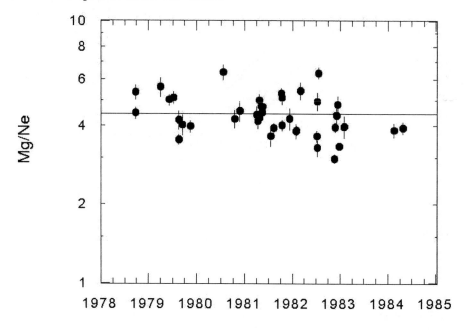

Fig. 7. Mg/Ne, relative to the corresponding photospheric ratio (Table 1), in gradual SEP events spanning 7 years.

3. Impulsive Solar-Flare Events

Energetic particles from impulsive flare events would not seem to be a good source of information on coronal abundances. Electromagnetic ion cyclotron

(EMIC) waves are produced between the gyrofrequencies of H and ^{4}He by streaming electrons in the flare plasma (Temerin and Roth 1992). The rare isotope ^{3}He, the only species whose gyro-frequency lies in this region, resonantly absorbs the waves, resulting in 1000-fold enhancements in ^{3}He/^{4}He ratios in the accelerated particles. Waves below the gyro-frequency of ^{4}He lead to enhancements in heavy elements (*e.g.* Miller and Reames 1996). Ne/O, Mg/O, and Si/O are enhanced by a factor of about 3 and Fe/O by a factor of about 7, relative to *coronal* abundances (Reames, Meyer, and von Rosenvinge, 1994; Reames 1995a).

Ionization state measurements of the energetic particles from impulsive events (Luhn *et al.* 1987) show the charge state of Fe to be 20.5±1.2 while Si and all elements below it are fully ionized. These charge states are appropriate to heating of the ions to a temperature ~10 MK in the flare. However, if the fully ionized ions have the same value of Q/A and resonate with the same waves, how can the enhancements in Ne/O, Mg/O, and Si/O occur? Reames, Meyer, and von Rosenvinge (1994) suggested that the heating and ionization occurs *after* acceleration. They interpreted the pattern of enhancements as evidence of acceleration of the ions from coronal plasma at a temperature 3-5 MK. At that temperature, the electron configurations of Ne, Mg, and Si are He-like and the three elements have similar values of Q/A, as shown in Figure 8, hence they are accelerated similarly. At that temperature, He and C are both fully ionized, so we expect He/C to have a value similar to that in the coronal source plasma. Thus, impulsive events confirm the SEP abundances for He/C, Mg/Ne and Si/Ne.

When averaged over many flares, the abundance enhancements show a pronounced dependence on Q/A when Q is taken as the value at ~3 MK as shown in Figure 9 (Reames 1995a). However, individual impulsive events do not show a systematic Q/A dependence as they do

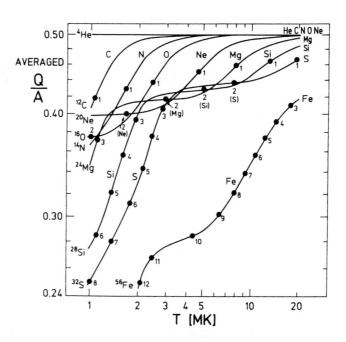

Fig. 8. Average Q/A as a function of plasma electron temperature based on Arnaud and Rothenflug (1985) and Arnaud and Raymond (1992) as plotted by Reames, Meyer, and von Rosenvinge (1994).

in gradual SEP events, although the statistics are not as good (Reames, Meyer, and von Rosenvinge 1994).

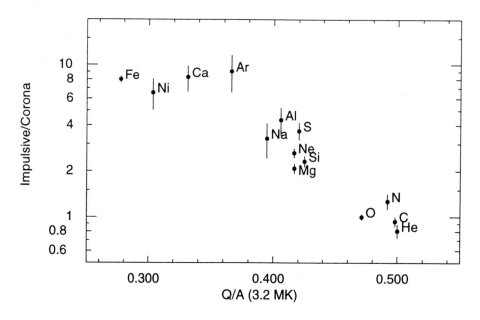

Fig. 9. The average abundance enhancement in impulsive-flare events vs. Q/A at 3.2 MK.

Finally, it is important to mention abundances obtained from γ-ray line measurements in solar flares. These measurements have now been made in several events and they also show the FIP-effect of coronal abundances (*e.g.* Ramaty, Mandzhavidze, and Kozlovsky 1996).

4. Co-rotating Interaction Regions

Co-rotating interaction regions (CIRs) are formed when high-speed solar-wind streams are emitted behind low-speed solar wind because of solar rotation. From the interaction a forward shock wave propagates outward into the low-speed wind and a reverse shock propagates sunward into the high-speed stream. Generally the shocks form outside 1 AU and strengthen out to several AU. While ions are accelerated at both shocks, those accelerated from the high-speed stream at the reverse shock predominate and are often observed streaming sunward at 1 AU.

The FIP dependence of these accelerated ions was first shown by Reames, Richardson, and Barbier (1991). Their FIP effect is similar to that seen in the fast solar wind itself (Gloeckler and Geiss 1989). Here the low-FIP elements are enhanced by a factor of only ~2 relative to the high-FIP elements, rather than the factor of ~4 we see in SEP events and in the low-speed wind. The presence of interstellar pickup ions in the solar wind also affects the abundances of H and He in the accelerated particles (Gloeckler *et al.*, 1993; Reames 1995).

Another interesting feature of energetic ions from CIRs is that a high value of C/O = 0.89±0.04 makes C move up to the level with the low-FIP ions. This is to be contrasted with a value of ~0.7 in the solar wind and 0.47±0.01 in the corona or photosphere. Richardson et al. (1993) found that the C/O ratio of the energetic ions was a function of the solar wind speed. This finding has been confirmed recently by Mason et al. (1997) using the high-resolution data from the Wind spacecraft. These results are shown in Figure 10.

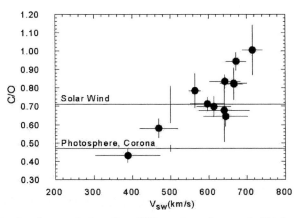

Fig. 10. C/O ratio of energetic ions from CIR events *vs.* the speed of high-speed stream.

5. Prospects

Energetic particles already provide information on the coronal abundances of a large number of elements. New observations on the Wind, SOHO, and ACE spacecraft can be expected to extend these measurements to heavier elements, to rarer elements and to isotopes.

Our greatest challenge now is to understand the physics of particle acceleration, both for its own sake and to improve our knowledge of the underlying source abundances. New measurements of abundance variations with improved statistics and higher time resolution may prove to be of great assistance here as well.

I would like to thank ISSI for their hospitality during this workshop and to thank C. K. Ng, A. J. Tylka, and T. T. von Rosenvinge for their comments on this manuscript. I also thank an unnamed referee for some helpful comments.

References

Arnaud, M., and Raymond, J.: 1992, *Astrophys. J.* **398**, 394.
Arnaud, M., and Rothenflug, R.: 1985, *Astron. and Astrophys. Suppl.* **60**, 425.
Bertsch, D.L., Fichtel, C.E., and Reames, D. V.:1969, *Astrophys. J. (Letters)*, **157**, L53.
Boberg, P. R., Tylka, A. J., and Adams, J. H.: 1996, *Astrophys. J. (Letters)* **471**, L65.
Breneman, H. H., and Stone, E. C.: 1985, *Astrophys. J. (Letters)*, **299**, L57.

Cliver, E. W.: 1996, in *High Energy Solar Physics*, edited by R. Ramaty, N, Mandzhavidze and X.-M. Hua, , AIP Conf. Proc **374**, (Woodbury, NY: AIP press), p. 45.
Fichtel, C. E., and Guss, D. E.: 1961, *Phys. Rev. Lett.* **6**, 495.
Garrard, T. L., and Stone, E. C.: 1994, *Adv. Space Res.* **14** (No. 10), 589.
Gloeckler, G., *et al.*: 1993, *Science* **261**, 70.
Gloeckler, G., and Geiss, J.: 1989, in *Cosmic Abundances of Matter*, edited by C. J. Waddington, A. I. P. Conf. Proc. **183**, (NY: AIP press) p.49.
Gosling, J. T.: 1993., *J. Geophys. Res.* **98**, 18949.
Grevesse, N., Noels, A., and Sauval, A. J.: 1996, in *Cosmic Abundances*, edited by S. Holt and G. Sonneborn, A. S. P. Conf. Series **99**, p. 117.
Kahler, S. W.: 1992, *Ann. Rev. Astron. Astrophys.* **30**, 113.
Kahler, S. W.: 1994, *Astrophys. J.* **428**, 837.
Lee, M. A.: 1983, *J. Geophys. Res.* **88**, 6109.
Luhn, A., *et al.*: 1985, *Proc. 19th Internat. Cosmic-Ray Conf.*, (La Jolla) **4**, 241.
Luhn, A., Klecker, B., Hovestadt, D., and Möbius, E.:1987, *Astrophys. J.* **317**, 951.
Marsch, E., von Steiger, R., and Bochsler, P.: 1995, *Astron. Astrophys.* **301**, 261.
Mason, G. M., Gloeckler, G., and Hovestadt, D. 1984, *Astrophys. J.* **280**, 902.
Mason, G. M., Mazur, J. E., Dwyer, J. R., Reames, D. V., and von Rosenvinge, T. T.: 1997, *Astrophys. J. Letters*, **486**, L149.
Mazur, J.E., Mason, G.M., Klecker, B., and McGuire, R.E.: 1992, *Astrophys. J.* **401**, 398.
Meyer, J. P.: 1985, *Astrophys. J. Suppl.* **57**, 151.
Meyer, J. P.: 1993, in *Origin and Evolution of the Elements*, edited by N Prantzos, E. Vangioni-Flam and M. Casse (Cambridge Univ. Press), p. 26.
Meyer, J. P.: 1996, in *Cosmic Abundances*, edited by S. Holt and G. Sonneborn, A. S. P. Conf. Series **99**, p. 127.
Miller, J. A., and Reames, D. V.: 1996, in *High Energy Solar Physics*, edited by R. Ramaty, N. Mandzhavidze, X.-M. Hua, AIP Conf. Proc. **374**, (Woodbury, NY: AIP press), p. 450.
Ramaty, R., Mandzhavidze, N. and Kozlovsky, B.: 1996, in *High Energy Solar Physics*, edited by R. Ramaty, N, Mandzhavidze and X.-M. Hua, , AIP Conf. Proc **374**, (Woodbury, NY: AIP press), p. 172.
Reames, D. V.: 1990, *Astrophys. J. Suppl.* **73**, 235.
Reames, D. V.: 1993, *Adv. Space Res.* **13** (No. 9), 331.
Reames, D. V.: 1995a, *Adv. Space Res.* **15** (No. 7), 41.
Reames, D. V.: 1995b, *Revs. Geophys. (Suppl.)* **33**, (U. S. National Report to the IUGG), 585.
Reames, D. V.: 1997, in: *Coronal Mass Ejections*, edited by N. Crooker, J. A. Jocelyn, J. Feynman, Geophys. Monograph **99**, (AGU press) p. 217.
Reames, D. V., Meyer, J. P., and von Rosenvinge, T. T.: 1994, *Astrophys. J. Suppl.*, **90**, 649.
Reames, D.V., Richardson, I.G., and Barbier, L.M.: 1991, *Astrophys. J. (Letters)* **382**, L43.
Richardson, I.G., Barbier, L.M., Reames, D.V., and von Rosenvinge, T.T.: 1993, *J. Geophys. Res.*, **98**, 13.
Temerin, M. and Roth, I.: 1992, *Astrophys. J. (Letters)* **391**, L105.
Tylka, A. J., Dietrich, W. F., and Boberg, P. R.: 1997, *Proc. 25th Internat. Cosmic-Ray Conf.*, (Durban) **1**, 101.
Tylka, A. J., Reames, D. V., and Ng, C. K.: 1998, in preparation.

UVCS/SOHO: THE FIRST TWO YEARS

S. R. CRANMER and J. L. KOHL
Harvard-Smithsonian Center for Astrophysics, 60 Garden St., Cambridge, MA, USA

G. NOCI
Università di Firenze, I-50125 Firenze, Italy

Abstract. The SOHO Ultraviolet Coronagraph Spectrometer (UVCS/SOHO) has observed the extended solar corona between 1 and 10 $R_\odot$ for more than two years. We review spectroscopic and polarimetric measurements made in coronal holes, equatorial streamers, and coronal mass ejections, as well as selected non-solar targets. UVCS/SOHO has provided a great amount of empirical information about the physical processes that heat and accelerate the solar wind, and about detailed coronal structure and evolution.

Key words: Corona, Solar Wind, UV Radiation, Spectroscopy, Line Profiles

1. Introduction

The Ultraviolet Coronagraph Spectrometer (UVCS) on the Solar and Heliospheric Observatory (SOHO) has made continuous spectroscopic observations of the extended solar corona since April 1996. The purpose of the UVCS/SOHO mission is to provide spectroscopic and polarimetric diagnostic measurements aimed at determining empirical values for coronal densities, outflow velocities, and other velocity distribution parameters (such as temperature) for electrons, hydrogen, and several minor ions (see Kohl *et al.*, 1995, 1997a, 1997b; Noci *et al.*, 1997a).

In its first two years of operation, UVCS/SOHO has observed the fast solar wind thought to emerge from coronal holes, the slow solar wind associated with bright equatorial streamers, and a number of dramatic transient events such as coronal mass ejections (CMEs). Measured profiles of at least 40 spectral lines (for H I and ions of C, N, O, Mg, Al, Si, S, Fe, and Ni) allow a detailed analysis of coronal abundances, ionization, and energy balance. Daily synoptic observations of the corona, combined with other special observations, have allowed the global three-dimensional structure of the solar-minimum corona to be derived tomographically. This paper reviews these many of observations, especially measurements of H I and O VI emission lines, and discusses the information learned about the physical processes controlling large-scale and small-scale structures in the extended corona. Where possible, we also discuss the implications of the UVCS/SOHO observations on theories which explain the heating, acceleration, and composition of the solar wind near the Sun.

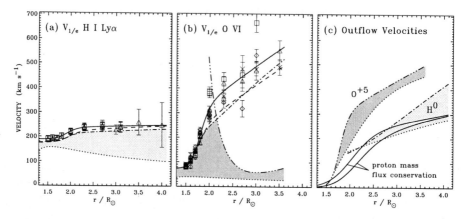

Figure 1. (a) Line widths and empirical most-probable speeds (parallel and perpendicular to the radial field) for H^0 atoms derived from H I Lyα profiles. Squares (triangles) denote $1/e$ half-widths over the north (south) polar holes, and the solid line is a fit to both. Also plotted: the most-probable speed corresponding to thermal distributions with temperature T_e (dotted line), and the perpendicular speeds speeds for two consistent models (dashed, dot-dashed lines). The shaded region denotes the possible values of the parallel most-probable speed. (b) Same as (a), but for O VI 1032, 1037 Å (squares [triangles] denote 1032 in the north [south]; diamonds [crosses] denote 1037 in the north [south]). The dash-triple-dotted line denotes an empirical upper limit on the parallel most-probable speed of O^{5+} ions. (c) Empirical outflow velocity over the poles, corresponding to the two consistent models in (a) and (b) for H^0 and O^{5+}. Solid lines denote proton mass flux conservation assuming superradial coronal-hole expansion (upper line) and radial expansion (lower line). Shaded regions correspond to those in (a) and (b). For more information, see Kohl *et al.* (1998a).

2. Coronal Holes and Fast Solar Wind

The most surprising initial results from UVCS/SOHO were the extremely wide line profiles of H I Lyα and O VI 1032, 1037 Å over the heliographic poles (see also the results of UVCS/Spartan; Kohl *et al.*, 1996). Resonantly scattered emission probes the velocity distributions along directions perpendicular (via the line widths) and parallel (via Doppler dimming) to the open magnetic field lines (e.g., Withbroe *et al.*, 1982; Noci *et al.*, 1987; Strachan *et al.*, 1993; Cranmer 1998).

H^0 atoms are closely coupled to the protons by charge transfer in the inner corona below about 3 $R_\odot$ (Olsen *et al.*, 1994; Allen *et al.*, 1998). H I Lyα observations have constrained the hydrogen velocity distributions to be mildly anisotropic (with $T_\perp > T_\parallel$) above heights of 2–3 $R_\odot$ (see Figure 1a). The O^{5+} ions, however, are strongly anisotropic, with perpendicular kinetic temperatures approaching 2×10^8 K at $3 R_\odot$ (Figure 1b; Kohl *et al.*, 1997a), and $T_\perp/T_\parallel \approx 10\text{--}100$. There is evidence that the outflow velocity for O^{5+} is larger than the outflow velocity for H^0 by as much as a factor of two (Figure 1c; see also Kohl *et al.*, 1998a; Li *et al.*, 1998a).

The UVCS measurements and the self-consistent empirical models for the plasma parameters imply that the extended corona is in a highly nonequilibrium thermodynamic state, and already exhibits similar anisotropic velocity distributions to those observed by *in situ* spacecraft in the solar wind (e.g., Feldman and Marsch, 1997). The kinetic temperatures (which by necessity contain both microscopic thermal motions and unresolved transverse wave motions) imply a strong preferential heating of heavy ions (Kohl *et al.*, 1998a; Cranmer *et al.*, 1999). Because the most-probable speed for O^{5+} is much larger than that of hydrogen above $\sim$ $2.2R_\odot$, the perpendicular temperatures must be *greater* than the approximate mass-proportionality often seen in the solar wind (see also Collier *et al.*, 1996, for *in situ* measurements beginning to indicate this preferential heating in the fast wind). UVCS has also measured broad profiles of Mg X 625 Å in coronal holes, but these lines are not as wide as O VI at comparable heights. Mg^{9+} ions may be more strongly dominated by proton collisions than O^{5+}, which may inhibit the ion heating and decrease the temperature of Mg^{9+} ions at low heights (Esser *et al.*, 1998), or the heating mechanisms may have a more complex mass or charge dependence (see also Kohl *et al.*, 1998b).

Coronal holes often exhibit fine structure in the form of ray-like polar plumes. UVCS has found narrower H I and O VI line widths in bright concentrations of plumes (Noci *et al.*, 1997a; Corti *et al.*, 1997; Giordano *et al.*, 1997a) and, when O VI line profiles are fitted with multiple Gaussians, more intense narrow components in plumes than out of plumes (Kohl *et al.*, 1997b, 1998b). This may be in agreement with theoretical models of plume energization (Wang, 1994, 1998). Temporal inhomogeneities in coronal holes are reported by Ofman *et al.* (1997), who used electron-scattered polarization measurements with the UVCS White Light Channel to infer the presence of compressive waves with periods of 6–50 minutes at $2R_\odot$.

3. Streamers and Slow Solar Wind

Raymond *et al.* (1998a), in this volume, review UVCS results for elemental abundances in bright coronal streamers, so we only provide a brief overview. At solar minimum, streamers occupy a stable and geometrically simple band around the solar equator, and are thought to be the source of the slow solar wind (e.g., Habbal *et al.*, 1997). The earliest UVCS/SOHO data showed a striking dichotomy between the "core" and bordering "legs" at low heights in the corona, presumably below any closed-field cusps (Noci *et al.*, 1997a). While H I Lyα and white light show a maximum intensity in the core, the O VI 1032, 1037 Å, Mg X 625 Å, Si XII 499 Å, and Fe XII 1242 Å lines show a strong core depletion and *edge brightening*. Noci *et al.* (1997b) investigated several mechanisms that could explain these observations, and concluded that there must be a strong gradient in the minor ion abundances between the streamer core and legs. This abundance variation may be explained by either the gravitational settling of heavy ions in the hydrostatic environment of closed loops (Raymond *et al.*, 1997), or by relative differences in the collision strength and outflow in flows between multiple current sheets (Noci *et al.*, 1997b).

Other UVCS observations of streamers have provided independent information about electron temperatures in the corona. Fineschi *et al.* (1997) made the first measurement of the electron-scattered component of H I Lyα, and found temperatures consistent with the inferred ionization balance (Raymond *et al.*, 1997) and intensities consistent with the observed white-light polarization brightness. Maccari *et al.* (1997) measured the electron temperature by comparing the intensities of H I Lyα and Lyβ, which have different radiative and collisional strengths. UVCS measurements have also helped constrain the plasma β (ratio of gas pressure to magnetic pressure) in streamers to be as high as 3–5 at low heights (Li *et al.*, 1998b), and found extremely high-temperature ions (e.g., Fe XIX, Ca XV, Ni XIV) over post-flare active-region loops (Raymond *et al.*, 1998c).

4. Coronal Mass Ejections

UVCS has observed several dozen coronal mass ejection (CME) transients in the extended corona. Spectroscopic measurements allow the *three-dimensional* velocity structure of the moving plasma to be probed: Doppler shifts give the line-of-sight velocity, Doppler dimming and time evolution gives the radial velocity, and motion along the slit gives the transverse (sun-tangential) velocity.

The CME on 6–7 June 1996 exhibited evidence for an untwisting magnetic field around an erupted flux tube (Antonucci *et al.*, 1997a). An event on 23 December 1996 showed emission lines from species with a large range of electron temperatures (from C III and N II to O VI), an increase in the H I Lyα emission by two orders of magnitude, and a flat emission measure distribution 3 or 4 orders of magnitude lower than typical emission measures for prominences (Ciaravella *et al.*, 1997a).

In Figure 2 we present UVCS data from an eruptive prominence CME on 11 December 1997, which indicates helical structure and an untwisting rate for the flux ropes which increases with radius (Ciaravella *et al.*, 1998). In general, UVCS has observed strong Doppler shifts of up to 400 km/s in magnitude in complex, small-scale "knots" of emission in CMEs. The relative intensity variations in different lines have been used to characterize CME origins (e.g., whether they are correlated with prominence eruptions or flares; Ciaravella *et al.*, 1997b).

5. Global Coronal Morphology

More than half of the observation time of UVCS/SOHO has been devoted to daily synoptic scans of the entire corona (see Strachan *et al.*, 1997; Romoli *et al.*, 1997). These measurements have been extremely valuable in efforts to model the

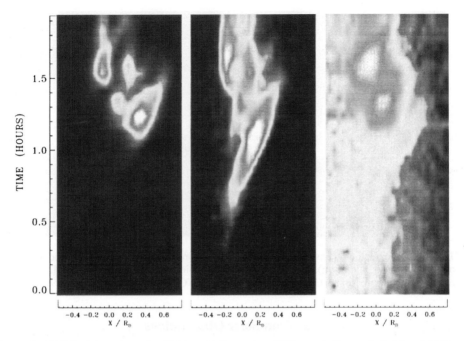

Figure 2. UVCS observation of a CME on 11 December 1997, at a heliocentric height of $1.7 R_\odot$. From left to right, panels denote H I Lyα intensity, O VI intensity, and H I Lyα Doppler shift (where -250 km s^{-1} is white). The x-coordinate is transverse distance along the spatial slit, and the y-coordinate is elapsed time from 23:25:31 UT.

three-dimensional structure and temporal evolution of the large-scale corona. The 27-day solar rotation provides a slow sampling of many lines of sight through the optically thin plasma, and it is possible to use these large data sets to *tomographically* reconstruct local line emissivities (e.g., Davila, 1994). Panasyuk (1998) has produced a global reconstruction of H I Lyα and O VI 1032, 1037 Å emissivities between 1.5 and $3 R_\odot$ during the SOHO Whole Sun Month campaign (10 August to 8 September 1996). Figure 3 shows two views of the reconstructed emission for O VI.

The UVCS synoptic program has often been extended to produce more detailed maps of coronal intensities and profile widths (Antonucci *et al.*, 1997b; Giordano *et al.*, 1997b). Extended scans in the boundaries between coronal holes and streamers have been studied extensively in the context of models of superradial (Dobrzycka *et al.*, 1998; Cranmer *et al.*, 1999) and radial (Habbal *et al.*, 1997) high-speed wind outflow in the corona.

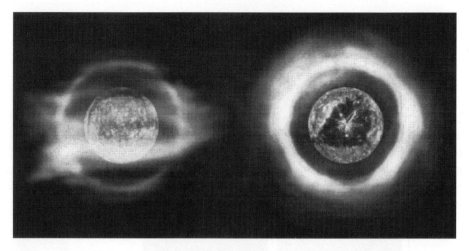

Figure 3. Tomographic reconstruction of the full solar corona as seen from the ecliptic plane (left) and the solar north pole (right). The inner solar-disk emission is Fe XII 195 Å from EIT/SOHO, and above $1.5 R_\odot$ the emission is O VI 1032, 1037 Å from UVCS/SOHO, during August 1996.

6. Non-Solar Observations

UVCS has observed many UV-bright stars which pass within $10 R_\odot$ of the sun. In addition to providing observations in a wavelength range extending below that covered by IUE or HST, stellar observations are useful for spectro-radiometric calibrations of the instrument. Stars observed in the first two years of operation include 38 Aqr, TT Ari, 53 Tau, 103 Tau, 121 Tau, τ Tau, ζ Tau, α Leo, ρ Leo, HD 142883, ω Sco, δ Sco, β Sco, HD 164794, HD 164492, and HD 164816. UVCS has also observed several sun-grazing comets (Raymond *et al.*, 1998b), the planets Venus and Jupiter, and the interplanetary emission of H^0 (backscattered from the local interstellar gas) and He^0 (enhanced in a heliospheric "gravitational focusing cone").

7. Summary

The first two years of observation with the UVCS instrument on SOHO have produced several significant advances in understanding the plasma conditions in the extended solar corona. Coordinated observations with other instruments on SOHO (e.g., CDS, CELIAS, LASCO, MDI, SUMER, and SWAN; see Fleck, Domingo, and Poland, 1995) and with other ground and space based instruments have also provided valuable insight. The empirical spectroscopic and polarimetric UVCS measurements act as important constraints of theories of coronal heating and solar wind acceleration. Even if no further observations are made after the mid-1998 recovery of SOHO, the massive data set gathered thus far is expected to yield additional discoveries as it is processed into a final archive and more in-depth analysis continues.

Acknowledgements

The authors wish to acknowledge the contributions of the many scientists and engineers who contributed to the work reported herein. A more complete list is presented in the author list and acknowledgements of Kohl *et al.* (1995). This work is supported by the National Aeronautics and Space Administration under grant NAG5-3192 to the Smithsonian Astrophysical Observatory, by Agenzia Spaziale Italiana, and by ESA's PRODEX program (Swiss contribution).

References

Allen, L. A., Habbal, S. R., and Hu, Y. Q.: 1998, *J. Geophys. Res.* **103**, 6551.
Antonucci, E., *et al.:* 1997a, *Ap. J. Letters* **490**, L183.
Antonucci, E., Giordano, S., Benna, C., Kohl, J. L., Noci, G., Michels, J., and Fineschi, S.: 1997b, in *The Corona and Solar Wind Near Minimum Activity, Fifth SOHO Workshop*, ed. A. Wilson (Noordwijk, The Netherlands: ESA), ESA SP-404, 175.
Ciaravella, A., *et al.:* 1997a, *Ap. J. Letters* **491**, L59.
Ciaravella, A., *et al.:* 1997b, in *Correlated Phenomena at the Sun, in the Heliosphere and in Geospace, 31st ESLAB Symposium*, ed. A. Wilson (Noordwijk, The Netherlands: ESA), ESA SP-415, 543.
Ciaravella, A., *et al.:* 1998, in preparation.
Collier, M. R., Hamilton, D. C., Gloeckler, G., Bochsler, P., and Sheldon, R. B.: 1996, *Geophys. Res. Letters* **23**, 1191.
Corti, G., Poletto, G., Romoli, M., Michels, J., Kohl, J., and Noci, G.: 1997, in *The Corona and Solar Wind Near Minimum Activity, Fifth SOHO Workshop*, ed. A. Wilson (Noordwijk, The Netherlands: ESA), ESA SP-404, 289.
Cranmer, S. R.: 1998, *Ap. J.* **508**, in press (1 Dec 1998).
Cranmer, S. R., *et al.:* 1999, *Ap. J.* **511,** in press (20 Jan 1999).
Davila, J. M.: 1994, *Ap. J.* **423**, 871.
Dobrzycka, D., Cranmer, S. R., Panasyuk, A. V., Strachan, L., and Kohl, J. L.: 1998, *J. Geophys. Res.*, submitted.
Esser, R., Fineschi, S., Dobrzycka, D., Habbal, S. R., Edgar, R. J., Raymond, J. C., Kohl, J. L., and Guhathakurta, M.: 1998, *Ap. J.*, submitted.
Feldman, W. C., and Marsch, E.: 1997, in *Cosmic Winds and the Heliosphere,* ed. J. R. Jokipii, C. P. Sonett, and M. S. Giampapa (Tucson: University of Arizona Press), 617.
Fineschi, S., Kohl, J. L., Strachan, L., Gardner, L. D., Romoli, M., Huber, M. C. E., and Noci, G.: 1997, presented at *The Corona and Solar Wind Near Minimum Activity, Fifth SOHO Workshop*, ed. A. Wilson (Noordwijk, The Netherlands: ESA), ESA SP-404.
Fleck, B., Domingo, V., and Poland, A., eds.: 1995, *The SOHO Mission, Solar Phys.* **162**, 1–531.
Giordano, S., Antonucci, E., Benna, C., Romoli, M., Noci, G., Kohl, J. L., Fineschi, S., Michels, J., and Naletto, G.: 1997a, in *The Corona and Solar Wind Near Minimum Activity, Fifth SOHO Workshop*, ed. A. Wilson (Noordwijk, The Netherlands: ESA), ESA SP-404, 413.
Giordano, S., Antonucci, E., Benna, C., Kohl, J. L., Noci, G., Michels, J., and Fineschi, S.: 1997b, in *Correlated Phenomena at the Sun, in the Heliosphere and in Geospace, 31st ESLAB Symposium*, ed. A. Wilson (Noordwijk, The Netherlands: ESA), ESA SP-415, 327.
Habbal, S. R., Woo, R., Fineschi, S., O'Neal, R., Kohl, J., and Korendyke, C.: 1997, *Ap. J. Letters* **489**, L103.

Kohl, J. L., et al.: 1995, Solar Phys. **162**, 313.
Kohl, J. L., Strachan, L., and Gardner, L. D.: 1996, Ap. J. Letters **465**, L141.
Kohl, J. L., et al.: 1997a, Solar Phys. **175**, 613.
Kohl, J. L., et al.: 1997b, Adv. Space Res. **20**, no. 1, 3.
Kohl, J. L., et al.: 1998a, Ap. J. Letters **501**, L127.
Kohl, J. L., et al.: 1998b, Ap. J. Letters, submitted.
Li, J., Raymond, J. C., Acton, L. W., Kohl, J. L., Romoli, M., Noci, G., and Naletto, G.: 1998b, Ap. J., submitted.
Li, X., Habbal, S. R., Kohl, J. L., and Noci, G.: 1998a, Ap. J. Letters **501**, L133.
Maccari, L., et al.: 1997, presented at *Correlated Phenomena at the Sun, in the Heliosphere and in Geospace, 31st ESLAB Symposium,* ed. A. Wilson (Noordwijk, The Netherlands: ESA), ESA SP-415.
Noci, G., Kohl, J. L., and Withbroe, G. L.: 1987, Ap. J. **315**, 706.
Noci, G., et al.: 1997a, Adv. Space Res. **20**, no. 12, 2219.
Noci, G., et al.: 1997b, in *The Corona and Solar Wind Near Minimum Activity, Fifth SOHO Workshop,* ed. A. Wilson (Noordwijk, The Netherlands: ESA), ESA SP-404, 75.
Ofman, L., Romoli, M., Poletto, G., Noci, G., and Kohl, J. L.: 1997, Ap. J. Letters **491**, L111.
Olsen, E. L., Leer, E., and Holzer, T. E.: 1994, Ap. J. **420**, 913.
Panasyuk, A. V.: 1998, J. Geophys. Res., in press.
Raymond, J. C., et al.: 1997, Solar Phys. **175**, 645.
Raymond, J. C., Suleiman, R., Kohl, J. L., and Noci, G.: 1998a, Space Sci. Rev., this volume.
Raymond, J. C., et al.: 1998b, Ap. J., in press.
Raymond, J. C., et al.: 1998c, in preparation.
Romoli, M., et al.: 1997, in *The Corona and Solar Wind Near Minimum Activity, Fifth SOHO Workshop,* ed. A. Wilson (Noordwijk, The Netherlands: ESA), ESA SP-404, 633.
Strachan, L., Kohl, J. L., Weiser, H., Withbroe, G. L., and Munro, R. H.: 1993, Ap. J. **412**, 410.
Strachan, L., et al.: 1997, in *The Corona and Solar Wind Near Minimum Activity, Fifth SOHO Workshop,* ed. A. Wilson (Noordwijk, The Netherlands: ESA), ESA SP-404, 691.
Wang, Y.-M.: 1994, Ap. J. Letters **435**, L153.
Wang, Y.-M.: 1998, Ap. J. Letters **501**, L145.
Withbroe, G. L., Kohl, J. L., Weiser, H., and Munro, R. H.: 1982, Space Sci. Rev. **33**, 17.

Address for correspondence: 60 Garden St., Mail Stop 50, Cambridge, MA 02138, USA

THE EXPANSION OF CORONAL PLUMES IN THE FAST SOLAR WIND

L. DEL ZANNA and R. VON STEIGER
International Space Science Institute, Hallerstrasse 6, CH-3012 Bern, Switzerland

M. VELLI
Dipartimento di Astronomia e Scienza dello Spazio, Università di Firenze,
Largo E. Fermi 5, I-50125 Firenze, Italy

Abstract. Coronal plumes are believed to be essentially magnetic features: they are rooted in magnetic flux concentrations at the photosphere and are observed to extend nearly radially above coronal holes out to at least 15 solar radii, probably tracing the open field lines. The formation of plumes itself seems to be due to the presence of reconnecting magnetic field lines and this is probably the cause of the observed extremely low values of the Ne/Mg abundance ratio.

In the inner corona, where the magnetic force is dominant, steady MHD models of coronal plumes deal essentially with quasi-potential magnetic fields but further out, where the gas pressure starts to be important, total pressure balance across the boundary of these dense structures must be considered.

In this paper, the expansion of plumes into the fast polar wind is studied by using a thin flux tube model with two *interacting* components, plume and interplume. Preliminary results are compared with both *remote sensing* and solar wind *in situ* observations and the possible connection between coronal plumes with pressure-balance structures (PBS) and microstreams is discussed.

Key words: MHD, Sun: Corona, Sun: Coronal Plumes, Solar Wind, FIP-effect

1. Introduction

Coronal plumes are dense structures located inside coronal holes and are mainly observed in white light or in UV and EUV lines (see Del Zanna et al., 1997, for a review). From these observations plumes appear to be from 2 to 5 times denser than the background coronal hole and to be roughly in hydrostatic equilibrium (at low altitudes). Plumes have an intrinsic magnetic nature, since they are observed to be rooted in strong magnetic field concentrations and, further out, they appear to trace the open field lines above coronal holes out to at least 15 solar radii (DeForest et al., 1997).

Among the open questions concerning the physics of plumes, there are two, strongly linked together: what exactly lies at their base and how are plumes formed? Before the Skylab mission plumes were believed to be rooted in unipolar flux concentrations, located at the vertices between supergranular cells. After the discovery of the presence of compact EUV enhancements at the base of the most bright plumes the attention shifted towards regions containing magnetic bipoles. These observations have suggested a possible explanation for plume formation (Wang and Sheeley, 1995; Velli et al., 1994): one or more bipoles are pushed by photospheric motions towards an open flux region located at a supergranular junction; eventually reconnection occurs, field lines open up and the required energy for plume formation is released. This picture seems to be confirmed also by recent observations that unipolar flux concentrations at the base of plumes have actually a complex geometry (DeForest *et al.*, 1997) and, when the magnetograms resolution is enhanced to $\sim 5 - 10\,\mathrm{G}$ (along the line of sight), that the flux concentration region indeed contains minority-polarity fields (Wang *et al.*, 1997).

A hint to answer these questions might be provided by optical *remote sensing* observations of polar plumes, that seem to indicate a very strong fractionation from the photosphere to the corona of the heavy ions abundances, with an enrichment of the low-FIP (First Ionization Potential) elements, e.g. Mg, with respect to the high-FIP elements, e.g. Ne. The ratios of the relative abundances normalized to the photospheric values, i.e. $[\mathrm{Mg/Ne}]_{\mathrm{cor}}/[\mathrm{Mg/Ne}]_{\mathrm{phot}}$, found in *some* coronal plumes can reach values of the order of 10 (Widing and Feldman, 1992; Sheeley, 1996). The current view on the elements fractionation is summarized, e.g., in Meyer (1993), von Steiger (1996) or von Steiger *et al.* (1997). Most fractionation models operate by atom-ion separation in the chromosphere, where the solar surface material on its way to the corona is partially ionized (e.g. von Steiger and Geiss, 1989). Another possibility is that this process occurs at the time of production of coronal material (Sheeley, 1996), which may be caused by chromospheric evaporation due to heating from the top: the induced electric current would produce a thermoelectric field with components perpendicular to the magnetic flux tube, dragging low-FIP elements into the tube where the material is evaporating upwards (Antiochos, 1994). In the case of polar plumes, where closed magnetic fields are thought to interact with unipolar open field lines, the source of heating could be easily provided by reconnection, which might also play a direct role in the ion/atom separation (Arge and Mullan, 1997).

On the other hand, the strong *FIP-effect* observed in coronal plumes poses a problem: since the fractionation measured *in situ* in the fast solar wind is definitely less than 2 (e.g. von Steiger *et al.*, 1997), what happens to plumes at large distances? Various possibilities have been suggested. Plumes could contain a static plasma that does not feed the solar wind, although this is difficult to explain in an open field geometry. Another possibility might be that the flux tubes containing the plumes disrupt somewhere between the Sun and the Earth, diluting their content in the fast wind. In any case, if the filling factor of plumes were small also in the solar wind, then it would be hard to measure plasma with strong fractionation and the only conclusion that we could draw would be that plumes are *not* the major source for the solar wind mass flux.

The aim of the present paper is precisely to study how dense structures (plumes) propagate in the fast solar wind. Both the plume and the interplume regions will be treated as thin magnetized flux-tubes (1-D) and the interaction between these two regions is provided by total lateral pressure balance (gas plus magnetic pressure). This picture is valid everywhere *but* near the coronal base, where a strong superradial expansion is observed (Ahmad and Withbroe, 1977; DeForest *et al.*, 1997). This is easily modeled by making use of a 2-D, quasi-potential field configuration in a low-beta plasma (Del Zanna et al., 1997; see also Del Zanna et al., 1998). In any case, after a few tens of thousand kilometers of height this expansion terminates and a 1-D approach becomes a very good approximation.

2. The pressure-balance model

The first model of this kind for the expansion of magnetically confined dense structures in the outer corona was an analytical study by Parker (1964), who considered a thin isothermal *streamer* expanding in an empty medium. He concluded that these dense structures should eventually fill the entire coronal hole, no matter what the initial filling factor, but (Casalbuoni et al., 1998) the expansion was predicted to occur too close to the Sun, essentially because the hydrostatic approximation used for the density profile broke down closer to the Sun than assessed. Also since an empty background was considered, the expansion of the denser streamer material was overestimated. These drawbacks were removed by Velli et al. (1994), who correctly solved the coupled equations, assuming no *a priori* given density profile, and allowing for the presence of plasma also in the interplume (or *interstream*) region. These models will be further improved here by allowing for a non-spherical overall coronal hole expansion, non-isothermal temperature profiles and mechanical momentum input by WKB Alfvén waves. The assumptions of this model are:

- The presence of two distinct regions nested one inside the other: interplume ($i = 1$) and plume ($i = 2$). The details of the spatial distribution of plumes in the coronal hole do not enter the model. Only the *filling factors* a_i are taken into account, which are normalized to their base values α_i, so that

$$\alpha_1 a_1 + \alpha_2 a_2 = 1. \qquad (1)$$

- The thin flux-tube approximation is employed: all quantities depend upon the heliocentric distance r alone and curvature terms are neglected (thus the plume's initial super-radial expansion modeled in the 2-D approach cannot be taken into account).
- Magnetic and mass fluxes are conserved separately in the two regions; the two regions interact through the continuity of total pressure across the plume–interplume boundary:

$$p_1 + \frac{B_1^2}{8\pi} = p_2 + \frac{B_2^2}{8\pi}. \qquad (2)$$

- The energy equation is not considered and the temperature is given *a priori*: this avoids the problem of having to choose *ad hoc* heating functions in order to yield reasonable temperature profiles.
- The overall coronal hole expansion is also given, in the form $r^2 f(r)$.
- Mechanical momentum input by WKB Alfvén waves is considered in the form of an additional wave pressure, both in the momentum equation and in Eq. (2). This may be written in terms of the Alfvénic Mach number $M_A =$

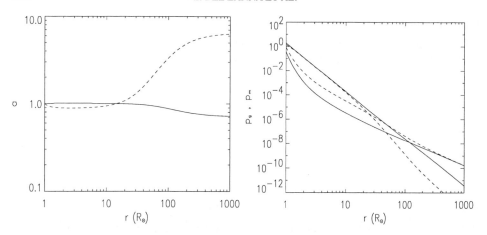

Figure 1. Left panel: non-radial expansion $a(r)$ (which is the normalized filling factor) for the inter-plume (solid line) and the plume (dashed line) regions. Here the initial filling factor of plumes is $\alpha_2 = 5\%$. The other values are $n_{10} = 10^8$ cm^{-3}, $n_{20} = 3 \times 10^8$ cm^{-3}, $T_1 = 10^6$ K, $T_2 = 1.2 \times 10^6$ K (hotter and denser plume). Right panel: the normalized magnetic and gas pressures for the same set of parameters (magnetic pressure curves are those starting with values larger than 1 at the base).

$\sqrt{4\pi\rho}u/B \sim \rho^{-1/2}$ (because from the conservation of magnetic and mass fluxes we have $\rho u/B = $ const):

$$p_w = p_{w0} \frac{M_{A0}}{M_A} \left(\frac{1 + M_{A0}}{1 + M_A} \right)^2. \tag{3}$$

The numerical solving procedure is the same employed in Velli et al. (1994): critical points in the momentum equation are found by means of a shooting method combined with an iterative technique (plume and interplume regions are coupled), while for larger distances the system of ODEs is integrated in the standard way.

To get a better physical understanding of the model consider first constant temperature profiles, a spherical overall expansion ($f = 1$) and no momentum addition by Alfvén waves ($p_w = 0$). The resulting filling factors for the two regions (a denser and hotter plume is considered) and the corresponding gas and magnetic pressure profiles are shown in Fig. 1 in logarithmic scale.

The plume expands radially within $\sim 10 R_\odot$, then a_2 starts to increase and, at the Earth orbit, the filling factor of plumes reaches a value as large as $\alpha_2 a_2 \sim 25\%$. Also note that the plasma beta reaches unity at $r \sim 20 R_\odot$ in plumes and at $r \sim 120 R_\odot$ outside. The effects of the coupling between the two regions clearly increase together with the plasma beta, confirming what was found in the low-β model. The initial plume contraction and further expansion are easily explained: near the base, where pressure balance is imposed, densities drop faster than B^2 (the decay is exponential) and the magnetic pressure thus dominates, so that there is a compression by the stronger external field. When the beta increases the situation is reversed: the higher gas pressure inside the plume pushes outwards the magnetic field lines and the plume fans out until this expansion (and the corresponding compression of the interplume region) smoothes the gas pressure differences in the two regions, as can be seen in the plot on the right hand side. Another interesting aspect is that the plume contains a faster plasma, and this is true *even* when its temperature is the same as outside. This result is clearly due to pressure-balance, which makes the ratio $n_2/n_1 \to 1$ at large distances, where magnetic pressure is negligible. A mathematical proof of this effect may be found in Casalbuoni *et al.* (1998).

In order to build a realistic fast solar wind model the isothermal assumption must be relaxed. While it used to be generally accepted that in the inner corona $T_e \approx T_p \approx 10^6$ K (our temperature T is defined as the average of the two temperatures), recent observations of Doppler broadenings of UV lines seem to indicate much higher ($\sim 10^7$ K) proton and heavier ions kinetic temperatures (e.g. Kohl *et al.*, 1996; Wilhelm *et al.*, 1998).

Another fundamental new result is that the acceleration to the final high speed flow velocities happens within the first 10-20 solar radii (Grall *et al.*, 1996) and that speeds as high as 150–200 km/s may be reached even at $2 - 3R_\odot$ (Strachan *et al.*, 1993; Corti *et al.*, 1997; Kohl *et al.*, 1997). Theoretical fast solar wind models which consider high proton temperatures have already been produced (McKenzie *et al.*, 1995; McKenzie *et al.*, 1997; Esser *et al.*, 1997; Hu *et al.*, 1997) and seen to succesfully yield a realistic high-speed wind already in the inner corona, whereas models where the wind is driven by Alfvén waves seem to fail to give the required strong acceleration near the coronal base (Wang, 1994; Habbal *et al.*, 1995).

The same results can be found with our model. Alfvén waves accelerate the wind only at large distances and the $\delta u \equiv \sqrt{<\delta u_w>^2}$ at 1AU required to give high speed flows is definitely too high (> 100 km/s). On the other hand, the presence of hot protons in the inner corona leads to very realistic outflow velocity profiles.

Since the energy equation is not considered in our model, the temperature profile must be provided *a priori*. We managed to come up with an analytical function which is able to simulate a *reasonable* temperature profile, with a peak in the inner corona caused by some form of heating (acting mainly on protons) and a given power law decay, and which yields realistic outflow velocities (see Fig. 2). Furthermore, an appropriate choice of the base number densities allows one to match also white light observational data (Fisher and Guhathakurta, 1995) in the inner corona (Fig. 3) and the high-speed solar wind properties at the Earth (Tab. 1).

All the values reported are well within the range of the corresponding observed values (the filling factors αa are basically unknown). The velocities in the range $r = 1-1.2R_\odot$ are very small (less than 5 km/s), in agreement with the observations. The average wave velocity fluctuation is taken to be larger in the plume by a factor 2 (yielding a factor 8 in the wave pressure) in order to simulate the possible effect of the reconnection dynamics at the base of the plume. This also enhances the velocity differences at 1 AU, yielding plumes that are faster than interplume regions. If on the other hand the same δu is assumed and this is larger than a critical value, the interplume region turns out to be faster than the plume at large distances (provided that the temperature profiles are also the same).

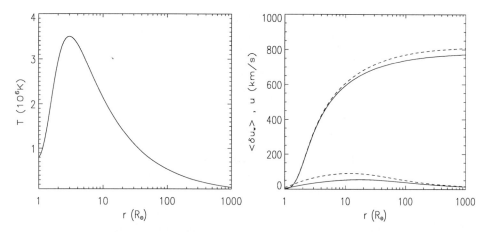

Figure 2. On the left, the temperature profile employed in our model. This is given by the following analytic function: $T = T_0[e^{c(1-1/x)}/x]^\delta$, where $x - 1 \sim (r - 1)^{n/\delta}$, T_0 is the base value, c is related to T_m, the maximum value at $r = r_m$, $-n$ is the power law index at great distances and δ is a parameter controlling the width of the maximum. Here T is the average between the proton and electron temperatures, and it is assumed to be the same in the two regions. The parameters used are $T_0 = 0.8$, $T_m = 3.5$ (both in units of 10^6 K), $r_m = 3R_\odot$, $\delta = 0.3$ and $n = 0.6$. The resulting value at the Earth is $T \approx 0.3$. On the right, the flow speed u and the average Alfvénic speed fluctuation δu. Solid (dashed) line is for the interplume (plume) region.

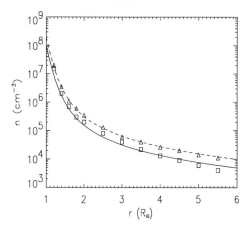

Figure 3. The resulting electron number densities in the inner corona compared to the white light data by the Spartan spacecraft (Fisher and Guhathakurta, 1995). The solid (dashed) line is the model prediction for the interplume (plume) region, whereas squares (triangles) are observational data for the corresponding region.

Note that the expansion of the dense structures is nearly radial when their temperature is the same as in the interplume region. This result seems to be consistent with recent observations by Woo and Habbal (1997), who also claim that the overall coronal hole expands basically radially. Anyway, similar results are found in the case of non-spherical coronal hole expansion ($f \neq 1$) *provided* the base mag-

Table I
Values of the various quantities, both in the interplume and plume regions, at the coronal base and at 1 AU.

	Interplume		Plume	
	$1\,R_\odot$	1 AU	$1\,R_\odot$	1 AU
αa	0.95	0.92	0.05	0.08
B (G)	1.39	$3.1\,10^{-5}$	1.28	$1.8\,10^{-5}$
n (cm^{-3})	$1\,10^8$	2.65	$2\,10^8$	3.09
u (km/s)	0.90	755	0.87	789
δu (km/s)	5	28	10	34

netic fields and densities are larger, thus giving values at 1 AU that are still in the observed range.

For a detailed study of the influence of the various parameters, especially those defining the temperature profiles, the reader is referred to Casalbuoni et al. (1998).

3. Conclusions

The inhomogeneities predicted by our model, remnants of coronal plumes signature, could have an observational counterpart in the so-called *pressure-balance structures* (PBS) (Thieme et al., 1990; McComas et al., 1996; Poletto et al., 1996) and/or in *microstreams* (Neugebauer et al., 1995). The former are defined by an anticorrelation between the gas and magnetic pressure, while the other plasma properties have average fast streams values; this is clearly in agreement with our model. On the other hand, microstreams, which can have both higher or lower velocity ($\sim$ 40 km/s), may be produced by different Alfvén waves fluxes. Larger velocity differences can be found *only* if different temperature profiles are assumed for the two regions.

However, the large variety of body and surface MHD waves that the dense filaments might carry can easily mask the intrinsic plasma properties and pressure balance, so that we do not believe that the absence in typical microstreams of such a balance is particularly significant to invalidate a possible interpretation in terms of our model.

Finally, our model predicts small filling factors for plumes at large distances, when the temperature in plumes is equal or less than the temperature in the background medium. As it was discussed in the introduction, this could explain why regions with strong FIP-effect are not observed in the fast solar wind. In order to find out whether PBSs are really the interplanetary signature of coronal plumes it would be interesting to measure their ion abundances.

References

Ahmad, I.A. and Withbroe, G.L.: 1977, *Sol. Phys.* **53**, 397.
Antiochos, S.K.: 1994, *Adv. Space Res.* **14**, (4)139.
Arge, C.N. and Mullan, D.J.: 1997, *BAAS, SPD meeting* **29**, 2.40.
Casalbuoni, S., Del Zanna, L., Habbal, S.R., and Velli, M.: 1998, *ApJ*, in preparation.
Corti, G., Poletto, G., Kohl, J.L., and Noci, G.: 1997, in *Proc. Fifth SoHO Workshop, ESA* **SP-404**.
DeForest, C.E., *et al.*: 1997, *Sol. Phys.* **175**, 393.
Del Zanna, L., Hood, A.W., and Longbottom, A.W.: 1997, *A&A* **318**, 963.
Del Zanna, L., Hood, A.W., Velli, M., and von Steiger, R.: 1998, in *Proc. Solar Jets and Coronal Plumes*, Guadeloupe, *ESA* **SP-421**, in press.
Esser, R., Habbal, S.R., Coles, W.A., and Hollweg, J.V.: 1997, *JGR* **102**, 7063.
Fisher, R. and Guhathakurta, M.: 1995, *ApJ* **447**, L139.
Grall, R.R., *et al.*: 1996, *Nature* **379**, 429.
Habbal, S.R., Esser, R., Guhathakurta, M., and Fisher R.: 1995, *GRL* **22**, 1465.
Hu, Y.Q., Esser, R., and Habbal, S.R.: 1997, *JGR* **102**, 14,661.
Kohl, J.L., Strachan, L., and Gardner, L.D.: 1996, *ApJ* **465**, L141.
Kohl, J.L., *et al.*: 1997, *Sol. Phys.* **175**, 613.
McComas, D.J., *et al.*: 1996, *A&A* **316**, 368.
McKenzie, J.F., Banaszkiewicz, M., and Axford, W.I.: 1995, *A&A* **303**, L45.
McKenzie, J.F., Axford, W.I., and Banaszkiewicz, M.: 1997, *GRL* **24**, 2877.
Meyer, J.-P.: 1993, in *Origin and Evolution of the Elements*, Cambridge Univ. Press, p. 26.
Neugebauer, M., *et al.*: 1995, *JGR* **100**, 23,389.
Parker, E.: 1964, *ApJ* **139**, 690.
Poletto, G., *et al.*: 1996, *A&A* **316**, 374.
Sheeley, N.R., Jr.: 1996, *ApJ* **469**, 423.
von Steiger, R. and Geiss, J.: 1989, *A&A* **225**, 222.
von Steiger, R.: 1996, in *Solar Wind Eight*, AIP 382, Woodbury, NY, p. 193.
von Steiger, R., Geiss, J., and Gloeckler, G.: 1997, in *Cosmic Winds and the Heliosphere*, J.R. Jokipii, C.P. Sonett, A.S. Giampapa (eds.), Tucson: University of Arizona Press, p. 581.
Strachan, L., Kohl, J.L., Weiser, H., and Withbroe, G.L.: 1993, *ApJ* **412**, 410.
Thieme, K.M., Marsch, E., and Schwenn, R.: 1990, *Ann. Geophys.* **8**, 713.
Velli, M., Habbal, S.R., and Esser, R.: 1994, *Space Sci. Rev.* **70**, 391.
Wang, Y.-M.: 1994, *ApJ* **435**, L153.
Wang, Y.-M. and Sheeley, N.R.: 1995, *ApJ* **452**, 457.
Wang, Y.-M., *et al.*: 1997, *ApJ* **484**, L75.
Widing, K.G. and Feldman, U.: 1992, *ApJ* **392**, 715.
Wilhelm, K., *et al.*: 1998, *ApJ*, in press.
Woo, R. and Habbal, S.R.: 1997, *GRL* **24**, 1159.

FRACTIONATION OF SI, NE, AND MG ISOTOPES IN THE SOLAR WIND AS MEASURED BY SOHO/CELIAS/MTOF

R. KALLENBACH[1], F.M. IPAVICH[2], H. KUCHAREK[3], P. BOCHSLER[4],
A.B. GALVIN[2], J. GEISS[1], F. GLIEM[5], G. GLOECKLER[2], H. GRÜNWALDT[6],
S. HEFTI[4], M. HILCHENBACH[6] and D. HOVESTADT[3]

[1] *International Space Science Institute, Hallerstrasse 6, CH-3012 Bern, Switzerland*

[2] *Space Physics Group, University of Maryland, College Park, MD 20742, USA*

[3] *Max-Planck-Institut für Extraterrestrische Physik, D-85740 Garching, Germany*

[4] *Physikalisches Institut, University of Bern, Sidlerstrasse 5, CH-3012 Bern, Switzerland*

[5] *Technische Universität, D-38106 Braunschweig, Germany*

[6] *Max-Planck-Institut für Aeronomie, D-37189 Katlenburg-Lindau, Germany*

Abstract. Using the high-resolution mass spectrometer CELIAS/MTOF on board SOHO we have measured the solar wind isotope abundance ratios of Si, Ne, and Mg and their variations in different solar wind regimes with bulk velocities ranging from 330 km/s to 650 km/s. Data indicate a small systematic depletion of the heavier isotopes in the slow solar wind on the order of $(1.4 \pm 1.3)\%$ per amu (2σ-error) compared to their abundances in the fast solar wind from coronal holes. These variations in the solar wind isotopic composition represent a pure mass-dependent effect because the different isotopes of an element pass the inner corona with the same charge state distribution. The influence of particle mass on the acceleration of minor solar wind ions is discussed in the context of theoretical models and recent optical observations with other SOHO instruments.

Key words: Solar Isotopic Composition, Solar Wind, SOHO

Abbreviations: SOHO – Solar and Heliospheric Observatory; CELIAS – Charge, Element, and Isotope Analysis System; MTOF – Mass Time-of-Flight Spectrometer

1. Introduction

The study of the solar isotopic composition is motivated by the interest in the composition of the early solar nebula which provides knowledge about the history of the solar system. The elements Si, Ne, and Mg are authentic witnesses for matter in the solar nebula because temperatures in the Sun never have been high enough to change their isotopic composition by nuclear burning. They represent two classes of elements: The solar abundances of the refractory elements Si and Mg are close to their terrestrial elemental abundances, whereas the solar abundance of the volatile element Ne is orders of magnitude higher than its terrestrial elemental abundance. Recent studies have shown that the solar wind isotopic compositions of Mg (Bochsler *et al.*, 1997; Kucharek *et al.*, 1998) and Si (Wimmer *et al.*, 1998) are very similar to their terrestrial isotopic compositions. However, the solar wind ^{20}Ne/^{22}Ne ratio (Geiss *et al.*, 1972; Kallenbach *et al.*, 1997a) is significantly higher than the terrestrial ratio (13.8 instead of 9.78). Interpreting these findings, the question arises how well the solar wind isotopic composition represents the solar source composition. In this paper we report on first experimental evidence that there is a small variation by $(1.4\pm1.3)\%$ per amu (2σ-error) of the isotope abundance ratios of Si, Ne, and Mg in different solar wind regimes. These results can be interpreted in the context of experimental (von Steiger *et al.*, 1995) and theoretical (Bodmer and Bochsler, 1996) studies on the Coulomb friction in the solar wind plasma, experimental results on the variation of the ^{4}He/^{1}H elemental abundance ratio (Borrini *et al.*, 1981) in the slow and fast solar wind, and recent measurements of ion velocity distributions in the inner corona by optical instruments on the SOHO spacecraft (Kohl *et al.*, 1997; Raymond *et al.*, 1997). Once the isotope abundance ratios can be corrected systematically for the variable fractionation processes in the solar wind, it finally may be possible to determine the solar source isotopic compositions with a precision of better than 0.5%, which would be relevant for cosmochemical considerations.

2. Instrumentation

The MTOF sensor (Figure 1) of the CELIAS experiment on board the SOHO spacecraft is an isochronous TOF mass spectrometer (Hovestadt *et al.*, 1995) with a resolution $M/\Delta M$ of about 100 which provides the possibility of resolving the different isotopes of almost all solar wind elements in the range from 3 to 60 amu. The instrument detects ions at solar wind bulk velocities of 300 to 1000 km/s corresponding to energies of about 0.3 to 3 keV/amu. The TOF of a single ion in a harmonic potential is measured between a pulse in a microchannel plate detector (Start MCP) triggered by secondary electrons released at its passage through a thin (2.1 μg/cm^2) carbon foil as start signal and the stop pulse of a position sensing ion microchannel plate detector (Ion MCP). This TOF is independent of the ion energy and simply proportional to the square root of mass per charge $(M/Q^*)^{1/2}$. The charge state distribution Q^* of the ions leaving the foil (Gonin *et al.*, 1995) is mainly a function of ion velocity and must be well understood for quantitative measurements of isotopic and elemental abundances. Typically, particles leave the foil as neutrals or singly charged ions independent of their incoming solar wind charge state. Details on the instrument calibration and the format of the data transferred by the telemetry can be found elsewhere (Kallenbach *et al.*, 1997a; Kallenbach *et al.*, 1998).

Isotope abundance measurements can be correlated with data from the proton monitor (PM) which is a subsystem of the CELIAS/MTOF sensor on board the SOHO spacecraft (Hovestadt *et al.*, 1995). The PM provides bulk velocity, kinetic temperature, and proton density of the solar wind each 5 min (Figure 2).

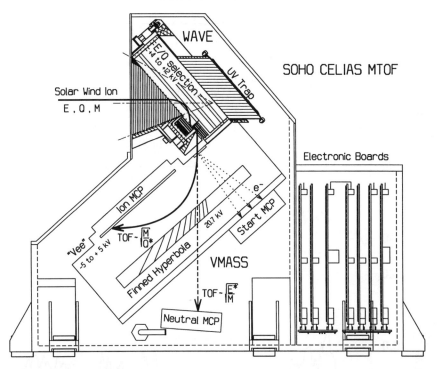

Figure 1. Schematic view of the SOHO/CELIAS/MTOF sensor. Almost the full velocity distribution of a given species of highly charged solar wind ions can be transmitted through the wide angle variable energy (WAVE) entrance system which has a conic field of view of $\pm 25°$ width (Hovestadt et al., 1995). After leaving the WAVE the solar wind particles can be accelerated or decelerated to optimize their detection efficiency in the isochronous VMASS time-of-flight (TOF) spectrometer where a V-shaped and a hyperbola deflection electrode generate an electrostatic harmonic potential.

3. Observations

TOF spectra of the time period from day 21 in 1996 to day 114 in 1997 have been accumulated. The data reduction for Ne and Mg solar wind isotopes as observed with the CELIAS/MTOF sensor is described in detail elsewhere (Kallenbach et al., 1997a; Kucharek et al., 1998). The Si solar wind isotopes have first been detected by the WIND/MASS sensor (Wimmer et al., 1998). In this work results from the SOHO/CELIAS/MTOF sensor are presented where it has been possible for the first time to analyze the variation of the isotope abundances of Si in different solar wind regimes. The mass spectra have been grouped into six classes of solar wind bulk velocities in the range from about 330 km/s to 650 km/s where the solar wind kinetic parameters have been derived from the PM data (Figure 2). Within these velocity classes the data have been filtered with respect to instrumental isotope fractionation. This fractionation arises from deviations in the solar wind energy-per-charge (E/Q) distributions of the different isotopic species which are passed through the WAVE electrostatic deflection system (Kallenbach et al.,

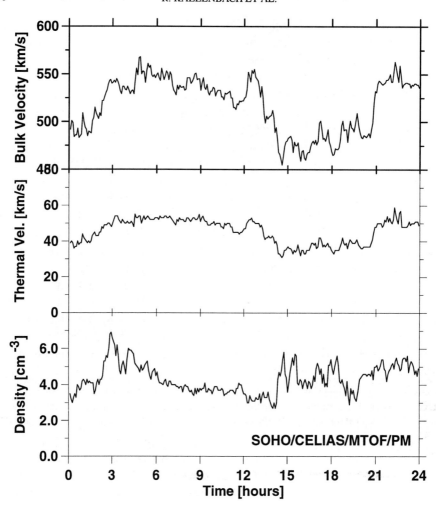

Figure 2. Solar wind bulk velocity, mean thermal velocity, and proton density on Aug. 1, 1996, as example for the measurements with the proton monitor which is a subsystem of the SOHO/CELIAS/MTOF sensor.

1997a). Minor fractionation is due to the fact that the VMASS detection efficiencies, the start efficiencies and the double coincidence efficiencies, are velocity dependent and at low velocities even slightly mass dependent. In this work data with instrument fractionations larger than ±15% have been disregarded. The remaining data have been corrected for the average instrument fractionation of typically a few percent within each velocity class. To calculate the instrument fractionation the entrance system voltage U_W, the post acceleration voltage V_f, the solar wind bulk velocity of Si $V_{bulk,Si}$, the kinetic temperature of Si $T_{kin,Si}$, and the charge state distribution of the solar wind Si ions are the variable input parameters that determine the folding of the solar wind distribution functions with the acceptance and efficiency functions of the MTOF sensor. U_W and V_f are available with 5 min time resolution from the housekeeping data. $V_{bulk,Si}$ is approximated by $1.02831 \times v_H - 14.4$ km/s (Hefti, 1997) where v_H is the solar wind proton bulk velocity (Figure 2) measured by the PM every 5 min. $T_{kin,Si}$ is assumed to be proportional to $T_{kin,H}$ where the proportionality factor is the particle mass in amu (Bochsler, 1984). $T_{kin,H}$ is also measured by the PM. The Si solar wind charge states are derived from the so-called freeze-in temperature assuming an ionization equilibrium in the corona and applying ionization and recombination rates for electronic collisions from Arnaud and co-workers (Arnaud and Rothenflug, 1985; Arnaud and Raymond, 1992). The relevant coronal freeze-in temperature is difficult to estimate because there are not yet explicit experimental studies on the Si charge states with the CELIAS experiment; however the Fe (Aellig et al., 1998) and the O (Hefti, 1997) freeze-in temperatures have been determined with high time resolution. In the latter work it has been found that the O freeze-in temperature correlates with the solar wind bulk velocity according to the relation $(1.99 - 0.001 \times v_H/(km/s))$ MK. The Si charge states have first been measured mainly in coronal holes at high latitudes by Ulysses/SWICS (Galvin et al., 1991). It has been found that the freeze-in temperatures of Si and O are not well correlated and that the abundance ratios of different charge state pairs such as Si^{11+}/Si^{10+} or Si^{9+}/Si^{8+} are not described by the same freeze-in temperature. This means that the formula for the freeze-in temperature of O does not apply for Si. On the other hand, during the time period and in the solar wind velocity ranges relevant for this work WIND/MASS has measured mean coronal freeze-in temperatures for Si (Wimmer et al., 1998) that agree with the mean value for O. Considering all studies of Ulysses/SWICS, WIND/MASS and SOHO/CELIAS/CTOF we conclude that the Si freeze-in temperatures in average do not deviate by much more than 0.1 MK from the O freeze-in temperatures during the time periods studied here. Simulations have been performed to estimate the variations in the instrument fractionation for Si isotopes depending on differences in the coronal freeze-in temperatures of O and Si. It turned out that these variations are not the dominating source of experimental uncertainties in the abundance ratios of Si isotopes. This is due to the fact that the E/Q acceptance of the WAVE is very wide, and only measurements are considered where the solar wind E/Q distributions of the Si isotopes fall into the plateau of the WAVE acceptance.

Figure 3 shows a mass spectrum of the MTOF sensor in the range of TOF channels where the Si isotopes are detected. The data displayed are selected from time intervals when the solar wind bulk velocity varied between 485 km/s and 535 km/s. The peaks originate from ^{28,29,30}Si particles that leave the carbon foil of the instrument singly charged. However, there is also instrumental interference from $^{56}Fe^{2+}$, $^{58}Ni^{2+}$, and $^{60}Ni^{2+}$ which are detected in the same TOF channels as $^{28,29,30}Si^+$ because the Fe and Ni particles do not only leave the thin carbon foil as neutrals or singly charged ions but also as doubly charged ions. The count numbers of the $^{56}Fe^{2+}$ and $^{58,60}Ni^{2+}$ instrumental interferences have been estimated from

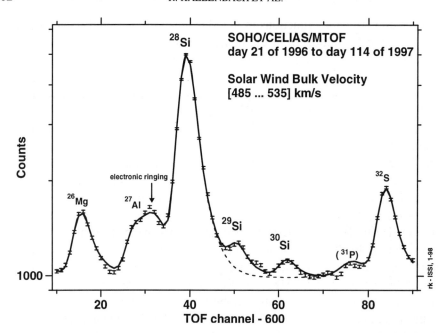

Figure 3. Mass spectrum of the MTOF sensor showing the signatures of the isotopes $^{28,29,30}\text{Si}^+$. In this spectrum the interferences from doubly charged Fe have been reduced to $^{56}\text{Fe}^{2+}/^{28}\text{Si}^+ = 0.0254$, $^{58}\text{Ni}^{2+}/^{29}\text{Si}^+ = 0.0164$, and $^{60}\text{Ni}^{2+}/^{30}\text{Si}^+ = 0.0059$ (see text). Data are selected from 5 min time intervals when the solar wind bulk velocity ranged between 485 km/s and 535 km/s. The "electronic ringing" peaks in the spectrum are a reproduceable instrumental effect arising from reflections in the fast signal lines. These ringing peaks are fixed in relative position at about seven channels to the left of the main peak. The line shape for the fit of the Si peaks has been derived from calibrations. The solid line represents the complete maximum likelihood fit whereas the dashed line excludes $^{29}\text{Si}^+$ and $^{30}\text{Si}^+$.

the actual spectrum of $^{56}\text{Fe}^+$ and $^{58,60}\text{Ni}^+$ applying the following model: The double coincidence efficiencies for $^{56}\text{Fe}^{2+}$ ($^{58,60}\text{Ni}^{2+}$ respectively) and $^{56}\text{Fe}^+$ ($^{58,60}\text{Ni}^+$ respectively) have been calculated using the calibrated instrument acceptance and efficiency functions. The incoming solar wind charge state distributions have been derived from the freeze-in temperatures for Fe measured by SOHO/CELIAS/CTOF (Aellig *et al.*, 1998) assuming that the Ni freeze-in temperatures do not differ too much from the Fe freeze-in temperatures. The charge exchange of Fe and Ni in thin carbon foils has been calibrated extensively (Gonin *et al.*, 1995). For most ranges of solar wind bulk velocities the $^{56}\text{Fe}^{2+}$ interference contributed about 15% to the $^{28}\text{Si}^+$ signature due to the wide E/Q acceptance of the entrance system. We have assumed that the uncertainty in the instrument model calculation is less than 20%. Therefore most abundance ratios have an additional experimental uncertainty due to the interferences of doubly charged Fe and Ni of about 3% that approximately equals the statistical uncertainty of the pure Si count numbers. At best, for the spectrum shown in Figure 3, the interferences could be reduced to $^{56}\text{Fe}^{2+}/^{28}\text{Si}^+ = 0.0254$, $^{58}\text{Ni}^{2+}/^{29}\text{Si}^+ = 0.0164$, and $^{60}\text{Ni}^{2+}/^{30}\text{Si}^+ = 0.0059$ by

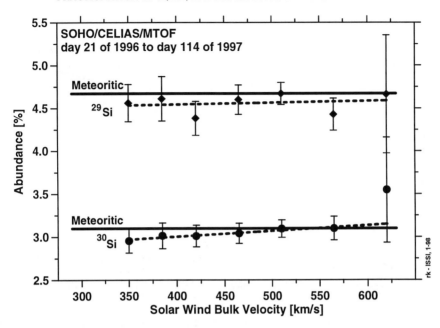

Figure 4. Solar wind ^{29}Si and ^{30}Si abundances versus the solar wind bulk velocity. For comparison the meteoritic abundances are shown. The abundances are in percent relative to the total number of Si isotopes ^{28}Si+^{29}Si+^{30}Si. The dashed lines show linear least-square fits to the measured data and indicate a weak correlation between the solar wind bulk velocity and the ^{29}Si and ^{30}Si abundances. The mean solar wind abundances are 4.47% for ^{29}Si and 3.04% for ^{30}Si.

selecting 5 min time intervals in the flight data where the voltage of the entrance system and the solar wind bulk velocity had values such that the solar wind ions ^{28,29,30}Si$^{8+...12+}$ are well transmitted and ^{56}Fe$^{9+...14+}$ and ^{58,60}Ni$^{9+...14+}$ are partly filtered out. This is possible because the Fe and Ni ions are approximately drifting at the same solar wind bulk velocity as the Si ions but have different mass-per-charge ratios and therefore different E/Q distributions. However, the method is not very efficient because both the solar wind E/Q distributions and the WAVE E/Q acceptance are very wide, and the Si E/Q distributions have to coincide with the plateau of the WAVE E/Q acceptance so that there is still considerable transmission for solar wind Fe and Ni.

Figure 4 shows the solar wind ^{29}Si and ^{30}Si abundances versus the solar wind bulk velocity compared to the meteoritic abundances (Anders and Grevesse, 1989). A trend can be seen in the sense that the heavier isotopes are depleted by a few percent in the slow solar wind. This trend has already been discussed for the solar wind Ne isotopic composition (Kallenbach *et al.*, 1997a), although statistics have not been sufficient to exclude the hypothesis that the isotopic ratio is independent of the solar wind bulk velocity. For Si the trend is also not significant within 95% confidence limits because of statistics. However, averaging the variations in the abundances of the Si, Ne, and Mg isotopes weighted by the inverse of their vari-

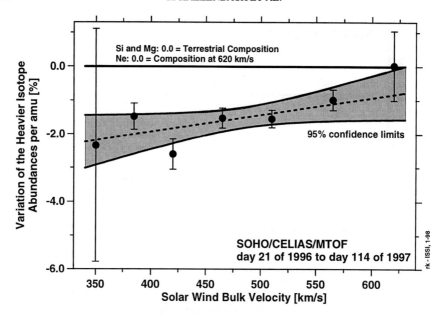

Figure 5. Weighted average of the variation in the abundances of the heavier Si, Ne, and Mg isotopes per unit mass compared to the abundances at 620 km/s. For Si and Mg the abundances at 620 km/s coincide with the terrestrial abundances. The abscissa of the plot may be offset by up to 1.5% per amu due to uncertainties in the instrument calibration (Kallenbach et al., 1997a).

ances a more significant result can be obtained. From Figure 5 it can be concluded that the depletion of the heavier isotopes per amu is $(1.4\pm1.3)\%$ (2σ-error) larger in the slow solar wind with a bulk velocity of about 350 km/s compared to the faster solar wind with a bulk velocity of 620 km/s. The zero level in Figure 5, indicating no depletion of the heavier isotopes, is defined by the terrestrial isotope abundances for Si and Mg, and by the isotopic composition of Ne in the solar wind with a bulk velocity of 620 km/s. The average overall depletion of the heavier isotopes independent of the solar wind bulk velocity may be interpreted as a signature of gravitational settling in the outer convective zone of the Sun. However, the absolute values on the abscissa of Figure 5 may not be calibrated well enough; overall they could be offset by up to 1.5% per amu (Kallenbach et al., 1997a).

The above described relative variation of the isotope abundance ratios with the solar wind bulk velocity has been derived based on instrument functions which correct for instrument efficiencies that are dependent on energy per mass. The efficiency ratio corrections for two different isotopes of Si, Ne or Mg are typically on the order of a few percent. The precision of the data reduction has carefully been studied by Kallenbach et al. (1997a). However, to eliminate any residual systematic uncertainty in the relative variation, the $^{24}Mg/^{26}Mg$ ratio has also been analyzed as a function of solar wind density in a narrow window of solar wind bulk velocities. According to Schwenn (1990) coronal-hole type solar wind typically has a density of a few particles per cm^3 whereas coronal-streamer type solar wind has

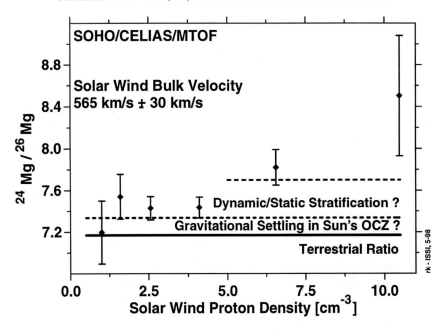

Figure 6. Solar wind $^{24}Mg/^{26}Mg$ ratio versus solar wind number density for solar wind bulk velocities between 535 km/s and 595 km/s.

densities of approximately 10 cm^{-3} and sometimes up to 100 cm^{-3}. In the range of solar wind bulk velocities between 535 km/s and 595 km/s considered here both types of solar wind can roughly be distinguished by their densities. The result of this study is shown in Figure 6: It clearly can be seen that at higher solar wind densities the heavier isotope is depleted compared to its abundance at lower solar wind densities.

4. Discussion

From a theoretical point of view several sites of isotope fractionation in the solar wind can be discussed: Gravitational settling in the outer convective zone, gravitational settling in closed magnetic structures of quiet helmet streamers, ion-neutral separation in the upper chromosphere and transition region (von Steiger and Geiss, 1989; Marsch *et al.*, 1995), inefficient Coulomb coupling of minor ions to protons and α-particles in the inner corona (Bodmer and Bochsler, 1996), coronal mass ejection (CME) events, corotating interaction region (CIR) events, and shocks in the interplanetary medium in general.

The effects of interplanetary shocks cancel out in long term averages.

The ion-neutral separation (von Steiger and Geiss, 1989; Marsch *et al.*, 1995) in the upper chromosphere and transition region is very important for elemental fractionation: Neutrals diffuse from the upper chromosphere into the transition region and are ionized within a time depending on their first ionization potential (FIP). Once the particles are ionized they are uplifted by the Coulomb drag. The varying FIPs lead to a strong elemental fractionation well known from solar wind elemental abundances. However, for Ne only 0.1% isotope fractionation and for Mg just 0.08% isotope fractionation are expected from this process.

The CME and CIR events connected to the SOHO spacecraft do not contribute very much to the data analyzed. Isotope fractionation in CMEs and CIRs is the topic of a pending study. Note, that there is an enrichment of heavy elements in CME related solar wind (Galvin, 1997; Wurz et al., 1998).

The effect of gravitational settling in the outer convective zone is estimated to be an enhancement of 2% for ^{20}Ne/^{22}Ne and ^{24}Mg/^{26}Mg and 1% for ^{28}Si/^{30}Si independent of the solar wind regime (Vauclair, 1981; Vauclair, 1998; Bochsler, 1998). Although data of Figure 5 indicate a weak overall depletion of the heavier isotopes of Si and Mg, independent of solar wind bulk velocity and compared to meteoritic and terrestrial abundances, it cannot be concluded definitely that this is an effect of gravitational settling in the outer convective zone because the whole abscissa of Figure 5 may be offset by up to 1.5% per amu due to systematic instrumental effects (Kallenbach et al., 1997a). The absolute values of the isotope abundance ratios cannot be determined with the same precision as their relative variations with solar wind bulk velocity due to systematic experimental uncertainties.

The main effects on the isotopic density and velocity profiles have to be expected from the gravitational settling in closed magnetic structures (static stratification) and from the Coulomb coupling of minor solar wind ions to the protons and α-particles in the inner corona (dynamic stratification).

Recent observations of the Ultraviolet Coronal Spectrograph (UVCS) on board SOHO (Raymond et al., 1997) have revealed that there is a reduction of O/H by about one order of magnitude in the cores of helmet streamers compared to O/H in their "legs". This is most likely due to static stratification by gravitational settling in the closed magnetic structure which also depletes the heavier isotopes of an element. However, elemental abundances in the slow solar wind (Geiss et al., 1995) are rather similar to the abundances observed spectroscopically in the "legs" of quiet streamers and in active region streamers (Raymond et al., 1997) which do not exhibit the strong depletion of O observed in the cores of quiet streamers. On the other hand, UVCS observations (Spadaro, 1998) indicate that about 10% of the slow solar wind may detach sporadically from closed magnetic loops. This is supported by a theory of Fisk (1998) in this issue stating that magnetic flux moves across coronal holes as a result of the interplay between the differential rotation of the photosphere and the non-radial expansion of the solar wind in more rigidly rotating coronal holes. This flux is deposited at low latitudes and should reconnect with closed magnetic loops, thereby releasing material from the loops to form the slow solar wind.

One of the largest variations in solar wind composition is the depletion of ^{4}He^{2+} in the slow solar wind (Borrini et al., 1981). Compared to its abundance in fast coronal-hole type solar wind $^4\text{He}^{2+}$ is depleted by up to a factor 3. In fact, in a preliminary study it has been found that the depletion of heavier isotopes is correlated with a low $^4\text{He}^{2+}/\text{H}^+$ abundance ratio (Kallenbach et al., 1997b). According to Borrini et al. (1981) the depletion of $^4\text{He}^{2+}$ is associated with sector boundaries that are coronal signatures of the slow interstream solar wind. Von Steiger et al. (1995) have shown that $^4\text{He}^{2+}$ with the least favourable drag factor (Burgers, 1969) of all minor solar wind ions must be depleted due to inefficient Coulomb coupling to the low proton flux in flow tubes with large geometrical expansion factors associated with sector boundaries. This has been proven by analyzing the solar wind O abundances associated with sector boundaries. O^{6+} and O^{7+} with their larger drag factors are depleted less compared to $^4\text{He}^{2+}$; in case of gravitational settling O should be depleted even more than He which has not been observed by Ulysses/SWICS. The variation of the Coulomb coupling in different solar wind regimes has been modeled by Bodmer and Bochsler (1996) and by Bochsler in this issue. According to their work, the fractionation of $^{24}\text{Mg}/^{26}\text{Mg}$ and $^{20}\text{Ne}/^{22}\text{Ne}$ is larger in the slow interstream solar wind with its strong superradial expansion than in the fast coronal-hole type solar wind in the sense that the heavier isotopes with their less favourable drag factors are depleted in the slow interstream solar wind.

Recently, it has been questioned whether the Coulomb drag is really necessary for the acceleration of minor solar wind ions. Optical observations with UVCS (Kohl et al., 1997) on polar coronal holes down to 1.5 solar radii have revealed that minor ion outflow velocities can be as high as the escape velocity from the solar surface. In this case, Coulomb collisions with protons and α-particles are not important for heavy solar wind ions to be carried away from the solar surface. However, preliminary results by Raymond et al. (1997) indicate that minor ion outflow velocities at the sources of the slow solar wind are much lower so that dynamic stratification by inefficient Coulomb coupling is a likely cause for isotope fractionation processes.

5. Conclusion

SOHO/CELIAS/MTOF delivers high quality data, which provide the possibility to observe the isotope abundance ratios in different solar wind regimes. From the results of the data analysis of the solar wind Si, Ne, and Mg isotopic compositions it can be concluded that the fractionation of isotopes in the solar wind is very small, especially when compared to fractionations observed in the isotopic compositions of solar energetic particles (Mason et al., 1994; Mewaldt et al., 1984; Selesnick et al., 1993). The variation of the isotope abundance ratios of Si, Ne, and Mg in the range of solar wind bulk velocities between 350 and 620 km/s is $(1.4\pm1.3)\%$ per amu (2σ-error). Rough estimates based on the results on the solar wind $^4\text{He}^{2+}$ depletion associated with sector boundaries as well as a theoretical model (Bodmer and Bochsler, 1996) which attributes the variable part of the depletion of the heavy isotopes in the solar wind to inefficient Coulomb coupling predict the observed trends with the correct orders of magnitude. A final statement may be possible after an analysis which includes the correlation of solar wind isotope abundance ratios with magnetic field data from the WIND spacecraft to identify the sector boundaries. Another possible explanation for the observed effect is the gravitational settling of material in closed magnetic structures which is released sporadically as a consequence of magnetic reconnection. The data on solar wind Si and Mg isotopes give no significant evidence for a general depletion of heavy isotopes in the solar wind independent of its bulk velocity which might be interpreted as the consequence of gravitational settling in the outer convective zone of the Sun but also do not rule out an effect of one or two percent per amu.

Correlated studies by remote particle sensing and spectroscopy on board SOHO may lead to a more detailed knowledge about structure and composition of the solar corona and about the sources and acceleration mechanisms of the slow interstream and the fast coronal-hole type solar wind.

Acknowledgements

This work was supported by the Swiss National Science Foundation, by the PRODEX program of ESA, by NASA grant NAG5-2754, and by DARA, Germany, with grants 50 OC 89056 and 50 OC 9605. The MTOF sensor has been developed in the CELIAS consortium (Hovestadt *et al.*, 1995) under the lead of the University of Maryland space physics group. The WAVE entrance system has been built under the guidance of the Physics Institute of the University of Bern at INTEC, Bern. The integrated sensor and the charge exchange of solar wind ions in thin carbon foils have been calibrated at the University of Bern, Switzerland, at the Strahlenzentrum of the University of Giessen, Germany, and at the Centre d'Etudes Nucleaires de Grenoble, France. The Technical University of Braunschweig, Germany, contributed the Data Processing Unit. The main author would like to thank for inspiring discussions with Rudolf von Steiger at the International Space Science Institute, Bern, with Matthias Aellig and Robert Wimmer-Schweingruber at the Physikalisches Institut, University of Bern, and with Thomas Zurbuchen, University of Michigan.

References

Aellig, M.R., *et al.*: 1998, 'Iron freeze-in temperatures measured by SOHO CELIAS CTOF', *J. Geophys. Res.*, in press.

Anders, E., and Grevesse, N.: 1989, 'Abundances of the elements: Meteoritic and solar', *Geochim. Cosmochim. Acta* **53**, 197–214.

Arnaud, M., and Rothenflug, R.: 1985, 'An updated evaluation of recombination and ionization rates', *Astron. Astrophys. Suppl. Ser.* **60**, 425–457.

Arnaud, M., and Raymond, J.: 1992, 'Iron ionization and recombination rates and ionization equilibrium', *Astrophys. J.* **398**, 394–406.

Bochsler, P., Gonin, M., Sheldon, R.B., Zurbuchen, T., Gloeckler, G., Hamilton, D.C., Collier, M.R., and Hovestadt, D.: 1997, 'Abundance of Solar Wind Magnesium Isotopes determined with WIND/MASS', *Proc. of the 8th International Solar Wind Conference, Dana Point, CA, USA*, 199–202.

Bochsler, P.: 1984, 'Helium and Oxygen in the Solar Wind: Dynamic properties and abundances of elements and Helium isotopes as observed with the ISEE-3 plasma composition experiment', Habilitation Thesis, University of Bern.

Bochsler, P.: 1998, 'Structure of the Solar Wind and Compositional Differences', *Space Sci. Rev.*, this volume.

Bodmer, R. and Bochsler, P.: 1996, 'Fractionation of minor ions in the solar wind acceleration process, paper presented at meeting, Eur. Geophys. Soc., The Hague, Netherlands.

Borrini, G., Wilcox, J.M., Gosling, J.T., Bame, S.J., and Feldman, W.C.: 1981, 'Solar wind helium and hydrogen structure near the heliospheric current sheet: A signal of coronal streamers at 1 AU', *J. Geophys. Res.* **86**, 4565–4573.

Burgers, J.M.: 1969, 'Flow Equations for Composite Gases', Academic, New York.

Fisk, L.A., Schwadron, N.A., and Zurbuchen, T.H.: 1998, 'On the slow solar wind', *Space Sci. Rev.*, in press.

Galvin, A.B., Ipavich, F.M., Gloeckler, G., von Steiger, R., and Wilken, B.: 1991, 'Silicon and oxygen charge state distributions and relative abundances in the solar wind measured by SWICS on Ulysses', *Proc. of the 7th International Solar Wind Conference, Goslar, Germany*, 337–340.

Galvin, A.B.: 1997, 'Minor Ion Composition in CME-Related Solar Wind', in: Coronal Mass Ejections, eds. Nancy Crooker, JoAnn Joselyn, and Joan Feynman, *Geophysical Monograph* **99**, 253–260.

Geiss, J., Bühler, F., Cerutti, H., Eberhardt, P., and Filleux, C.: 1972, 'Solar Wind Composition Experiment', *Apollo 16 Preliminary Scientific Report, NASA Spec. Publ.* **SP-315**, 14.1.

Geiss, J., Gloeckler, G., von Steiger, R., Balsiger, H., Fisk, L.A., Galvin, A.B., Ipavich, F.M., Livi, S., McKenzie, J.F., Ogilvie, K.W., and Wilken, B.: 1995, 'The Southern High-Speed Stream: Results from the SWICS instrument on ULYSSES', *Science* **268**, 1033–1036.

Gonin, M., Kallenbach, R., Bochsler, P., and Bürgi, A.: 1995, 'Charge exchange of low energy particles passing through thin carbon foils: Dependence on foil thickness and charge state yields of Mg, Ca, Ti, Cr and Ni', *Nucl. Instrum. Methods Phys. Res.* **B101**, 313–320.

Hefti, S.: 1997, 'Solar Wind Freeze-in Temperatures and Fluxes Measured with SOHO/CELIAS/CTOF and Calibration of the CELIAS Sensors', PhD Thesis, University of Bern.

Hovestadt, D., et al.: 1995, 'Charge, Element, and Isotope Analysis System onboard SOHO', *Sol. Phys.* **162**, 441–481.

Kallenbach, R., et al.: 1997, 'Isotopic composition of solar wind neon measured by CELIAS/MTOF on board SOHO', *J. Geophys. Res.* **102**, 26895–26904.

Kallenbach, R., et al.: 1997, 'Limits to the fractionation of isotopes in the solar wind as observed with SOHO/CELIAS/MTOF', *ESA* **SP-415**, 33–37.

Kallenbach, R., et al.: 1998, 'Isotopic Composition of Solar Wind Calcium: First in situ Measurement by CELIAS/MTOF on board SOHO', *Astrophys. J. Lett.* **498**, L75–L78.

Kohl, J.L., et al. 1997, 'First results from the SOHO ultraviolet coronagraph spectrometer', *Solar Physics* **175**, 613–644.

Kucharek, H., et al.: 1998, 'Magnesium isotopic composition as observed with the MTOF sensor on the CELIAS experiment on the SOHO spacecraft', *J. Geophys. Res.*, submitted.

Marsch, E., von Steiger, R., and Bochsler, P.: 1995, 'Element fractionation by diffusion in the solar chromosphere', *Astron. Astrophys.* **301**, 261–276.

Mason, G.M., Mazur, J.E., and Hamilton, D.C.: 1994, 'Heavy-ion isotopic anomalies in ^{3}He-rich solar particle events', *Astrophys. J.* **425**, 843–848.

Mewaldt, R.A., Spalding, J.D., and Stone, E.C.: 1984, 'A high-resolution study of the isotopes of solar flare nuclei', *Astrophys. J.* **280**, 892–901.

Raymond, J.C., et al.: 1997, 'Composition of coronal streamers from the SOHO ultraviolet coronagraph spectrometer', *Solar Physics* **175**, 645–665.

Schwenn, R.: 1990, 'Large-Scale Structure of the Interplanetary Medium', in: Physics of the Inner Heliosphere I, eds. R. Schwenn and E. Marsch, *Physics and Chemistry in Space - Space and Solar Physics* **20**, 99–181.

Selesnick, R.S., Cummings, A.C., Cummings, J.R., Leske, R.A., Mewaldt, R.A., Stone, E.C., and von Rosenvinge, T.: 1993, 'Coronal abundances of neon and magnesium isotopes from solar energetic particles', *Astrophys. J.* **418**, L45–L48.

Spadaro, D.: 1998, 'SOHO observations of the outer solar atmosphere: status and prospects', *Annales Geophysicae* **16, suppl. III**, C915.

Vauclair, S.: 1981, *Astrophys. J.* **86**, 513, and this issue.

Vauclair, S.: 1998, *Space Sci. Rev.*, this issue.

von Steiger, R., and Geiss, J.: 1989, 'Supply of fractionated gases to the corona', *Astron. Astrophys.* **225**, 222–238.

von Steiger, R., Wimmer-Schweingruber, R.F., Geiss, J., and Gloeckler, G.: 1995, 'Abundance variations in the solar wind', *Adv. Space Res.* **15**, (7)3–(7)12.

Wimmer-Schweingruber, R.F., Bochsler, P., Kern, O., Gloeckler, G., and Hamilton, D.C.: 1998, 'First in situ detection of solar wind silicon isotopes: WIND/MASS results', *J. Geophys. Res.*, submitted.

Wurz, P., *et al.*: 1998, 'Elemental Composition of the January 6, 1997, CME', *Geophys. Res. Lett.*, in press.

Address for correspondence: International Space Science Institute, Hallerstrasse 6, CH-3012 Bern, Switzerland, e-mail: reinald.kallenbach@issi.unibe.ch.

SOLAR EUV AND UV EMISSION LINE OBSERVATIONS ABOVE A POLAR CORONAL HOLE

K. WILHELM and R. BODMER
Max-Planck-Institut für Aeronomie, D-37191 Katlenburg-Lindau, Germany

Abstract. The roll manoeuvre of SOHO on September 3, 1997 provided the opportunity to study the northern coronal hole with SUMER slits in east-west orientation, so that polar plumes and inter-plume lanes could be observed simultaneously. A preliminary analysis of the observations shows that lines emitted by ions with the lowest formation temperatures (with the exceptions of Ne^{7+} and Ar^{7+}) have the largest ratios of plume to lane radiances at heights between 35 000 km and 70 000 km above the photosphere. All lines have narrower widths inside plumes than outside. Electron densities have been deduced in plumes and lanes from Si VIII and Mg VIII line radiance ratios. The Mg IX pair was used to determine the corresponding electron temperatures. Neon (with a high first-ionization potential) is found to be less abundant relative to magnesium (with low FIP) in a plume compared to an inter-plume lane, but the variation is smaller than previously determined Ne/Mg abundance ratios in a plume relative to the photosphere.

1. Introduction

The plasma conditions above solar polar coronal holes are of considerable interest, because it is well established that the high-speed solar wind streams originate in these regions (see, e.g., Krieger *et al.*, 1973; Geiss *et al.*, 1995a; Woch *et al.*, 1997), and the acceleration processes of the fast solar wind thus can be studied there. The most prominent features above polar coronal holes are polar plumes or rays (van de Hulst, 1950a, 1950b; Saito, 1965; Newkirk and Harvey, 1968; Ahmad and Withbroe, 1977; Antonucci *et al.*, 1997; Wang *et al.*, 1997; DeForest *et al.*, 1997; Wilhelm *et al.*, 1998). Widing and Feldman (1989, 1992, 1993) found a dependence of the neon-to-magnesium abundance ratio for different structures of the solar atmosphere, and, in particular, a very low ratio in a bright plume. As, on the other hand, abundances measured in fast solar wind streams by *in situ* observations are close to those of the photosphere (e.g., Geiss *et al.*, 1995b), any differences observed in plumes and inter-plume lanes are of great importance. We report here on such observations obtained on September 3, 1997.

2. Instrumentation and Observations

The SUMER instrument and its radiometric calibration have been described elsewhere (Wilhelm *et al.*, 1995; Hollandt *et al.*, 1996; Lemaire *et al.*, 1997; Wilhelm *et al.*, 1997a; Wilhelm *et al.*, 1997b). SUMER can provide plume and inter-plume lane observations simultaneously during a roll manoeuvre of the spacecraft about

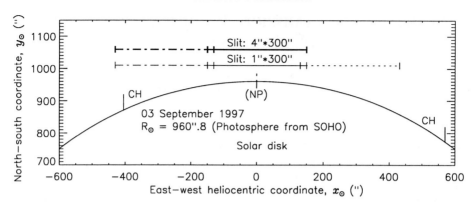

Figure 1. Slit positions during the September 3, 1997 observations. EIT/SOHO images were inspected to determine the extensions of the polar coronal hole. The northern hole was well developed during the spacecraft roll.

its Sun-pointing axis by 90°. The pointing positions of the slits are shown in Figure 1. They were chosen so as to intersect plumes and inter-plume lanes at two heights above the limb. The exposure time for each spectrum was 500 s. The spectral resolution element in first order is $\approx$ 44 mÅ. A selection of eleven wavelength intervals of 40 Å each had to be made due to time constraints. Even then, the sequence could not be completed in the 7 hours available, and electron temperatures and some density values cannot be deduced for the high slit position at $x_\odot = 281''$.

3. Data Analysis

All individual spectra have been visually inspected for data quality and prominent coronal lines. The lines discussed in this contribution are listed in Table I. They have been selected to cover a wide range of temperatures, T_M – the temperature of maximum ionic fraction in ionization equilibrium (Arnaud and Rothenflug, 1985) –, but also to include elements with high and low first-ionization potentials (FIP) (7.6 eV for Mg; 21.5 eV for Ne). It was also important to have density and temperature sensitive line pairs available for plasma diagnostics.

The radiation, even in "bright" plumes, is rather faint, and does not allow us to employ the full spatial resolution of SUMER ($\approx 1''$) with reasonable integration times. By binning 30 to 32 spatial elements (detector pixels) into one element of $\approx 31''$, adequate line profiles for most of the lines could be obtained for all but the darkest regions. Flat-field corrections and rectification of the geometric distortion were not required with this low spatial resolution. Gauss fits for all lines were then computed and formed the data base for the following studies. The line widths (in Doppler velocity, $v_{1/e}$) were corrected for instrumental effects.

Table I

Observational Parameters, Radiance Ratios and Line Width Correlation

Position[a] $y_\odot$ (″)	Line	Wavelength (Å)	Temperature T_M (K)	Ratio L_P/L_I	Ratio L_{P0}/L_{P1}	Correlation $\rho(L, v_{1/e})$
1010	Mg X[b]	624.94	$1.1 \cdot 10^6$	1.60		−0.78
1059				1.54	2.1	−0.83
1010	Na IX[b]	681.66[c]	$8.5 \cdot 10^5$	1.82		−0.77
1059				2.11	3.1	−0.55
1010	Ar VIII	700.24[c]	$4.3 \cdot 10^5$	2.29		−0.82
1059				2.05	4.0	−0.35
1010	Mg IX[d]	706.07	$9.5 \cdot 10^5$	2.17		−0.91
1059				2.03	3.1	−0.91
1010	Ne VIII	770.41	$6.3 \cdot 10^5$	1.64		−0.91
		780.33		1.69		−0.93
1059		770.41		1.76	4.5	−0.89
		780.33		1.84	4.3	−0.89
1010	Mg VIII	772.28	$8.1 \cdot 10^5$	2.78		−0.75
		782.36		2.70		−0.92
1059		772.28		2.70	4.7	−0.73
		782.36		2.49	4.4	−0.58
1010	Fe XII	1242.00	$1.4 \cdot 10^6$	1.52		−0.79
1059				1.25	2.8	−0.88
1010	Si VIII[e]	1445.75	$8.5 \cdot 10^5$	2.26		−0.80
1059				2.75	3.1	−0.78
1010	Fe X	1463.48	$9.5 \cdot 10^5$	1.87		−0.53
1059				1.66	2.9	−0.83
1010	Fe XI	1467.08	$1.1 \cdot 10^6$	1.65		−0.66
1059				1.47	2.6	−0.84

[a] Centre of slit in heliocentric coordinates; all $x_\odot = 0''$.
[b] Observed in second order.
[c] Wavelengths from Curdt et al. (1997); all other wavelengths from Feldman et al. (1997).
[d] Temperature sensitive line pair together with Mg IX (λ 749.55 Å).
[e] Density sensitive line pair together with Si VIII (λ 1440.49 Å).

4. Results

Some of the results are compiled in Table I. The table refers to the central portion from $x_\odot = -140''$ to $+140''$ only, where plumes and inter-plume lanes can be clearly distinguished. It lists the radiance ratio, L_P/L_I, between the brightest plume and the first lane towards the west for heights of $\approx 35\,000$ km and $\approx 70\,000$ km ($y_\odot = 1059''$), and the ratio, L_{P0}/L_{P1}, indicating the altitude dependence inside the central plume. Also given are the correlation coefficients, $\rho(L, v_{1/e})$. With with a few exceptions, they indicate a strong anti-correlation between the width of a

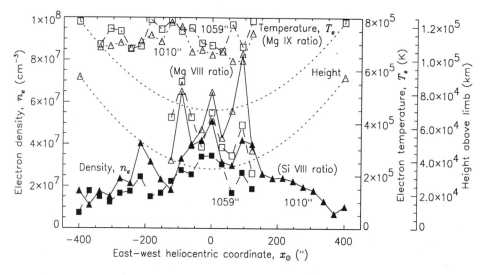

Figure 2. Electron densities (Si VIII and Mg VIII line pairs; left scale) and electron temperatures (Mg IX line pair; right scale) are plotted at the slit positions in Figure 1. The actual altitudes above the limb are indicated by dotted lines with a scale at the extreme right. The Mg VIII measurements have been shown only for the central region and are less reliable than the Si VIII data (see text).

line and its radiance. This confirms and extends UVCS/SOHO and earlier SUMER observations made in O VI (λ 1032 Å) (Antonucci *et al.*, 1997; Kohl *et al.*, 1997; Hassler *et al.*, 1997; Wilhelm, 1998).

The Si VIII line pair at 1440 Å and 1445 Å can be used for electron density diagnostics (Feldman *et al.*, 1978; Dwivedi, 1991; Doschek *et al.*, 1997; Laming *et al.*, 1997). The intensity ratio of the Mg VIII inter-combination lines at 772 Å and 782 Å is also density sensitive (Laming *et al.*, 1997) and is available for our analysis, albeit with the complication that the line at 772 Å is blended by two lines (S VIII? and ?). We estimated the blend contributions from the line shape to be 20 %, but have to accept large uncertainties for the critical electron density determinations. The intensity considerations, on the other hand, will not be significantly influenced. In Figure 2, the results for the plume and lane regions above the north pole are shown for $y_\odot = 1010''$ and $1059''$. The variations of the electron density, n_e, between plumes and lanes are quite pronounced and amount to about a factor of two for the central region. This can be seen both from the Si VIII line ratio and the Mg VIII line ratio. The absolute value obtained from the Mg VIII lines should, with the complications mentioned above, not be considered to be very accurate. Note also that the spatial resolution of $\approx 31''$ employed is not sufficient for resolving some of the narrow structures, for which the variations in n_e will be underestimated.

The Mg IX line at 750 Å and the inter-combination line at 706 Å provide electron temperature diagnostics (Keenan *et al.*, 1984; Wilhelm *et al.*, 1998). The temperature, T_e, also shown in Figure 2, varies between 630 000 K and 780 000 K at

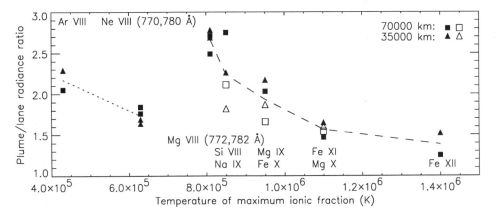

Figure 3. The ratios of the plume to inter-plume lane radiances are plotted versus the formation temperature of the ions emitting the lines. The open symbols refer to Na IX, Fe X, and Mg X. Two lines are shown both for Ne VIII and Mg VIII. The mean values for each temperature level are connected by a dotted line for high FIP elements and by dashed lines for low FIP elements.

$\approx 35\,000$ km ($y_\odot = 1010''$) altitude and between 670 000 K and 790 000 K at $\approx 70\,000$ km ($y_\odot = 1059''$). The temperature is anti-correlated with density for $x_\odot$ from $-160''$ to $140''$ with correlation coefficients of $\rho(n_e, T_e) = -0.51$ at the lower height and -0.77 higher up. Similar results were obtained with SUMER earlier (Wilhelm *et al.*, 1998).

In the context of abundance discussion, the radiance ratios of plumes to inter-plume lanes are of particular relevance. Using data of Table I, the ratios are plotted in Figure 3 as a function of T_M. They exhibit a discontinuity between Ne VIII and Mg VIII, with low FIP elements on the left and high FIP elements on the right side. It might be of interest to compare the radiances of the Ne VIII ($\lambda\lambda$ 770, 780 Å) and the Mg VIII ($\lambda\lambda$ 772, 782 Å) lines for plumes and inter-plume lanes. We find $L_P(770\,\text{Å})/L_P(772\,\text{Å}) = 10$ and $L_P(780\,\text{Å})/L_P(782\,\text{Å}) = 7.7$ in the central plume at $T_e \approx 660\,000$ K. In the inter-plume lane, we measure $L_I(770\,\text{Å})/L_I(772\,\text{Å}) = 17$ and $L_I(780\,\text{Å})/L_I(782\,\text{Å}) = 12.3$ at an electron temperature of $T_e \approx 750\,000$ K.

5. Discussion

The line width results will only briefly be discussed here, by noting that the narrower line widths inside plumes compared to those in inter-plume lanes are a characteristic feature of all lines analysed in this study.

We will concentrate on the question whether we can deduce any information on the FIP effect from our observations. From Figure 2, we find at 35 000 km (with values for 70 000 km in parentheses) electron densities of $n_e = 5 \cdot 10^7$ cm^{-3} ($3.4 \cdot 10^7$) for the central plume and $3 \cdot 10^7$ cm^{-3} ($1.7 \cdot 10^7$) for the first lane towards the west. For allowed lines formed by electron impact excitation from the ground state, the radiance varies as $L \sim n_e^2$ if the abundance and the ionic fractions are constant. For such lines we would expect $L_P/L_I = 5^2/3^2 = 2.78$ ($3.4^2/1.7^2 = 4.0$) for the plume-to-lane radiance ratio. With the Atomic Data and Analysis Structure (ADAS) software (Summers et al., 1996), we find a density dependence of $L(772 \text{ Å}) \sim n_e^{2.11}$ and $L(782 \text{ Å}) \sim n_e^{1.95}$ for the Mg VIII inter-combination lines. Again from Figure 2, we get $T_e \approx 660\,000$ K at 35 000 km inside and outside the plume; at 70 000 km we have $T_e \approx 670\,000$ K for the plume and $T_e \approx 770\,000$ K outside. Consequently, there is no temperature effect to be expected at the low altitude, but the lower temperature inside plumes reduces the contribution function by $\approx 33\%$ in ionization equilibrium at 70 000 km. The expected intensity ratios would be $L_P/L_I = 2.94$ (2.89) and 2.71 (2.59), respectively – very close to the observed values (see Table I). For ions with higher T_M, the ratios L_P/L_I in Figure 3, decrease with respect to Mg VIII. It is unlikely that abundance variations are of major importance, because the Mg IX, X lines and the Fe X, XI lines fall off relative to Mg VIII in a very similar manner. The ratios might be explained by assuming that the low plume temperature has an attenuating effect on high-temperature lines inside plumes and that hotter regions are surrounding the plumes.

However, the small ratios of L_P/L_I for the Ne VIII lines ($T_M \approx 630\,000$ K) cannot be understood in this way. In fact, from temperature considerations, we would even predict an increase of the Ne VIII ratios relative to Mg VIII (λ 772 Å), in line with the trend in Figure 3. Assuming again electron impact excitation for the Ne VIII resonance lines, which can be justified by the factor of ≈ 2 we find for the ratio, $L(770 \text{ Å})/L(780 \text{ Å})$, of the line radiances (see, e.g., Noci et al., 1987) inside and outside the plume, we can write $L_P \sim n_P(\text{Ne}^{7+}) \cdot n_{eP}$ and $L_I \sim n_I(\text{Ne}^{7+}) \cdot n_{eI}$, where n_P, n_I are the densities of Ne^{7+} and n_{eP}, n_{eI} the electron densities. We have to compare a radiance ratio of $L_P/L_I = 2.78$ (4.0) to an observed mean ratio of 1.67 (1.80), and finally obtain that, inside the central plume, the Ne^{7+} density is reduced to $1.67/2.78 = 0.60$ ($1.8/4.0 = 0.45$) of the value one would deduce from the Mg^{7+} (and the electron density) data. Incidentally, this means that the Ne^{7+} density is the same inside and outside of the plume at 35 000 km. Taking into account the close match of the plume electron temperature at 70 000 km and the formation temperature of Ne^{7+}, the Ne VIII lines should be emitted with higher efficiency inside plumes than outside (we estimate a factor of 1.6 from the contribution function). Hence, the neon abundance in a plume is reduced relative to magnesium by ≈ 1.7 at 35 000 km and ≈ 3.5 at 70 000 km altitude.

The values are smaller than the factor of about 10 (relative to photospheric levels) found by Widing and Feldman (1992) for a strong plume. Consequently it must be concluded that the missing enrichment of low FIP elements in the fast solar wind cannot be fully reconciled with strongly enriched plumes by equally strong plume-to-lane variations. It should be mentioned, however, that it is by no means certain that the lane observations represent measurements without plume "contamination", especially at the low altitude, and thus the depletion factors found have to be considered as lower limits. In particular, we do not claim that we have established an altitude dependence of the depletion.

6. Conclusions

Much more work is required for Ar^{7+} and the other ions than could be done in this preliminary analysis, in which we could confirm higher electron densities and lower electron temperatures inside plumes than in inter-plume lanes. Using these plasma parameters, we found neon-to-magnesium depletion factors in plumes of 1.7 and 3.5 relative to inter-plume lanes.

Acknowledgements

The SUMER project is financially supported by DLR, CNES, NASA, and the ESA PRODEX programme (Swiss contribution). SUMER is part of *SOHO*, the *Solar and Heliospheric Observatory*, of ESA and NASA.

References

Ahmad, I. A. and Withbroe, G. L.: 1977, 'EUV Analysis of Polar Plumes', *Solar Phys.* **53**, 397.
Antonucci, E., *et al.*: 1997, in B. Schmieder, J. C. del Toro Iniesta, and M. Vazquez (eds.), 'First Results from UVCS: Dynamics of the Extended Corona', *Advances in the Physics of Sunspots*, ASP Conference Series 118, 273.
Arnaud, M. and Rothenflug, R.: 1985, 'An Updated Evaluation of Recombination and Ionization Rates', *Astron. Astrophys. Suppl. Ser.* **60**, 425.
Curdt, W., *et al.*: 1997, 'The Solar Disk Spectrum Between 660 and 1175 Å (First Order) Obtained by SUMER on SOHO', *Astron. Astrophys. Suppl. Ser.* **126**, 281.
DeForest, C. E., *et al.*: 1997, 'Polar Plume Anatomy: Results of a Co-ordinated Observation', *Solar Phys.* **175**, 375.
Doschek, G. A., *et al.*: 1997, 'Electron Densities in the Solar Polar Coronal Holes from Density-Sensitive Line Ratios of Si VIII and S X', *Astrophys. J.* **482**, L109.
Dwivedi, B. N.: 1991, 'Forbidden Line Ratios From Si VIII and S X Coronal Ions', *Solar Phys.* **131**, 49.
Feldman, U., Doschek, G. A., Mariska, J. T., Bhatia, A. K., and Mason, H. E.: 1978, 'Electron Densities in the Solar Corona From Density-Sensitive Line Ratios in the N I Isoelectronic Sequence', *Astrophys. J.* **226**, 674.
Feldman, U., *et al.*: 1997, 'A Coronal Spectrum in the 500–1610 Å Wavelength Range Recorded at a Height of 21 000 Kilometers Above the West Solar Limb by the SUMER Instrument on Solar and Heliospheric Observatory', *Astrophys. J. Suppl. Ser.* **113**, 195.
Geiss, J., *et al.*: 1995a, 'The Southern High-Speed Stream: Results from the SWICS Instrument on Ulysses', *Science* **268**, 1033.
Geiss, J., Gloeckler, G., and von Steiger, R.: 1995b, 'Origin of the Solar Wind From Composition Data', *Space Sci. Rev.* **72**, 49.
Hassler, D. M., Wilhelm, K., Lemaire, P., and Schühle, U. 1997, 'Observations of Polar Plumes with the SUMER Instrument on SOHO', *Solar Phys.* **175**, 375.
Hollandt, J., *et al.*: 1996, 'Radiometric Calibration of the Telescope and Ultraviolet Spectrometer SUMER on SOHO', *Appl. Opt.* **35**, 5125.

Keenan, F. P., Kingston, A. E., Dufton, P. L., Doyle, J. G., and Widing, K. G.: 1984, 'Mg IX and Si XI Line Ratios in the Sun', *Solar Phys.* **94**, 91.

Kohl, J. L., *et al.*: 1997, 'Measurements of H I and O VI Velocity Distributions in the Extended Solar Corona With UVCS/SOHO and UVCS/SPARTAN 201', *Adv. Space Res.* **20**, 3.

Krieger, A. S., Timothy, A. F., and Roelof, E. C.: 1973, 'A Coronal Hole and Its Identification as the Source of a High Velocity Solar Wind Stream', *Solar Phys.* **29**, 505.

Laming, J. M., *et al.*: 1997, 'Electron Density Diagnostics for the Solar Upper Atmosphere From Spectra Obtained by SUMER/SOHO', *Astrophys. J.* **485**, 911.

Lemaire, P., *et al.*: 1997, 'First Results of the SUMER Telescope and Spectrometer on SOHO, II. Imagery and Data Management', *Solar Phys.* **170**, 105.

Newkirk, Jr., G. and Harvey, J.: 1968, 'Coronal Polar Plumes', *Solar Phys.* **3**, 321.

Noci, G., Kohl, J. L., and Withbroe, G. L.: 1987, 'Solar Wind Diagnostics From Doppler-Enhanced Scattering', *Astrophys. J.* **315**, 706.

Saito, K.: 1965, 'Polar Rays of the Solar Corona', *Publications of the Astronomical Society of Japan* **17**, 1.

Summers, H. P., Brooks, D. H., Hammond, T. J., and Lanzafame, A. C.: 1996, Atomic Data and Analysis Structure (University of Strathclyde).

Van de Hulst, H. C.: 1950a, 'The Electron Density of the Solar Corona', *Bull. Astron. Inst. Netherlands* **11**, 135.

Van de Hulst, H. C.: 1950b, 'On the Polar Rays of the Corona', *Bull. Astron. Inst. Netherlands* **11**, 150.

Wang, Y.-M., *et al.*: 1997, 'Association of Extreme-Ultraviolet Imaging Telescope (EIT) Polar Plumes with Mixed-Polarity Magnetic Network', *Astrophys. J.* **484**, L75.

Widing, K. G. and Feldman, U.: 1989, 'Abundance Variations in the Outer Solar Atmosphere Observed in Skylab Spectroheliograms', *Astrophys. J.* **344**, 1046.

Widing, K. G. and Feldman, U.: 1992, 'Element Abundances and Plasma Properties in a Coronal Polar Plume', *Astrophys. J.* **392**, 715.

Widing, K. G. and Feldman, U.: 1993, 'Nonphotospheric Abundances in a Solar Active Region', *Astrophys. J.* **416**, 392.

Wilhelm, K., *et al.*: 1995, 'SUMER – Solar Ultraviolet Measurements of Emitted Radiation', *Solar Phys.* **162**, 189.

Wilhelm, K., *et al.*: 1997a, 'First Results of the SUMER Telescope and Spectrometer on SOHO, I. Spectra and Spectroradiometry', *Solar Phys.* **170**, 75.

Wilhelm, K., et al: 1997b, 'Radiometric Calibration of SUMER: Refinement of the Laboratory Results under Operational Conditions on SOHO', *Appl. Optics* **36**, 6416.

Wilhelm, K.: 1998, 'Results from the SUMER Telescope and Spectrometer – Solar Ultraviolet Measurements of Emitted Radiation – on SOHO', in *Proceedings of the SOLTIP III Symposium*, Beijing, October 14-18, 1996, in press.

Wilhelm, K., *et al.*: 1998, 'The Solar Corona Above Polar Coronal Holes as Seen by SUMER on SOHO', *Astrophys. J.* **500**, in press.

Woch, J., *et al.*: 1997, 'SWICS/Ulysses Observations: The Three-Dimensional Structure of the Heliosphere in the Declining/Minimum Phase of the Solar Cycle', *Geophys. Res. Lett.* **24**, 2885.

SOLAR ENERGETIC PARTICLE ISOTOPIC COMPOSITION

D. L. WILLIAMS, R. A. LESKE, R. A. MEWALDT and E. C. STONE
California Institute of Technology, Pasadena, CA, 91125.

Abstract. We discuss isotopic abundance measurements of heavy ($6 \leq Z \leq 14$) solar energetic particles with energies from ~15 to 70 MeV/nucleon, focusing on new measurements made on SAMPEX during two large solar particle events in late 1992. These measurements are corrected for charge/mass dependent acceleration effects to obtain estimates of coronal isotopic abundances and are compared with terrestrial and solar wind isotope abundances. An example of new results from the Advanced Composition Explorer is included.

1. Introduction

Solar energetic particles (SEP) provide a sample of solar material that can be used (1) to determine the elemental and isotopic composition of the Sun, the dominant reservoir of material in the solar system, and (2) to study particle acceleration. However, such studies must take into account that the SEP elemental composition can vary greatly from one event to another. Previous studies have shown that there are two classes of SEP events (Reames 1990, 1998). Impulsive events are enriched in ^{3}He and heavy elements, and are associated with impulsive solar flares. Gradual events, which we consider here, are accelerated by shocks associated with coronal mass ejections, and have, on average, a composition similar to that of the corona.

There are two established differences between the elemental composition of gradual events and that of the photosphere. (1) The abundance of elements with high (>10 eV) first ionization potential (FIP) is depleted in SEPs by a factor of ~4 (Cook *et al.*, 1984). This FIP-fractionation is apparently introduced as material passes from the photosphere to the corona (see e.g., Hénoux, 1998; Geiss, 1998). (2) There are event-to-event variations in SEP composition for $Z \geq 6$ that are well-organized by the ionic charge-to-mass ratio (Q/M) of the elements (Breneman and Stone, 1985). Coronal abundances can be derived by correcting for such variations.

In this analysis we combine the observed variations in the heavy element composition with ionic charge state data to correct for the expected Q/M-dependent isotope fractionation and to obtain SEP-based coronal isotope abundances. These coronal isotope abundances can be compared to estimates of the bulk composition of solar system matter (e.g., Anders and Grevesse, 1989), which are based primarily on terrestrial material.

2. Isotopic Composition and Fractionation

Although heavy SEP isotopes were first measured in the 1970's, there have been only a few measurements to date (see Table I). Of these, only the ISEE-3 and

Table I

Summary of published SEP isotope measurements. The symbols correspond to the data shown in Figures 1 and 2.

Instrument	Time Period	Number of SEP Events	Symbol and Reference
IMP-8	1974 to 1979	7,10	△ Dietrich and Simpson (1979, 1981)
ISEE-3	1978 23–27 Sept	1	□ Mewaldt et al. (1979, 1984)
S81-1	1982 May–Dec	4	◊ Simpson et al. (1984)
SAMPEX	1992 30 Oct–2 Nov	1	▼ Williams (1998)
SAMPEX	1992 2–7 Nov	1	◆ Williams (1998)

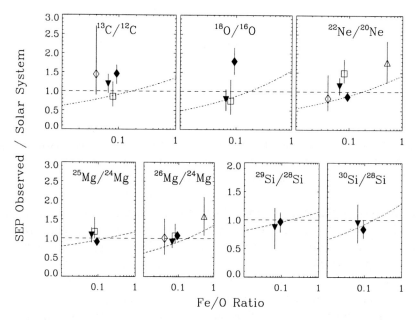

Figure 1. Comparison of SEP isotopic ratios observed in interplanetary space as a function of the ratio of iron to oxygen for the SEP events. The dotted lines show the expected Q/M fractionation, based on the Fe/O ratio of each SEP event. For definition of the symbols, see Table I.

SAMPEX results come from observations of individual SEP events. These new SAMPEX results supersede those in Selesnick et al. (1993).

Measured values for seven isotope ratios are shown in Figure 1, normalized to the solar system abundances of Anders and Grevesse (1989). The dotted lines show an estimate of the expected isotope fractionation, as discussed below. The ratios ^{22}Ne/^{20}Ne and ^{26}Mg/^{24}Mg both appear to follow this fractionation, although the statistical uncertainties are fairly large.

To derive coronal abundances using the approach of Breneman and Stone it is necessary to know the ionic charge states of the SEPs. Breneman and Stone used measurements of the mean charge states of 0.5 to 2.5 MeV/nucleon SEPs (Luhn

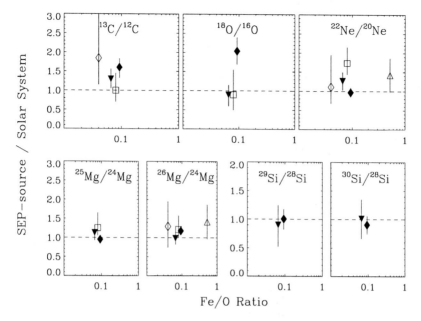

Figure 2. The SEP isotopic abundance ratios corrected for acceleration and propagation effects, normalized by the solar system isotopic ratios (Anders and Grevesse, 1989). These are plotted versus the iron to oxygen ratio in each SEP event. For definition of the symbols, see Table I.

et al., 1984) to correct their ~5 to 30 MeV/nucleon data and found that variations in the elemental composition of $6 \leq Z \leq 28$ nuclei could be parameterized by a power law in particle charge-to-mass ratio, Q/M. The power law index has been found to range from extremes of –4 to +5 (Garrard and Stone, 1994).

The degree of Q/M-dependent fractionation for $6 \leq Z \leq 28$ isotopes in an SEP event can be estimated from the Fe/O ratio (Mewaldt and Stone, 1989), since there is a large difference in the average Q/M ratio of Fe and O, and since both elements are usually well-measured in SEP events. In Figure 1, the dotted line shows the expected isotope fractionation based on the Fe/O ratio in each study. Mewaldt and Stone (1989) used this approach to correct the isotope measurements of Dietrich and Simpson (1979) and Simpson *et al.* (1984). There were no charge state measurements in most of those events, and so the average charge state values from Luhn *et al.* (1984) were assumed. There are, however, two possible pitfalls in this approach. First, the Fe/O ratio in SEP events can vary even after the Q/M fractionation is taken into account, due to variations in the magnitude of the FIP-dependent fractionation (Garrard and Stone, 1994). Thus, to find Q/M-dependent variations independent of FIP-dependent effects, it is best to use only elements with low FIP, as in Williams (1998).

The second pitfall arises from assuming that the same average charge states apply to all isotope measurements. If the charge states vary from one event to another, or vary with energy, then the calculated fractionation can be incorrect. Indeed,

Table II

SEP isotopic ratios observed in two SEP events with the MAST instrument on SAMPEX (Williams, 1998) corrected to the SEP source. Also shown are the "solar system" ratios of Anders and Grevesse (1989).

Mass Ratio	Energy Interval[1] (MeV/nucleon)	Isotopic Ratio (percent)		
		Oct 30, 1992 Event	Nov 2, 1992 Event	"Solar System"
$^{13}C/^{12}C$	14 – 70	$1.44^{+0.28}_{-0.27}$	$1.79^{+0.25}_{-0.31}$	1.11
$^{14}C/^{12}C$	14 – 70	< 0.25	< 0.15	0
$^{15}N/^{14}N$	15 – 70	< 0.59	< 1.31	0.369
$^{17}O/^{16}O$	17 – 70	< 0.12	< 0.16	0.038
$^{18}O/^{16}O$	17 – 70	$0.18^{+0.05}_{-0.06}$	$0.42^{+0.07}_{-0.07}$	0.201
$^{21}Ne/^{20}Ne$	19 – 70	< 0.67	< 1.40	0.243
$^{22}Ne/^{20}Ne$	19 – 70	$9.27^{+1.58}_{-1.73}$	$7.09^{+0.88}_{-0.85}$	7.31
$^{25}Mg/^{24}Mg$	20 – 70	$14.24^{+2.35}_{-2.49}$	$12.01^{+1.11}_{-1.24}$	12.6
$^{26}Mg/^{24}Mg$	20 – 70	$13.74^{+2.17}_{-2.28}$	$16.41^{+1.53}_{-1.53}$	13.9
$^{29}Si/^{28}Si$	22 – 70	$4.56^{+1.70}_{-1.93}$	$5.10^{+0.83}_{-0.93}$	5.07
$^{30}Si/^{28}Si$	22 – 70	$3.40^{+1.11}_{-1.17}$	$3.04^{+0.51}_{-0.54}$	3.36

[1] The upper limit to the energy interval was chosen so that the observed particles are not contaminated by galactic cosmic rays.

there is recent evidence for both energy-dependent and event-to-event variations in the charge state of Fe (Oetliker et al., 1997; Mazur et al., 1998; Moebius et al., 1998). Only the SAMPEX results have ionic charge state measurements based on the same particles used to determine isotope abundances (Leske et al., 1995).

The derived coronal isotope ratios are shown in Figure 2. These have been corrected for Q/M-dependent effects within the limitations discussed above. Although the ratios are plotted against the Fe/O ratio, as in Figure 1, after the Q/M correction there is no apparent trend with respect to Fe/O. Table II lists the new isotope measurements from SAMPEX (Williams, 1998). There is fair agreement between the measurements and the solar system values. However, in the 11/02/92 SEP event the $^{13}C/^{12}C$ ratio is ~2 standard deviations greater than the solar system value, and the $^{18}O/^{16}O$ ratio is ~3 standard deviations greater. This suggests that there could be further differences between the SEP-source material and the standard solar system abundances or that additional fractionation effects may have occurred during the acceleration process.

3. Neon Isotope Abundances

Neon is a volatile element that is depleted in terrestrial and meteoritic material. Studies of meteoritic neon have identified a number of isotopically distinct components (see Figure 3 and the review by Black, 1983). Neon-A, thought to be a primordial component, was chosen by Cameron (1982) for his table of solar sys-

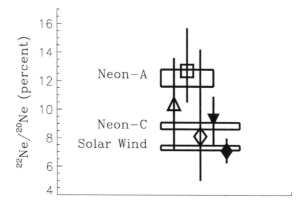

Figure 3. The SEP-source ratio of ^{22}Ne/^{20}Ne compared to other samples of solar system neon. SEP measurements: △ Dietrich and Simpson (1979); □ Mewaldt *et al.* (1984); ◊ Simpson *et al.* (1984); ▼ and ◆ Williams (1998). Neon-A and Neon-C are discussed in the text. The solar wind value is from Geiss *et al.* (1972).

tem abundances. Neon-B is thought to represent implanted solar wind, since its composition is close to the ^{22}Ne/^{20}Ne=0.073 value measured directly in the solar wind (Geiss *et al.*, 1972). Neon-C, with ^{22}Ne/^{20}Ne=0.089±0.002, is thought to be due to implanted energetic particles with energies <0.1 MeV/nucleon (Wieler *et al.*, 1986; Wieler, 1998). However, the particle fluxes required to account for the observed Neon-C density appear to be greater than the present day SEP flux (Humbert *et al.*, 1998), and so the source of Neon-C is as yet unknown.

Early measurements of the ^{22}Ne/^{20}Ne ratio in SEPs were more consistent with Neon-A than with solar wind neon, apparently supporting Cameron's choice of Neon-A as representative of solar system material. However, the new, more precise SAMPEX results favor the solar wind value (or Neon-C) rather than Neon-A. Thus, the ISEE-3 ratio, which is ~2.5 standard deviations higher, suggests that an additional fractionation process may have occurred, as discussed above for ^{13}C and ^{18}O.

4. SEP Isotope Studies on ACE

The Advanced Composition Explorer (ACE), launched in August, 1997, is making coordinated measurements of the elemental, isotopic, and ionic charge state composition of nuclei from H to Ni ($1 \geq Z \geq 28$), from solar wind to galactic cosmic ray energies. In the ~10 to 100 MeV/nucleon energy range, SEP isotope measurements are being carried out by the Solar Isotope Spectrometer (SIS; Stone *et al.*, 1998), which has improved resolution and collecting power over earlier SEP isotope spectrometers.

Shortly after the launch of ACE, two large SEP events in November, 1997 heralded the onset of activity leading to the next solar maximum. The Nov. 6 event

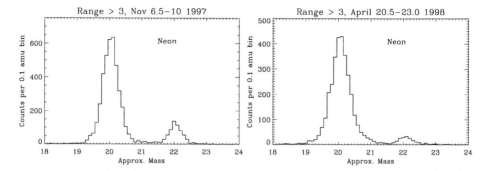

Figure 4. Neon histograms from the SIS instrument on ACE. *Left:* Particles at energies ∼25 to 100 MeV/nucleon from the SEP event which started on 6 November 1997. *Right:* Particles at energies ∼25 to 50 MeV/nucleon from the SEP event which started on 20 April 1998. These preliminary results show a large variation in the ^{22}Ne/^{20}Ne ratio between these two SEP events.

was unusually enriched in heavy nuclei (Fe/O≃1), with a very hard energy spectrum detected by ground-level neutron monitors. Figure 4 shows mass histograms of Ne nuclei from SEP events on 6 Nov. 1997 and 20 April 1998. The ^{22}Ne/^{20}Ne ratio in the Nov. 6 event is ∼0.15, considerably greater than that observed in earlier events. Indeed, there is also evidence for similar mass-dependent fractionation in the isotopes of C, O, Mg, Si, S, Ca, and Fe (Leske *et al.*, 1998). The 20 April 1998 event was Fe-poor and has a ^{22}Ne/^{20}Ne ratio of ∼0.05 (Leske *et al.*, 1998).

If plotted in Figure 1 at Fe/O≃1, the ^{22}Ne/^{20}Ne ratio for the Nov. 6 event would be greater than that expected from the simple Q/M dependent fractionation model described here. However, the curves in Figure 1 assume an Fe charge state of +15, and there are indications that the charge state of >15 MeV/nucleon Fe in this event may be greater than this (Cohen *et al.*, 1998; Mazur *et al.*, 1998; Moebius *et al.*, 1998). If so, it is possible that the observed enrichments in Fe/O and ^{22}Ne/^{20}Ne can be accounted for with the same Q/M dependent fractionation, although this hypothesis has yet to be tested. With a large geometry factor and excellent mass resolution, SIS promises to usher in a new era of SEP isotope studies that will extend to a wide range of elements with excellent statistical accuracy.

5. Discussion

The results presented here support the assumption that the Q/M dependent fractionation observed in SEP elemental abundances also acts to fractionate isotopic abundances. After correcting for this fractionation, most of the available SEP isotope measurements are consistent with the solar system abundances of Anders and Grevesse (1989). Although there remain isolated isotopic anomalies in some SEP events (e.g., ^{13}C/^{12}C and ^{18}O/^{16}O observed by SAMPEX in the 11/02/92 event and ^{22}Ne/^{20}Ne observed by ISEE-3 in the 9/23/78 event), there is no unequivocal evidence for additional fractionation effects beyond those considered here. However, the recent evidence for energy-dependent and event-to-event variations in SEP charge states emphasizes the need for simultaneous measurements of both the mass and ionic charge distributions of SEPs if fractionation effects are to be reliably evaluated. The SAMPEX results presented here are the only measurements for which this is possible at present.

It is now clear that the ^{22}Ne/^{20}Ne ratio in SEPs can be greater than the corresponding ratio in the solar wind in individual SEP events. While this provides possible support for the interpretation of Neon-C as implanted SEPs, there is not yet sufficient data to estimate the long-term average ^{22}Ne/^{20}Ne ratio in SEPs, and there remains a large uncertainty in the absolute abundance of Neon-C in lunar material (Wieler, 1998).

The SEP isotope measurements to date generally support the assumption that SEPs and solar wind come from the same coronal reservoir. However, these studies are still at an exploratory stage: there are observations in only a limited number of events, and the uncertainties are large. It is hoped that in the near future, the combination of improved SEP isotope measurements from ACE, charge state measurements from ACE and SAMPEX, and solar wind isotope studies by WIND, SOHO, and ACE, will shed new light on this subject.

Acknowledgements

We appreciate the contributions of C. M. S. Cohen, A. C. Cummings, T. T. von Rosenvinge, and M. E. Wiedenbeck to the development and data analysis of the SIS instrument on ACE. This work was also funded by NASA grants NAG5-2963, and NAG5-6912. One of us (D. L. Williams) would like to acknowledge support from the NASA Graduate Student Researchers Program, grant number NGT-51156.

References

Anders, E. and Grevesse, N.: 1989, 'Abundances of the Elements: Meteoritic and Solar', *Geochim. Cosmochim. Acta* **53**, 197–214.
Black, D. C.: 1983, 'The Isotopic Composition of Solar Flare Noble Gases', *Ap. J.* **266**, 889–894.
Breneman, H. H. and Stone, E. C.: 1985, 'Solar Coronal and Photospheric Abundances from Solar Energetic Particle Measurements', *Astrophys. J. Lett.* **199**, L57–L61.
Cameron, A. G. W.: 1982, 'Elemental and Nuclidic Abundances in the Solar System', in C. A. Barnes, D. D. Clayton and D. N. Schramm (eds), *Essays in Nuclear Astrophysics*, Cambridge Univ. Press, Cambridge, pp. 23–43.
Cohen, C. M. S., Cummings, A. C., Leske, R. A., Mewaldt, R. A., Stone, E. C., Christian, E. R., von Rosenvinge, T. T., Dougherty, B. L., Wiedenbeck, M. E.: 1998, 'Inferred Charge States of High Energy Solar Particles from the Solar Isotope Spectrometer on ACE', *Geophys. Res. Lett.*, submitted.
Cook, W. R., Stone, E. C. and Vogt, R. E.: 1984, 'Elemental Composition of Solar Energetic Particles', *Astrophys. J.* **279**, 827–838.
Dietrich, W. F. and Simpson, J. A.: 1979, 'The Isotopic and Elemental Abundances of Neon Nuclei Accelerated in Solar Flares', *Astrophys. J. Lett.* **231**, L91–L95.
Dietrich, W. F. and Simpson, J. A.: 1981, 'The Isotopic Composition of Magnesium Nuclei in Solar Flares', *Astrophys. J. Lett.* **245**, L41–L44.

Garrard, T. L. and Stone, E. C.: 1994, 'Composition of Energetic Particles from Solar Flares', *Adv. Space Res.* **14**(10), 589–598.

Geiss, J., Bühler, F., Cerutti, H., Eberhardt, P., and Filleux, Ch.: 1972, 'Solar Wind Composition Experiment', *Apollo 16 Prel. Sci. Report*, NASA SP-**315**, Sect. 14, 3375–3398.

Geiss, J.: 1998, 'Solar Wind Abundance Measurements: Constraints on the FIP Mechanism', *Space Science Reviews*, this volume.

Hénoux, J.-C.: 1998, 'FIP Fractionation: Theory', *Space Science Reviews*, this volume.

Humbert, F., Wieler, R. and Marty, B.: 1998, 'Nitrogen and Argon in Individual Mineral Grains of a Lunar Soil: Evidence for a Major Non-Solar Nitrogen Component', *Lunar and Planetary Science Conference*. Abstract no. 1034.

Leske, R. A., Christian, E. R., Cohen, C. M. S., Cummings, A. C., Dougherty, B. L., Mewaldt, R. A., Stone, E. C., von Rosenvinge, T. T. and Wiedenbeck, M. E.: 1998, 'First Results on the Isotopic Composition of Solar Energetic Particles From the Solar Isotope Spectrometer on ACE', *Trans. Am. Geophys. Union (Suppl.)* **79**, S256.

Leske, R. A., Cummings, J. R., Mewaldt, R. A., Stone, E. C. and von Rosenvinge, T. T.: 1995, 'Measurements of the Ionic Charge States of Solar Energetic Particles Using the Geomagnetic Field', *Astrophys. J. Lett.* **452**, L149–L152.

Luhn, A., Klecker, B., Hovestadt, D., Gloeckler, G., Ipavich, F. M., Scholer, M., Fan, C. Y. and Fisk, L. A.: 1984, 'Ionic Charge States of N, Ne, Mg, Si, and S in Solar Energetic Particle Events', *Adv. Space Res.* **4**(2), 161–164.

Mazur, J. E., Mason, G. M., Looper, M. D., Leske, R. A. and Mewaldt, R. A.: 1998, 'Charge States of Solar Energetic Particles Using the Geomagnetic Cutoff Technique: SAMPEX Measurements in the 1997 November Solar Particle Event', *Geophys. Res. Lett.*, submitted.

Mewaldt, R. A., Spalding, J. D. and Stone, E. C.: 1984, 'A High-Resolution Study of the Isotopes of Solar Flare Nuclei', *Astrophys. J.* **280**, 892–901.

Mewaldt, R. A., Spalding, J. D., Stone, E. C. and Vogt, R. E.: 1979, 'The Isotopic Composition of Solar Flare Accelerated Neon', *Astrophys. J. Lett.* **231**, 97–100.

Mewaldt, R. A. and Stone, E. C.: 1989, 'Isotope Abundances of Solar Coronal Material Derived from Solar Energetic Particle Measurements', *Astrophys. J.* **337**, 959.

Moebius, E., Popecki, M., Klecker, B., Kistler, L. M., Bogdanov, A., Galvin, A. B., Heirtzler, D., Hovestadt, D., Lund, E. J., Morris, D. and Scmidt, W. K. H.: 1998, 'Energy Dependence of the Ionic Charge State Distribution During the November 7 – 9, 1997 Solar Energetic Particle Event as Observed with ACE SEPICA', *Geophys. Res. Lett.*, submitted.

Oetliker, M., Klecker, B., Hovestadt, D., Mason, G. M., Mazur, J. E., Leske, R. A., Mewaldt, R. A., Blake, J. B. and Looper, M. D.: 1997, 'The Ionic Charge State of Solar Energetic Particles with Energies of 0.3-70 MeV per Nucleon', *Astrophys. J.* **477**, 495–501.

Reames, D. V.: 1990, 'Energetic Particles from Impulsive Solar Flares', *Ap. J. Supp.* **73**, 235–251.

Reames, D. V.: 1998, 'Solar Energetic Particles: Sampling Coronal Abundances', *Space Science Reviews*, this volume.

Selesnick, R. S., Cummings, A. C., Cummings, J. R., Leske, R. A., Mewaldt, R. A. and Stone, E. C.: 1993, 'Coronal Abundances of Neon and Magnesium Isotopes from Solar Energetic Particles', *Astrophys. J. Lett.* **418**, L45–L48.

Simpson, J. A., Wefel, J. P. and Zamow, R.: 1984, 'Isotopic and Elemental Composition of Solar Energetic Particles', *Proc. 18th Internat. Cosmic Ray Conf.*, Vol. 10, Bangalore, pp. 322–325.

Stone, E. C., Cohen, C. M. S., Cook, W. R., Cummings, A. C., Gauld, B., Kecman, B., Leske, R. A., Mewaldt, R. A., Thayer, M. R., Dougherty, B. L., Grumm, R. L., Milliken, B. D., Radocinski, R. G., Wiedenbeck, M. E., Christian, E. R., Shuman, S. and von Rosenvinge, T. T.: 1998, 'The Solar Isotope Spectrometer for the Advanced Composition Explorer', *Space Sci. Rev.*, in press.

Wieler, R.: 1998, 'The Solar Noble Gas Record in Lunar Samples and Meteorites', *Space Science Reviews*, this volume.

Wieler, R., Baur, H. and Signer, P.: 1986, 'Noble Gases From SEPs Revealed by Closed System Stepwise Etching of Lunar Soil Materials', *Geochim. Cosmochim. Acta* **50**, 1997–2017.

Williams, D. L.: 1998, *Measurements of the Isotopic Composition of Solar Energetic Particles with the MAST Instrument onboard SAMPEX*, PhD thesis, California Institute of Technology.

O^{5+} IN HIGH SPEED SOLAR WIND STREAMS: SWICS/ULYSSES RESULTS

ROBERT F. WIMMER-SCHWEINGRUBER
Physikalisches Institut, Universität Bern, Switzerland

RUDOLF VON STEIGER and JOHANNES GEISS
International Space Science Institute, Bern, Switzerland

GEORGE GLOECKLER and FRED M. IPAVICH
Dept. of Physics, University of Maryland, Maryland, USA

BEREND WILKEN
Max-Planck-Institut für Aeronomie, Katlenburg-Lindau, Germany

Abstract. Recent observations with UVCS on SOHO of high outflow velocities of O^{5+} at low coronal heights have spurred much discussion about the dynamics of solar wind acceleration. On the other hand, O^{6+} is the most abundant oxygen charge state in the solar wind, but is not observed by UVCS or by SUMER because this helium-like ion has no emission lines falling in the wave lengths observable by these instruments. Therefore, there is considerable interest in observing O^{5+} *in situ* in order to understand the relative importance of O^{5+} with respect to the much more abundant O^{6+}. High speed streams are the prime candidates for the search for O^{5+} because all elements exhibit lower freezing-in temperatures in high speed streams than in the slow solar wind. The Ulysses spacecraft was exposed to long time periods of high speed streams during its passage over the polar regions of the Sun. The Solar Wind Ion Composition Spectrometer (SWICS) on Ulysses is capable of resolving this rare oxygen charge state. We present the first measurement of O^{5+} in the solar wind and compare these data with those of the more abundant oxygen species O^{6+} and O^{7+}. We find that our observations of the oxygen charge states can be fitted with a single coronal electron temperature in the range of 1.0 to 1.2 MK assuming collisional ionization/recombination equilibrium with an ambient Maxwellian electron gas.

Key words: Solar wind, solar wind acceleration, corona

1. Introduction

The derivation of a coronal temperature profile is of considerable interest for an understanding of the mechanisms heating the corona and accelerating the solar wind. Ions of different elements form at different coronal temperatures and can be detected optically at different distances from the Sun. However, quantification of the results is rather complicated and strongly dependent on model assumptions and atomic data (e. g. Esser *et al.*, 1997). O^{5+} has long been known to be a strong emitter in the coronal EUV (e. g. Withbroe, 1971). Recently, O^{5+} has also been detected with optical instruments aboard the SOHO space craft (e. g. Kohl *et al.*, 1997, or Warren *et al.*, 1997). There are three points about these observations that have spurred discussion. The O^{5+} lines are very broad, which is attributed to wide velocity distributions along the line of sight (e. g. Kohl *et al.*, 1997). O^{5+} seems to flow outwards at velocities considerably higher than previously thought and these

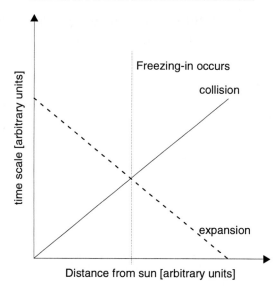

Figure 1. Schematic of the freezing in process. See text for discussion.

high velocities are observed unexpectedly close to the solar surface (e. g. Kohl *et al.*, 1997). Finally, the fact that no O^{5+} had been reported from *in situ* measurements in the solar wind made the measurements of this ion so far out in the solar corona seem to be surprising. Why was this ion not seen in the solar wind? We will address this last question in this work. O^{6+} forms at a higher temperature than the detected O^{5+} and is much more abundant in the solar wind. However, the emission lines of helium-like O^{6+} do not lie in the acceptance range of the UVCS and SUMER instruments and therefore it is not to be expected to observe this ion.

The relative abundances of the charge states of an element are generally believed to result from a "freezing-in" process (Hundhausen *et al.*, 1968). This can be viewed as a competition between two time scales, expansion and ionization/recombination (see Figure 1). The hot electron plasma in the outer corona interacts through collisions with the ions which are accelerated into the solar wind. As the collision time scale exceeds the expansion time scale the charge state of the ions is "frozen-in". Under the assumption that the ions are in collisional equilibrium with the ambient electron gas it is possible to derive a coronal temperature from the relative abundances of the charge states. Furthermore, using the charge states of different elements measured *in situ* it is, in principle, possible to derive constraints on the electron temperature and outflow velocities of ions in the corona (e. g. Geiss *et al.*, 1995; Ko *et al.*, 1997; Esser *et al.*, 1997).

With the Solar Wind Ion Composition Spectrometer (SWICS) aboard the Ulysses spacecraft it is possible to determine simultaneously the energy, mass, and mass per charge of an ion of the solar wind. The SWICS instrument (Gloeckler *et al.*, 1992) is a time of flight mass spectrometer with a stepped E/q entrance system/energy analyzer. In addition to measurement of these two quantities (time of flight and E/q) the energy of the solar wind ion is measured in a solid state detector. This results in three measurements which allows simultaneous determination of the three quantities energy, mass, and mass per charge of a solar wind ion. It is thus possible to determine the relative abundances of even the rare charge states of ions, in particular also of O^{5+}, which is the subject of this work. It is the aim of this paper to report on the relative abundance of O^{5+} in high speed streams which originate in coronal holes (Krieger et al., 1973), regions which have been well studied with the UVCS and SUMER instruments, and in which O^{5+} has been detected by these instruments. The out-of-ecliptic orbit of Ulysses over the polar regions of the Sun is ideally suited for this study.

2. Data Selection and Analysis

For this work we selected the first 150 days of 1996 when Ulysses was fully immersed in the high speed stream which originated in the northern polar coronal hole. Two hour averages of some solar wind parameters for this time period are plotted in Figure 2. From top to bottom we have the α/p ratio, oxygen and carbon freezing-in temperatures, and finally bulk solar wind (proton) speed. We observe a constant α/p value of $4.7 \pm .1\%$ which is typical of high speed streams (Bame et al., 1977), as well as constant oxygen and carbon freezing-in temperatures of 1.2 MK and 0.9 MK respectively, which are the values typically measured in this type of solar wind (Geiss et al., 1995). Bulk solar wind speed varied between ~ 710 km/s and ~ 820 km/s with an average of $\sim 760 \pm 20$ km/s. Oxygen and carbon freezing-in temperatures were derived from the number density ratios of O^{7+} to O^{6+} and C^{6+} to C^{5+} respectivley using the ionization and recombination rates of Arnaud and Rothenflug (1985). This time period shows no unusual characteristics and the source region of this wind lies well within the boundaries of the northern polar coronal hole.

In Figure 3 we show a contour plot of measured energy versus measured time of flight accumulated in an E/q-step in which O^{5+} is expected to enter given the average bulk solar wind alpha particle speed of 760 km/s. The principal ions O^{6+}, C^{5+}, and C^{4+} are identified, as are some other, less abundant, heavy ions. The thick solid hyperbola-like curve is the theoretical curve in $E - T$ space along which oxygen ions need to fall. It includes energy losses in the carbon foil and is thus not just a hyperbola as would be the case if the ions lost no energy during their passage through the carbon foil. O^{5+} is visible as an extension of the C^{4+} peak towards higher energy and lesser time of flight. We see that O^{5+} lies on the theoretical curve for oxygen. If the O^{5+} peak were a tail of the C^{4+} peak due to energy loss in the carbon foil, it would have to lie to the lower right of the C^{4+} peak, along a similar curve for carbon, and not to the upper right, as it is observed to lie.

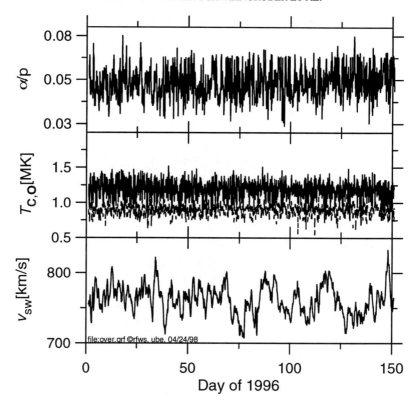

Figure 2. Solar wind parameters for the time period analysed. From top to bottom: α/p ratio, oxygen (solid) and carbon (dashed) freezing-in temperatures, and bulk alpha particle speed.

Clearly, O^{5+} is a very rare charge state of oxygen, and therefore contamination from neighboring ions needs to be carefully considered. In our analysis phase space densities of ions are extracted from $E - T$ matrices (such as shown in Figure 3) which are accumulated from the measured E and T over a given period of time. For each E/q step the procedure consists of fitting a sum of model peaks to such an $E-T$ matrix. The positions and widths of the peaks are kept fixed for each E/q step, only the abundances are allowed to vary. Ideally, this procedure should remove all contamination from the peak under investigation. However, our knowledge of the exact shapes of the peaks is incomplete and this results in some relic contamination.

In order to further reduce this contamination of the O^{5+} peak, we exploited the fact that all heavy ions in the solar wind and especially in high speed streams flow at the same speed (von Steiger et al., 1995, and references therein) although alpha particles and protons generally exhibit differential streaming (Neugebauer, 1981). Thus, due to its larger mass per charge, O^{5+} will enter the instrument at a higher value of E/q than O^{6+} or C^{4+}. For each step of the E/q analyzer, we derived the differential flux for each oxygen charge state using the fitting procedure just described. For each charge state and each E/q step we divided the corresponding

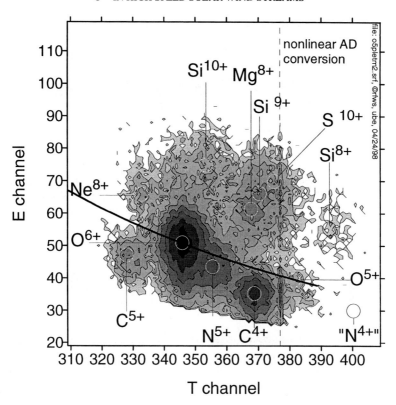

Figure 3. Contour plot of counts per $E - T$ bin accumulated in the principal E/q step for O^{5+} over the time period investigated. Contour levels start at 2 and are stepped by factors of 2.

speed by the bulk alpha particle speed. Together with the values of differential flux this division of speeds results in the reduced distribution function, $f(w)$, where $w \doteq v/v_\alpha$. Figure 4 shows such a reduced distribution function for O^{6+}. The data (solid circles) were accumulated and averaged over the entire period investigated. The distribution is well approximated by a κ-function (solid line) with a κ typical of high speed streams (Collier *et al.*, 1996). The distribution peaks at unity, reflecting the fact that the bulk heavy ions flow at the bulk alpha particle speed. Therefore, if one plots the reduced distribution function of O^{5+} vs. w, contamination of O^{5+} by O^{6+} should show up as a hump at $w = \sqrt{5/6}$. Similarly, contaminations of other ions are manifested as humps at characteristic values of $w = \sqrt{\frac{m/q}{16/5}}$.

3. Results

In the manner described in the previous section we obtained the data plotted in Figure 5 for O^{5+} (solid squares). In order to obtain the relative importances of the contaminations, we fitted a sum of four κ-functions (one for O^{5+}, O^{6+}, C^{4+},

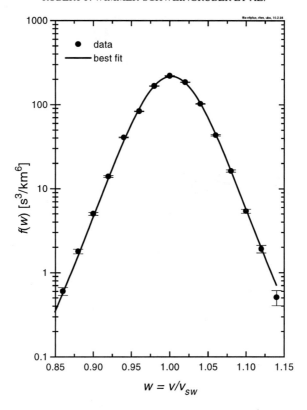

Figure 4. Reduced distribution function for O^{6+}.

and He^{2+}) to the data. The widths and the κ were held fixed at the value derived from O^{6+} (see Figure 4). In other words, all ions were assumed to have the same κ and thermal speeds. The positions in w were fixed at $w = \sqrt{\frac{m/q}{16/5}}$, and thus there were only the four amplitudes of the peaks left as fitting parameters. The various contributions to the data are shown by a dashed line (O^{5+}), a dotted line (O^{6+}), a dashed-dotted line (C^{4+}), and a dot-dot-dash line (He^{2+}). The solid line shows the sum of all the contributions.

The reduced distribution function of O^{5+} can thus be viewed as a sum of the dominant contribution from O^{5+} plus contributions from other ions. That O^{5+} is the dominant contribution can be seen from the fact that the distribution function peaks at $w = 1$. The only other solar wind ion which could peak at $w = 1$ in this plot is S^{10+} which has the same m/q value as O^{5+}. However, S^{10+} is well removed from O^{5+} in an $E - T$ plot (see Figure 3). A similar procedure can be applied to the O^{7+} data. O^{7+} is contaminated by O^{6+} and He^{2+}.

Figure 6 shows the data for O^{6+} (circles), O^{7+} (triangles), and O^{5+} (squares). The solid lines going through the data are the contributions of the main ions to the

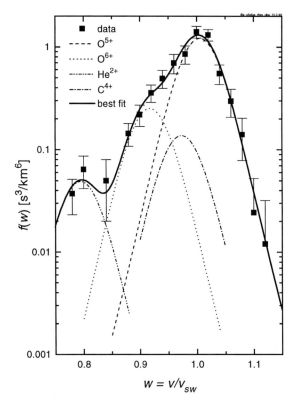

Figure 5. Reduced distribution function for O^{5+} with contributions from O^{6+}, He^{2+}, and C^{4+}.

respective data series. The dashed line shows the best fit to all the O^{5+} data (the solid line of Figure 3), the dotted line the best fit to all the O^{7+} data.

Thus the relative abundances of O^{5+}, O^{6+}, and O^{7+} are given by the relative amplitudes of the peaks for the main ions.

4. Discussion and Conclusions

In the previous section we have shown that O^{5+} exists in high speed streams and that is is more than two orders of magnitude rarer than O^{6+}. In Figure 7 we have plotted the values of the amplitudes derived in the previous section vs. the charge of the ion. The filled circles are the measured values, while the open diamonds are theoretical equilibrium values normalized to O^{6+} and based on the ionization/recombination rates of Arnaud and Rothenflug, (1985) assuming an ambient electron gas temperature of 1.075 MK. Thus for high speed streams it seems to be possible to explain the charge state distribution of oxygen by one equilibrium electron temperature of about 1.1 MK. We can interpret this as the local thermodynamical equilibrium (LTE) temperature at the radius where freezing in of oxygen

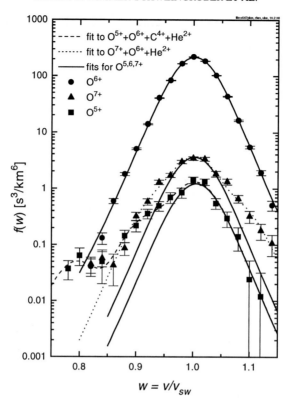

Figure 6. Reduced distribution functions for O^{6+} (circles), O^{7+} (triangles), and O^{5+} (squares). See text for discussion.

occurs, i. e. where the ionization/recombination timescale exceeds the expansion time scale. The high outflow velocities measured by UVCS (Kohl *et al.*, 1997) do not invalidate the concept of LTE, in fact, the results presented in this work seem to constrain any deviation from LTE to be small. The observation that the charge state composition of oxygen in high speed streams can be explained by one single electron temperature lends strong support to the observations of UVCS of high outflow velocities near the Sun. The ratio of O^{5+}/O^{6+} freezes in considerably further out than the ratio of O^{7+}/O^{6+}. In order for the two ratios to result from approximately the same coronal electron temperature, the expansion time needs to be falling steeply in the distance range where oxygen freezes in (see Figure 1). The drop in expansion time is driven by the acceleration of the plasma out of the solar gravitational well. This is equivalent to the steep speed gradients measured by the UVCS team. In fact measurements by Klinglesmith (1997) show outflow speeds with an average of approximately 800 km/s at a heliocentric distance of 2.4 solar radii. According to that work high speed streams would already have attained their speed very near the solar surface.

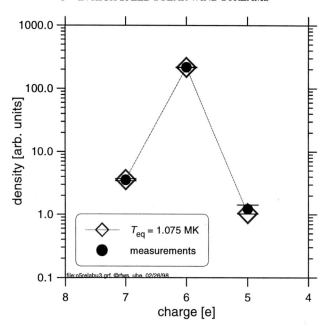

Figure 7. Relative abundances of O^{7+}, O^{6+}, and O^{5+} (circles) and calculated relative abundances (open diamonds) using the ionization and recombination rates of Arnaud and Rothenflug (1985).

Our observation that the oxygen charge state composition can be explained by one electron temperature is reminiscent of the same observation for silicon and iron (Geiss *et al.*, 1995). However, these elements freeze in after the coronal electron temperature maximum. In this region, where electron temperature is decreasing, the charge state distribution is governed mainly by recombination. Of course, these elements have many more charge states and thus are a more effective tracers of the coronal electron temperature profile.

The observations of O^{5+} by the UVCS and SUMER instruments aboard the SOHO spacecraft are not surprising, O^{5+} does exist in small, but detectable quantities in high speed streams.

Acknowledgements

We are very grateful to the many individuals of the SWICS team without whose work this investigation would not have been possible. We gratefully acknowledge helpful discussions with M. R. Aellig, H. Balsiger, P. Bochsler, R. Bodmer, and R. O. Neukomm.

References

Arnaud, M. and Rothenflug, R.: 1985, 'An updated evaluation of recombination and ionization rates', *Astron. Astrophys. Suppl. Ser.* **60**, 425–457.

Bame, S. J., Asbridge, J. R., Feldman, W. C. and Gosling, J. T.: 1977, 'Evidence for a Structure-Free State at High Solar Wind Speeds', *J. Geophys. Res.* **82**, 1487–1492.

Collier, M., Hamilton, D. C., Gloeckler, G., Bochsler, P. and v. Steiger, R.: 1996, 'Neon-20, Oxygen-16, and Helium-4 densities, temperatures, and suprathermal tails in the solar wind determined by WIND/MASS', *Geophys. Res. Lett.* **23**, 1191–1194.

Esser, R., Edgar, R. J. and Brickhouse, N. S.: 1997, 'High Minor Ion Outflow Speeds in the Inner corona and Observed Ion Charge States in Interplanetary Space', *Astrophys. J.*, submitted.

Geiss, J., Gloeckler, G. and von Steiger, R.: 1995, 'Origin of the solar wind from composition measurements', *Space Sci. Rev.* **72**, 49–60.

Geiss, J., Gloeckler, G., von Steiger, R., Balsiger, H., Fisk, L. A., Galvin, A. B., Ipavich, F. M., Livi, S., McKenzie, J. F., Ogilvie, K. W. and Wilken, B.: 1995, 'The Southern High-Speed Stream: Results From the SWICS Instrument on Ulysses', *Science* **268**, 1033–1036.

Gloeckler, G., Geiss, J., Balsiger, H., Bedini, P., Cain, J. C., Fischer, J., Fisk, L. A., Galvin, A. B., Gliem, F., Hamilton, D. C., Hollweg, J. V., Ipavich, F. M., Joos, R., Livi, S., Lundgren, R., Mall, U., McKenzie, J. F., Ogilvie, K. W., Ottens, F., Rieck, W., Tums, E. O., von Steiger, R., Weiss, W. and Wilken, B.: 1992, 'The Solar Wind Ion Composition Spectrometer', *Astron. Astrophys. Suppl. Ser.* **92**, 267–289.

Hundhausen, A., Gilbert, H. and Bame, S.: 1968, 'Ionization State of the Interplanetary Plasma', *J. Geophys. Res.* **73**, 5485–5493.

Klinglesmith, M.: 1997, *The Polar Solar Wind from 2.5 to 40 Solar Radii: Results of Intensity Scintillation Measurements*, PhD thesis, University of California, San Diego.

Ko, Y. K., Fisk, L. A., Geiss, J., Gloeckler, G. and Guhathakurta, M.: 1997, 'An empirical study of the electron temperature and heavy ion velocities in the south polar coronal hole', *Sol. Phys.* **171**, 345–361.

Kohl, J., Noci, G., Antonucci, A., Tondello, G., Huber, M. C. E., Gardner, L. D., Nicolosi, P., Fineschi, S., Raymond, J. C., Romoli, M., Spadaro, D., Siegmund, O. H. W., Benna, C., Ciaravella, A., Cranmer, S. R., Giordano, S., Karovska, M., Martin, R., Michels, J., Modigliani, A., Naletto, G., Panasyuk, A., Pernechele, C., Poletto, G., Smith, P. L. and Strachan, L.: 1997, 'Measurements of H I and O VI velocity distributions in the extended solar corona with UVCS/SOHO and UVCS/SPARTAN 201', *Adv. Space Res.* **20**, 3–14.

Kohl, J., Noci, G., Antonucci, A., Tondello, G., Huber, M. C. E., Gardner, L. D., Nicolosi, P., Strachan, L., Fineschi, S., Raymond, J. C., Romoli, M., Spadaro, D., Panasyuk, A., Siegmund, O. H. W., Benna, C., Ciaravella, A., Cranmer, S. R., Giordano, S., Karovska, M., Martin, R., Michels, J., Modigliani, A., Naletto, G., Pernechele, C., Poletto, G. and Smith, P. L.: 1997, 'First results from the SOHO ultraviolet coronagraph spectrometer', *Sol. Phys.* **175**, 613–644.

Krieger, A. S., Timothy, A. F. and Roelof, E. C.: 1973, 'A coronal hole and its identification as the source of a high velocity solar wind stream', *Sol. Phys.* **29**, 505–525.

Neugebauer, M.: 1981, 'Observations of Solar Wind Helium', *Fundam. Cosmic Phys.* **7**, 131–199.

von Steiger, R., Geiss, J., Gloeckler, G. and Galvin, A. B.: 1995, 'Kinetic Properties of heavy ions in the solar wind from SWICS/ULYSSES', *Space Sci. Rev.* **72**, 71–76.

Warren, H. P., Mariska, J. T. and Wilhelm, K.: 1997, 'Observations of doppler shifts in a solar polar coronal hole', *Astrophys. J.* **490**, L187–L190.

Withbroe, G. L.: 1971, 'A comparison of solar EUV intensities and K-coronameter measurements', *Sol. Phys.* **18**, 458–473.

Address for correspondence: Robert F. Wimmer-Schweingruber, Physikalisches Institut, University of Bern, Sidlerstrasse 5, CH-3012 Bern, Switzerland

ELEMENT AND ISOTOPIC FRACTIONATION IN CLOSED MAGNETIC STRUCTURES

T. H. ZURBUCHEN, L. A. FISK, G. GLOECKLER* and N. A. SCHWADRON
Department of Atmospheric, Oceanic and Space Sciences, University of Michigan, Ann Arbor, USA

Abstract. Recent papers have suggested that the slow solar wind is a super-position of material which is released by reconnection from large coronal loops. This reconnection process is driven by large-scale motions of solar magnetic flux driven by the non-radial expansion of the solar wind from the differentially rotating photosphere into more rigidly rotating coronal holes.

The elemental composition of the slow solar wind material is observed to be fractionated and more variable than the fast solar wind from coronal holes. Recently, it has also been reported that fractionation also occurs in ^{3}He/^{4}He. This may be interpreted in the frame-work of an existing model for fractionation on large coronal loops in which wave-particle interactions preferentially heat ions thereby modifying their scale-heights.

Key words: Solar wind, solar wind composition, FIP fractionation, ^{3}He/^{4}He fractionation, solar magnetic field

1. Introduction

In this paper we discuss a recent model for fractionation processes occurring in the low corona. This model is strongly motivated by recent *in situ* and remote sensing observations. The *in situ* observations have demonstrated that the elemental and isotopic composition in the solar wind strongly depend on the actual solar wind regime observed. Remote solar observations, e.g., from the SOlar and Heliospheric Observatory (SOHO), show that these compositional differences are also seen low in the solar atmosphere. It has been suggested that the elemental fractionation process occurs in the chromosphere (see, e.g., Geiss, 1982) where the solar material is being ionized. Judge *et al.* (1998) discuss recent composition observations in the chromosphere. Abundances in the solar wind are observed to be highly variable in time and they exhibit structures on very small spatial scales. This renders some of the time-stationary chromospheric fractionation mechanisms (see, e.g., Marsch *et al.*, 1995; Peter, 1998) unlikely candidates. The first of these models also does not naturally imply a difference in composition patterns in the fast, coronal hole associated wind compared to slow solar wind. A fractionation theory is needed which does not strongly depend on the geometrical properties of the separating region and can naturally account for composition differences between fast and slow wind.

We introduce a new concept by first considering the overall picture of fast and slow solar wind in the heliosphere, at least near solar minimum, when the concepts for the overall structure are well developed: At high heliographic latitudes, the polar coronal holes give rise to a fast, ~ 750 km/s flow, which is remarkably steady

---

* Also at Department of Physics and IPST, University of Maryland, College Park, USA

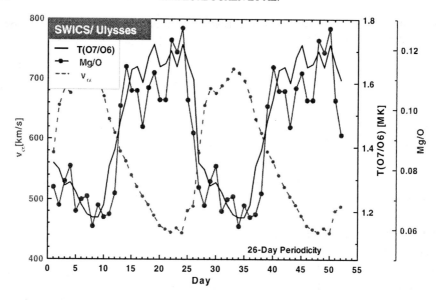

Figure 1. Superposed epoch analysis of Ulysses data showing the systematic variation in the Mg/O ratio, the solar wind speed, and the coronal temperature inferred from the O^{7+}/O^{6+} ratio during the (sidereal) solar rotation period. The abrupt transition of freeze-in temperatures and composition indicate a different origin for fast and slow wind. Figure adapted from von Geiss *et al.* (1995b).

(e.g., Phillips *et al.*, 1995). However, at low latitudes, surrounding the streamer belt, the flow is slower, ~ 400 km/s, but also more variable in density and speed (e.g. Gosling, 1996).

The most interesting difference between fast and slow wind is clearly associated with the elemental abundances and charge state ratios as for example measured by Ulysses-SWICS. A key observation from a paper by von Geiss *et al.* (1995b) is reproduced in Figure 1. Shown here are the results of a superposed epoch analysis of Ulysses-SWICS data showing the solar wind speed (plotted here is the speed the α-particles which is the same as the proton speed to within an Alfven speed) and the Mg/O ratio. Mg has a low first ionization potential (FIP) and is therefore easier to ionize than O. Also plotted is the so-called *freeze-in* temperature T_O of the solar wind which is a measure for the electron temperature about $> 1 - 2$ solar radii from the photosphere. This temperature is typically around $1.5 \cdot 10^6$ K and highly variable when the solar wind is slow, and around $1.2 \cdot 10^6$ K in fast solar wind. Equivalently, the Mg/O abundance ratio is fractionated in slow solar wind to become ~ 0.11, and close to photospheric in the fast solar wind. The steep transitions between the two solar wind types has been interpreted by Geiss *et al.* (1995a) to indicate a clear difference in the origin of fast and slow wind streams. These observations would be very hard to explain in a theory where coronal holes were the source of both fast and slow solar wind (see, e.g., Bravo *et al.*, 1997).

We will first introduce a global theory for the magnetic configuration of the low corona and then relate slow solar wind observations to loop properties in the low latitude corona. A recent observational result will then be described which shows the spatial dependence of the ^{3}He/^{4}He-ratio in the solar wind. In section 3, this will then be interpreted in the framework of the theory described.

2. Theory for the Slow Solar Wind and its Element Fractionation

First, consider a theory for the origin and of the slow solar wind and its relation to the solar global magnetic field configuration. This theory is a natural consequence of the new concept for the heliospheric magnetic field in fast solar wind proposed by Fisk (1996), which is, that the footpoints of the heliospheric magnetic field move on the solar wind source surface. This motion results from an interplay between the differential rotation of the photosphere and the non-radial expansion of the solar wind into more rigidly rotating coronal holes.

The theory was motivated by the Ulysses observation of low-energy CIR-modulated particles which were observed up to very high heliospheric latitudes (Simnett et al., 1995). Fisk (1996) interpreted these observations to be clear indications for direct magnetic connections from low to high heliospheric latitudes. Such connections are not possible in a standard Parker magnetic field configuration, but they are a natural consequence of the new field configuration which results in magnetic field transport in the corona, leading to magnetic field connections of different heliospheric latitudes.

In Figure 2 the characteristics of this magnetic field transport from high to low latitude are shown (for details, see Fisk et al., 1998b, or Schwadron et al., 1998). The time-scale associated with this transport is of the order $\tau \simeq 1/\omega \sim 100$ days, since $\omega = \Omega_{equ} - \Omega_{pole}$, and Ω_{equ} (Ω_{pole}) is the equatorial (polar) rotation rate. On this time-scale the entire high-latitude magnetic flux is dumped into the low latitude regions. In Zurbuchen et al. (1997) direct observational evidence in support of this theory was presented. It is pointed out that the high-latitude magnetic field as measured by Ulysses shows signatures of this transport, which are very hard to explain with a standard solar magnetic field model. If this high-latitude transport is really present, Maxwell's equations imply a transport at low latitudes. A natural mechanism for such a transport is reconnection, as is illustrated conceptually in Figure 2. This reconnection process releases material from previously closed magnetic structures such as loops. It can readily be demonstrated, as is discussed in Fisk et al. (1998b), that the typical properties of this transport is fully determined by the high latitude magnetic field configuration. For details concerning this magnetic field transport and reconnection scenario refer to Fisk et al. (1998b).

In this theory, there are therefore distinct regions from where solar wind can emerge (see Fisk et al., 1998a): First, it can be emitted along continuously open magnetic field lines in coronal holes, yielding the fast steady wind. On the other hand, slow solar wind emerges from closed magnetic field regions, which are continuously opened by reconnection of what had been closed magnetic structures.

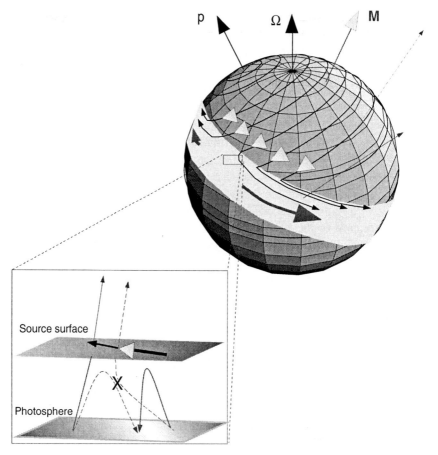

Figure 2. Reconnection scenario as described in the text. Shown is the source surface of the solar wind at about 2 solar radii in a frame co-rotating with the equatorial rotation rate. In open magnetic field regions, close to the pole, footpoints move in latitude. They are eventually convected into the band of closed magnetic fields at low latitudes. Due to sub-source surface reconnection events, a diffusive transport in longitude closes the footpoint curves on the solar wind source surface.

This makes the slow solar wind a time-dependent flow of plasma which is fed from many distinct sources. The compositional signatures of the slow solar wind are therefore closely related to the physical properties of the magnetic loops from where the wind emerges.

Recently, data from the SWICS-ACE instrument (for instrument descriptions see, Gloeckler *et al.*, 1998b) has been presented by Hefti *et al.* (1998) which is consistent with the framework described above. Figure 3 shows composition measurements of solar wind with a time-resolution of 13 minutes during a period of five days. During day 92 there are signatures of a coronal mass ejection event, afterwards there is a period of standard slow solar wind. Large fluctuations of the O charge state ratio n_7/n_6 and also of the Fe/O-ratio are visible through out the flux as described by Fisk (1996), reconnection is systematically forced at low latitudes. These reconnection events result in an intermittent source of plasma which was previously confined in closed loops, the slow solar wind.

The elemental composition of solar wind plasma is then strongly dependent on the actual source in the low corona, which is of course intimately linked to the chromosphere, where the first ionization occurs. The focus of the study of fractionation effects should therefore be not only on the chromosphere, the actual location of the first ionization, but also on the transition region and the low corona, where the plasma is undergoing wave-heating and gravitational settling. Unfortunately, the direct experimental test of these effects is very difficult, since all optical observations of line-of-sight integrals are dominated by structures of high number density. Large, low density loops would probably be more likely candidates for the source of the slow solar wind.

We should also remember that the theory described here applies only in the years around solar minimum, when there are well-developed polar coronal holes and a very distinct region of magnetically closed topology. It is not clear how this mechanism will apply near solar maximum. There are certainly coronal holes on the Sun nearer to solar maximum, but they are short lived, and the concept of field line motion across them, with resulting reconnection in closed loops, may be quite different.

Acknowledgements

The work was supported, in part, by NASA contract NAS5-32626 and NASA/JPL contract 955460. T.H.Z. and N.A.S. were also supported, in part, by an NFS grant ATM 9714070. We thank S. Hefti, R. von Steiger, E. H. Avrett, and C. Pei for many useful discussions.

References

Bravo, S., and Stewart, G. A.: 1997, 'Fast and slow solar wind from solar coronal holes', *Astrophys. J.* **489**, 992.
Fisk, L. A.: 1978, '^{3}He-rich flares: A possible explanation', *Astrophys. J.* **224**, 1048.
Fisk, L. A.: 1996, 'Motion of the footpoints of heliospheric magnetic field lines at the sun: Implications for recurrent energetic particle events at high heliographic latitudes', *J. Geophys. Res.* **101**, 15547.
Fisk, L. A., Schwadron, N. A., and Zurbuchen, T. H.: 1998a, 'On the slow solar wind', *Space Sci. Rev.*, in press.
Fisk, L. A., Zurbuchen, T. H., and Schwadron, N. A.: 1998b, 'On the slow solar wind: I. Origin in the coronal magnetic field', *Astrophys. J.*, in press.
Geiss, J.: 1982, 'Processes affecting abundances in the solar wind', *Space Sci. Rev.* **33**, 201.
Geiss, J., et al.: 1995a, 'The southern high-speed stream: Results from the SWICS instrument on Ulysses', *Science* **268**, 1033.
Geiss, J., Gloeckler, G., and von Steiger, R.: 1995b, 'Origin of the solar wind from composition data', *Space Sci. Rev.* **72**, 49.

Gloeckler, G., and Geiss, J.: 1998, 'Measurement of the abundance of helium-3 in the Sun and in the local interstellar cloud with SWICS on Ulysses', in *Primordial Nuclei and Their Galactic Evolution*, Eds. N. Prantzos, M. Tosi and R. von Steiger, Kluwer, 275.

Gloeckler, G, Bedini, P., Fisk, L. A., Zurbuchen, T. H., Ipavich, F. M., Cain, J., Tums, E. O., Bochsler, P., Fischer, J., Wimmer-Schweingruber, R. F., Geiss, J., and Kallenbach, R.: 1998, 'Investigation of the composition of solar and interstellar matter using solar wind and pickup ion measurements with SWICS and SWIMS on the ACE spacecraft', *Space Sci. Rev.*, in press.

Gosling, J. T., 1996: 'Physical nature of low-speed solar wind', in *Robotic Exploration close to the Sun: Scientific Basis*, AIP Conference Proc., **385**.

Hefti, S., Zurbuchen, T. H., Fisk, L. A., Gloeckler, and G., Schwadron, N. A., 1998: 'Compositional variations in the slow solar wind: ACE/SWICS results', *Eos Trans. AGU*, Spring Meet. Suppl., S259.

Judge, P. G., and Peter, H.: 1998, 'The structure of the chromosphere', *Space Sci. Rev.*, this volume.

Kohl, J. L. *et al.*: 1997, 'First results from the solar ultraviolet coronagraph spectrometer', *Solar Phys.* **175**, 613.

Marsch, E., von Steiger, R., and Bochsler, P.: 1995, 'Element fractionation by diffusion in the solar chromosphere', *Astron. Astrophys.* **301**, 261.

Peter, H.: 1998, 'Element separation in the chromosphere', *Space Sci. Rev.*, this volume.

Phillips, J. L., *et al.*: 1995, 'Ulysses solar wind plasma observations at high southerly latitudes', *Science* **268**, 1030.

Roth, I., and Temerin, M.: 1997, 'Enrichment of ^{3}He and heavy ions in impulsive solar flares', *Astrophys. J.* **477**, 940.

Schwadron, N. A., Fisk, L. A., and Zurbuchen, T. H.: 1998, 'On the slow solar wind: II. Element fractionation', *Astrophys. J.*, in press.

Simnett, G. M., Sayle, K. A., Tappin, S. J., and Roelof, E. C.: 1995, 'Corotating particle enhancements out of the ecliptic plane', *Space Sci. Rev.* **72**, 327.

Vernazza, J., Avrett, E. H., and Loeser, R.: 1981: *ApJS* **45**, 635.

Zurbuchen, T. H., Schwadron, N. A., and Fisk, L. A.: 1997, 'Direct observational evidence for a heliospheric magnetic field with large excursions in latitude', *J. Geophys. Res.* **102**, 24175.

Address for correspondence: Thomas H. Zurbuchen, Space Research Laboratory, University of Michigan, 2455 Hayward Street, Ann Arbor, MI 48109-2143

COMPOSITION ASPECTS OF THE UPPER SOLAR ATMOSPHERE
Rapporteur Paper III

R. VON STEIGER

International Space Science Institute, Hallerstrasse 6, CH-3012 Bern, Switzerland

Abstract. This rapporteur paper discusses the solar corona and the solar wind in the context of their chemical composition. The abundances of elements, both obtained by optical and by in situ observations, are used to infer the sources of the slow solar wind and of the fast streams. The first ionisation potential (FIP) fractionation effect is also discussed, in particular the agreed basics and the open questions.

1. Introduction

This rapporteur paper is an attempt to summarise what we know – and do not know – about the composition of the upper solar atmosphere from remote observations and about the composition of the solar wind from in situ measurements, and how the two relate to each other. Specifically, the following questions will be addressed:
- How does the composition of the slow solar wind, which is enhanced in elements with a low first ionisation potential (low-FIP) by a factor of 3–5, relate to the remote observations of coronal streamers?
- How does the composition of the fast solar wind streams, which are little enhanced in low-FIP elements, relate to the remote observations of coronal holes?
- What can be said about the boundary between slow wind and fast streams from composition measurements?
- What can we deduce from the charge state distributions of elements in the solar wind about the coronal temperature and its profile?
- Is the FIP fractionation effect appropriately named, or should it rather be called "FIT" (first ionisation time), "FRT" (first recombination time), "SIP" (second ionisation potential), or even "R/V" (refractivity/volatility) effect?

2. Coronal Streamers and Slow Solar Wind

Figure 1 shows the picture of a coronal streamer taken by SOHO/UVCS in the light of O VI. Streamers are generally thought to be (associated with) the source of slow solar wind. However, it is not clear how the material escapes from the obviously closed magnetic field, which constitutes the streamer, out into interplanetary space. Two possibilities come to mind: (a) quasi-stationary reconnection at the top of the

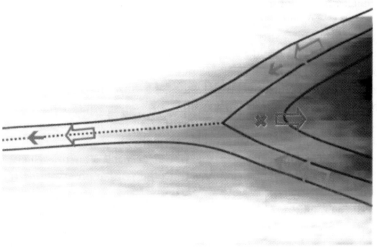

Figure 1. Upper panel: Coronal streamer in O VI 1032 Å as recorded by UVCS on 29 January 1996. (Courtesy of SOHO/UVCS consortium. SOHO is a project of international cooperation between ESA and NASA.) *Lower panel:* Schematic of the solar wind expansion geometry, superimposed on the negative of the image above. Blue lines indicate the magnetic field with the current sheet represented by a dotted line, green arrows indicate "open lanes" for the solar wind expansion, and red crosses indicate "closed lanes" that do not contribute to the solar wind.

streamer near the base of the current sheet, or (b) flow from the roots of the streamer along the open field lines at its periphery.

The two scenarios can be tested by looking at the composition of the slow solar wind, in particular at the times of current sheet crossings which map back to the tip of the streamer. It has been known for a long time that the He/H ratio is significantly lower there than anywhere else in the solar wind (Borrini *et al.*,

1981). However, this result alone cannot resolve the issue: it could be a consequence of gravitational settling in scenario (a) or of the fact that He has the least favourable drag factor with protons (as defined by Geiss *et al.*, 1970) in scenario (b). Only when heavy ions such as oxygen could routinely be measured in the solar wind by SWICS on Ulysses (Gloeckler *et al.*, 1992) it became possible to discriminate between the two scenarios. These heavy ions are observed to be present with close to their normal abundances at times of sector boundary crossings. This clearly favours scenario (b) over (a), since the heavy ions would be depleted even more strongly by gravitational settling than helium, but no strong depletion is expected due to insufficient drag with protons because of their high drag factors, which in turn are due to their high charge states (Wimmer-Schweingruber, 1994; von Steiger *et al.*, 1995).

This conclusion was recently confirmed directly by Raymond *et al.* (1997), who observed with SOHO/UVCS that the abundances of oxygen (cf. Fig. 1) and other high-FIP elements are depleted by an order of magnitude relative to photospheric values in the core of a quiescent equatorial streamer (by gravitational settling – scenario (a)) but resemble abundances measured in the slow wind along its periphery. When looking at Fig. 1, two caveats should be kept in mind: O^{5+} contributes less than a percent to the solar wind oxygen (Wimmer-Schweingruber *et al.*, 1998a), which in turn contributes less than 0.1% of the solar wind flow; we are thus looking at a very dilute tracer. Moreover, the observed emission is dominated by features of high density, whereas low-density regions (which might be of different composition) contribute little to it, so therefore line-of-sight effects may be important. Nevertheless, the large compositional difference between the core and the periphery of the streamer must be real, particularly since it is also observed in other elements. We conclude that scenario (b) is indeed correct, i.e. that the slow solar wind originates from the streamer's roots and flows *around* it, whereby the protons drag the heavy ions along by Coulomb friction.

3. Coronal Holes and High Speed Streams

Since the Skylab era coronal holes are known to be the source region of high speed streams (Krieger *et al.*, 1973). Figure 2 shows the picture of a coronal hole taken by SOHO/EIT in the light of Fe IX/X. The expansion geometry appears to be much simpler in this case as compared to the slow wind. The fields are obviously open, the acceleration occurs rapidly (Kohl *et al.*, 1997), the charge states of heavy ions indicate a rapid freezing-in process (see Sect. 5) and the abundances are close to photospheric values (see Sect. 6.1). So where is the problem?

Polar plumes, which are easily spotted as bright features in Fig. 2, have been reported to be composed differently than the fast solar wind streams. Skylab observations indicated that they appear to be enriched in low-FIP elements (such as in the Mg/Ne ratio) even more strongly than the slow solar wind is, at least in a promi-

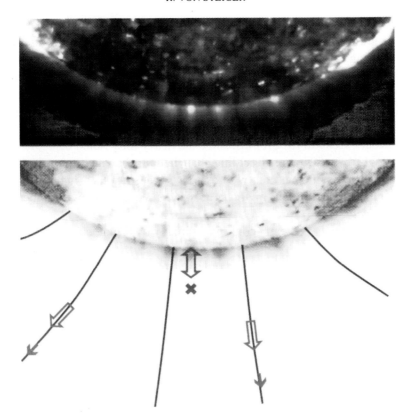

Figure 2. Upper panel: South polar region with polar coronal hole and polar plumes in Fe IX/X 171 Å as recorded by EIT on 8 May 1996. (Courtesy of SOHO/EIT consortium. SOHO is a project of international cooperation between ESA and NASA.) *Lower panel:* Schematic of the solar wind expansion geometry, superimposed on the negative of the image above. The meaning of the symbols is as in Fig. 1.

nently visible plume observed by Widing and Feldman (1992). This is surprising since plumes are likely to be open fields amidst the coronal holes (Del Zanna *et al.*, 1998), from where we know the fast streams emanate. These are not only little biased in composition, but also very uniform, showing little variations in composition, if any. Recent observations of SOHO/SUMER indeed indicate that the abundances observed in the interplume coronal hole are compatible with those in fast streams (Feldman, 1998).

This makes it clear that polar plumes are not a major source of the fast streams, and the absence of compositional variability in fast streams makes it questionable whether they are a solar wind source at all. Plumes could be essentially static features not contributing to the outward mass flow. In fact, the sign of the speed within plumes is not known with certainty. Alternatively (and perhaps more likely), it is simply the low filling factor of plumes within the coronal holes which makes it unlikely to encounter any anomalous composition signature by the time the plasma is observed at ≥ 1 AU. Judging from Fig. 2, the filling factor at the surface is 10% at most, and recent models of polar plumes indicate that it does not strongly increase outwards (Del Zanna *et al.*, 1998). [A third possibility could be that the apparently anomalous composition observations were obtained in atypical, particularly strong and bright plumes (Feldman, 1998, personal communication)]. Nevertheless, a systematic search for compositional fine structures as possible interplanetary remnants of polar plumes within the largely uniform fast streams, e. g. in microstreams or in pressure-balanced structures, should be undertaken.

4. The Coronal Hole – Streamer Belt Boundary

We have seen that both the expansion geometry as well as the element and charge-state composition of the two fundamental solar-wind types are significantly different from each other. On Fig. 2 a rather sharp boundary can be seen to separate the two source regions. Nevertheless, this does not guarantee that the boundary is equally sharp in interplanetary space. One might expect a mixing layer between the two stream types, particularly in corotating interaction regions (CIRs) where a fast stream rams into the previously emitted slow wind from a more easterly longitude.

Composition measurements have shown that this is clearly not the case. Wimmer-Schweingruber *et al.* (1997; 1998b) have analyzed the composition changes across CIRs and showed that the changes in charge state and in elemental composition exactly match with the location of the stream interface as inferred from the kinetic parameters to within the best time resolution possible with SWICS (13 minutes). This was confirmed during Ulysses' fast latitude scan, where the changes in charge state composition at the boundaries of the polar streams occurred almost from one hourly value to the next and coincided with the stream boundary as inferred from other (kinetic and magnetic) parameters (Balogh and von Steiger, 1998). Solar wind charge states thus are a reliable dye marking the solar-wind flow-type out to at least 5 AU. We conclude that this provides evidence against the view that the slow wind merely is a "tired fast stream" from the coronal hole boundaries (Bravo and Stewart, 1997), but rather constitutes a fundamentally different type of solar wind (cf. von Steiger *et al.*, 1997; Neugebauer *et al.*, 1998).

5. The Coronal Temperature and Temperature Profile

It is well known that the charge state distribution of an element in the solar wind freezes in at some altitude in the corona, typically at $1.2 - 3 R_\odot$, and thus can serve as a proxy for the coronal *electron* temperature at or near the altitude where the expansion time scale overcomes the ionisation/recombination time scale (Hundhausen, 1972; Wimmer-Schweingruber *et al.*, 1998a).

In fast streams, all charge states of an element indicate approximately the same temperature: C at 1 MK, O at 1.2 MK, Fe at 1.25 MK, and Si at 1.45 MK (Geiss et al., 1995; Ko et al., 1997). Therefore the freezing-in temperatures may, in fact, be used to obtain a rough coronal temperature profile. It is sufficient to assume a monotonously decreasing density as a function of heliocentric distance in order to show that the electron temperature must have a maximum of about 1.5 MK at a distance of a few solar radii. The freezing-in temperatures derived from O and Fe ionisation states are about equal, but their ionisation/recombination coefficients differ by about two orders of magnitude, which means that Fe can keep in collisional equilibrium with the electrons to much lower densities – and thus to larger distances – than O can. The coefficients of Si are intermediate between O and Fe, and since it freezes-in at a higher temperature than both of them, the temperature profile must have a maximum between the freezing-in altitudes of O and Fe (Geiss et al., 1995; see Ko et al., 1997, for quantitative details).

In the slow solar wind the picture is less clear. Contrary to what is observed in fast streams, the freezing-in temperatures of the charge states of any one element do not yield a single temperature there (e. g., von Steiger et al., 1997). There is a clear tendency towards higher charge states, caused by the higher temperatures in the streamer belt as compared to coronal holes. The charge state spectrum is not simply shifted, though, but low charge states remain clearly present, and the distribution appears thus broadened (von Steiger et al., 1997; Ko et al., 1997). However, the latter observation may well be an artifact caused by the limited time resolution of SWICS/Ulysses of at least a day for obtaining a Fe charge state spectrum with reasonable statistics. When observed with the high time resolution of SOHO/CELIAS or SWICS/ACE, it seems that the individual charge state distributions are roughly isothermal, but the associated temperature is highly variable on short time scales of hours to minutes (Aellig et al., 1998; Hefti and Zurbuchen, 1998). This may be interpreted in the framework of a new model of the slow solar wind (Fisk et al., 1998): When open field lines – which are transported across the coronal hole due to the differential rotation of the photosphere and the fact that the magnetic axis does not coincide with the rotation axis – hit the streamer belt region, they reconnect with the closed loops present there and thus are transported back to the other side of the Sun where their next migration across the coronal hole can start. During that transport along the rim of the streamer belt the field lines "hop" from one loop to the next and thus are being fed with plasma that may well have quite different temperatures, e. g. depending on the size of the loop.

The observed freezing-in temperatures, both in fast streams and in the slow wind, appear to be compatible with coronal electron temperatures clearly in excess of 1 MK. On the other hand, SUMER observations suggest significantly lower temperatures of < 1 MK. This discrepancy has not been resolved as yet, but it should be noted that the concept of freezing-in temperature is an indirect one. Certainly the freezing-in process does not occur abruptly at a sharp altitude (and temperature), but may be more or less gradual. Moreover, the freezing-in temperatures are normally defined as collision equilibrium with electrons *of a Maxwellian velocity distribution*. The presence of suprathermal tails on the electron distribution function may well have a significant influence on the observed charge states and thus on the inferred freezing-in temperatures (Owocki and Scudder, 1983; Bürgi, 1987), although recent studies indicate that the temperature estimates obtained from charge states are rather robust and independent of the details of the electron distribution function (Ko *et al.*, 1996; Aellig, 1998).

6. The FIP Effect

The First Ionisation Potential (FIP) fractionation effect has been presented and discussed extensively at this workshop and elsewhere in this volume (Del Zanna *et al.*, 1998; Feldman, 1998; Geiss, 1998; Hénoux, 1998; Judge and Peter, 1998; Peter, 1998; Raymond *et al.*, 1998; Reames, 1998; Rutten, 1998; Wieler, 1998; Wilhelm and Bodmer, 1998; Zurbuchen *et al.*, 1998). I therefore concentrate here on the known basics and on the open issues.

6.1. FIP OBSERVATIONS

Observations of the FIP fractionation, i. e. the enrichment of elements with a FIP $\lesssim$ 10 eV by a factor of 3–5 in the slow solar wind relative to the photosphere, and of its reduced strength to a factor of only 1.5–2 in fast streams, to date mainly come from the SWICS/Ulysses sensor (Geiss, 1998; cf. von Steiger *et al.*, 1997, for a review; note that the latter paper had essentially been written in 1994 but was delayed by the publisher!). Earlier missions such as ISEE-3 and AMPTE gave similar, but nevertheless not quite as trustworthy results, either because the sensor lacked SWICS' post-acceleration and mass separation, or because the results were obtained in the shocked solar wind behind the Earth's bow shock, in the magnetosheath. On the other hand, newer sensors such as WIND/MASS or SOHO/CELIAS/MTOF have a fantastic mass resolution and detected several new elements such as Na, Al, P, etc. (Bochsler *et al.*, 1997). No abundance measurements of these have been published as yet, but it is hoped that this may change in the future. Abundances of low-FIP elements other than the classical trio FeSiMg would be essential to decide whether the low-FIP plateau is indeed a plateau (as is the case for solar energetic particles, cf. Reames, 1998), or whether there may be a tendency to stronger enrichments of elements with lower FIPs.

6.2. FIP MODELS

A broad variety of models of the FIP effect have been proposed and evaluated to varying degrees of detail (cf. Hénoux, 1998; von Steiger, 1996). Since the paper of Geiss (1982) it is generally accepted that the FIP fractionation occurs by atom-ion separation at an altitude where the solar atmosphere is partially neutral, i. e. in the upper chromosphere and the lower transition region. The models may be classified according to their modes of ionisation and of separation. While most models take the mode of ionisation to be by UV and EUV photons (even though none of them has gone as far as treating the radiative transfer self-consistently), little consensus exists about the mode of separation*. Atom-ion separation across magnetic field lines seems a popular possibility, but there again several different ideas exist regarding the driving force (e. g. von Steiger and Geiss, 1989; Vauclair, 1996). Other models use the magnetic field to enclose the ions and preferentially heat them by waves (Tagger et al., 1995; Zurbuchen et al., 1998). Still others use the magnetic field solely as a guide for the ions, which permits the problem to be reduced to one dimension, and obtain the fractionation by diffusion (Marsch et al., 1995; Peter, 1998). These models have been criticised for merely shifting the problem to the boundary conditions, which themselves might even be unphysical (Geiss, 1998; McKenzie et al., 1998). Nevertheless, these models have given a remarkably simple fractionation formula between any two elements i and k, $f_{ik} = (r_{kH}/r_{iH})\sqrt{\tau_k/\tau_i}\sqrt[4]{(A_k/A_i)(A_i+1)/(A_k+1)}$ which only involves atomic parameters (the ionisation time, τ_i, the atom radius for collisions with H, r_{iH}, and the mass number, A_i), and which was obtained under very simple and clearly stated conditions (cf. Sect. 3 of Marsch et al., 1995). The formula thus rests on robust atomic parameters (and, via τ_i, on the solar spectrum), and it correctly gives the very weak mass dependency. On the other hand, it seems to predict a stronger FIP enrichment for elements with very short ionisation times. However, it should be kept in mind that these elements are unlikely to get neutral anywhere in the solar atmosphere and, in these cases, the formula obviously cannot apply since the conditions for which it was derived do not occur. It would be somewhat surprising that the diffusive fractionation models correctly yield the fractionation for all elements observed so far if they were based on erroneous physics.

6.3. THE NAME OF THE FIP EFFECT

From the section above it might seem appropriate, as it has been proposed before, to rename the FIP effect to "FIT"-effect (for first ionisation time, the most important parameter in the fractionation formula above). This choice is further motivated by the observations of Wieler (1998), who showed that the abundances of Kr (FIP = 14.0 eV) and Xe (FIP = 12.1 eV), obtained from inclusions of solar wind in the lunar regolith, do not fall on the plateau together with the other high-FIP element, but are enriched by an intermediate factor, i. e., less than the low-FIP elements are. When representing the same data as a function of FIT rather than FIP, these elements nicely fall in line with the general pattern (von Steiger et al., 1997). Nevertheless, I think that FIP is the right term to be used, because it is a fundamental

---

* The situation is somewhat reminiscent of the galactic evolution of the primordial D/H abundance ratio, for which different models yield D depletion factors of 1 (no depletion) to ∞ (total destruction) (Tosi, 1998). The situation for the FIP fractionation is somewhat more fortunate, though, because it is known from observations what the models ought to predict.

atomic parameter, whereas the FIT is not an unambiguously defined term. It is obtained by folding the atom's ionisation cross section with the solar spectrum, which is strongly variable over the solar cycle, particularly in the EUV (traditionally, a solar minimum spectrum has been used in the FIT calculations). Moreover, ionisation cross sections may become outdated by better measurements or theoretical calculations. A reevaluation of the FITs, which are still mainly based on Geiss and Bochsler (1985) and von Steiger (1988), would certainly be worthwhile.

Other names for the FIP effect were also discussed at this workshop, e. g. "FRT" (first recombination time) or "SIP" (second ionisation potential), but were quickly discarded because they would not order the data nearly as well. One other possibility, however, has been put forward by Meyer *et al.* (1997): Based on the observation that the galactic cosmic rays (GCRs) also show an overabundance of low-FIP elements, they argue that this is caused by the differences in refractivity and volatility of the observed elements, an "R/V"-effect. Of course, the FIP is closely connected with the R/V, but in the case of GCR this is indeed the better term since the fractionation is probably caused by condensation of refractive material in dusty stellar envelopes and subsequent acceleration of the volatile fraction. Because this is an entirely different process than the one occurring in the solar atmosphere (whatever it may be in detail), the term "FIP" remains preferable in the case at hand.

6.4. THE FIP EFFECT IN OTHER STARS

In the light of the previous section, it should not be surprising that the Sun is not the only star showing a FIP effect. It has even been observed directly, by comparison of chromospheric and coronal line ratios of different elements, in αCen AB (Drake *et al.*, 1997) and in ϵEri (Laming *et al.*, 1996), but not in αCMi, Procyon (Drake *et al.*, 1995). Given the fact that αCMi is hotter than the Sun (spectral class F5 IV vs. G2 V) whereas αCen AB and ϵEri are cooler (G2 V, K0 V, and K2 V), I should like to conclude with a brave (and likely wrong) speculation: There is a dividing line running vertically across the HR diagram somewhere between F5 and G2, to the right of which all stars show the FIP effect while those to the left do not.

7. Conclusions

Going back to the questions listed in the Introduction, the following answers can now be given:
- The slow solar wind flows around streamers, from their roots and along the periphery.
- The fast solar wind flows freely out of the open field in the coronal holes. Polar plumes are not likely to play a major role in its generation.
- The boundary between slow wind and fast streams is exceedingly sharp, indicating that the two (quasi-)stationary solar wind types are fundamentally different and keep their identity out to at least 5 AU.

- In the coronal holes, the electron temperature rises to a maximum of ~ 1.5 MK at $\sim 1.5 R_\odot$ and decreases from there further out. No single temperature profile can be given for the streamer belt region, as the slow solar wind probably consist of a mixture of many contributions at different temperatures.
- The FIP effect is indeed appropriately named, since the ionisation potential is a fundamental atomic property while the first ionisation times depend on the assumed solar spectrum. The evaluation of FITs should receive a second look in the future.

Finally, we should not forget that all these results were obtained in the declining and minimum phase of the solar activity cycle or Schwabe cycle (Schwabe, 1838). It is not clear how well, if at all, they will hold during times of maximum activity. We might well be in for a few nice surprises.

Acknowledgements

I wish to thank all the participants of this workshop for making it such an interesting experience which was well worth all the preparatory and editing work. In particular, I have benefited from discussions with M. Aellig, A. Balogh, P. Bochsler, L. Del Zanna, J. Geiss, J.-C. Hénoux, P. Judge, R. Kallenbach, J. Kohl, H. Peter, J. Raymond, R. Rutten, K. Wilhelm, and R. Wimmer. I thank M. C. E. Huber, T. Zurbuchen, and S. Hefti for many useful remarks on the manuscript. Support from the University of Michigan, where the work on this paper was completed, is gratefully acknowledged.

References

Aellig, M. R.: 1998, *Freeze-in Temperatures and Relative Abundances of Iron Ions in the Solar Wind Measured with SOHO/CELIAS/CTOF*, Ph. D. thesis, University of Bern.
Aellig, M. R., Grünwaldt, H., Bochsler, P., Wurz, P., Hefti, S., Kallenbach, R., Ipavich, F. M. *et al.*: 1998, 'Iron Freeze-in Temperatures Measured by SOHO/CELIAS/MTOF', *J. Geophys. Res.* **103**, 17,215–17,222.
Balogh, A. and von Steiger, R.: 1998, 'The Heliosphere at Solar Minimum: Ulysses Observations During Its Fast Latitude Scan in 1994–95', *Rev. Geophys.*, submitted.
Bochsler, P., Hovestadt, D., Gruenwaldt, H., Hilchenbach, M., Ipavich, F. *et al.*: 1997, 'The Sun at Minimum Activity: Results from the CELIAS Experiment on SOHO', *Proc. 5th SOHO Workshop*, Vol. 404 of *ESA SP*, Oslo, Norway, 17–20 June, pp. 37–43.
Borrini, G., Gosling, J. T., Bame, S. J., Feldman, W. C. and Wilcox, J. M.: 1981, 'Solar Wind Helium and Hydrogen Structure Near the Heliospheric Current Sheet: A Signal of Coronal Streamers at 1 AU', *J. Geophys. Res.* **86**, 4565–4573.
Bravo, S. and Stewart, G. A.: 1997, 'Fast and Slow Wind from Solar Coronal Holes', *Astrophys. J.* **489**, 992–999.
Bürgi, A.: 1987, 'Effects of Non-Maxwellian Electron Velocity Distribution Functions and Non-spherical Geometry on Minor Ions in the Solar Wind', *J. Geophys. Res.* **92**, 1057–1066.
Del Zanna, L., von Steiger, R. and Velli, M.: 1998, 'The Expansion of Coronal Plumes in the Fast Solar Wind', *Space Sci. Rev.*, this volume.

Drake, J. J., Laming, J. M. and Widing, K. G.: 1995, 'Stellar Coronal Abundances II: The FIP Effect and its Absence in the Corona of Procyon', *Astrophys. J.* **443**, 393–415.
Drake, J. J., Laming, J. M. and Widing, K. G.: 1997, 'Stellar Coronal Abundances. V. Evidence for the First Ionization Potential Effect in alpha Centauri', *Astrophys. J.* **478**, 403.
Feldman, U.: 1998, 'FIP Effect in the Upper Solar Atmosphere: Spectroscopic Results', *Space Sci. Rev.*, this volume.
Fisk, L. A., Schwadron, N. and Zurbuchen, T. H.: 1998, 'On the Slow Solar Wind', *Space Sci. Rev.* in press.
Geiss, J.: 1982, 'Processes Affecting Abundances in the Solar Wind', *Space Sci. Rev.* **33**, 201–217.
Geiss, J.: 1998, 'Solar Wind Abundance Measurements: Constraints on the FIP Mechanism', *Space Sci. Rev.*, this volume.
Geiss, J. and Bochsler, P.: 1985, 'Ion Composition in the Solar Wind in Relation to Solar Abundances', *Proc. Rapports Isotopiques dans le Système Solaire*, Cepadues-Editions, Toulouse, Paris, pp. 213–228.
Geiss, J., Hirt, P. and Leutwyler, H.: 1970, 'On Acceleration and Motion of Ions in Corona and Solar Wind', *Solar Phys.* **12**, 458–483.
Geiss, J., Gloeckler, G., von Steiger, R., Balsiger, H., Fisk, L. A., Galvin, A. B., Ipavich, F. M., Livi, S., McKenzie, J. F., Ogilvie, K. W. and Wilken, B.: 1995, 'The Southern High Speed Stream: Results from the SWICS Instrument on Ulysses', *Science* **268**, 1033–1036.
Gloeckler, G., Geiss, J., Balsiger, H., Bedini, P., Cain, J. C., Fischer, J., Fisk, L. A., Galvin, A. B., Gliem, F., Hamilton, D. C., Hollweg, J. V., Ipavich, F. M., Joos, R., Livi, S., Lundgren, R., Mall, U., McKenzie, J. F., Ogilvie, K. W., Ottens, F., Rieck, W., Tums, E. O., von Steiger, R., Weiss, W. and Wilken, B.: 1992, 'The Solar Wind Ion Composition Spectrometer', *Astron. Astrophys. Suppl.* **92**, 267–289.
Hefti, S. and Zurbuchen, T. H.: 1998, http://solar-heliospheric.engin.umich.edu/ace.html.
Hénoux, J.-C.: 1998, 'FIP Fractionation: Theory', *Space Sci. Rev.*, this volume.
Hundhausen, A. J.: 1972, *Coronal Expansion and Solar Wind*, Physics and Chemistry in Space, Vol. 5, Springer-Verlag.
Judge, P. G. and Peter, H.: 1998, 'The Structure of the Chromosphere', *Space Sci. Rev.*, this volume.
Ko, Y.-K., Fisk, L. A., Gloeckler, G. and Geiss, J.: 1996, 'Limitations on Suprathermal Tails of Electrons in the Lower Solar Corona', *Geophys. Res. Lett.* **23**, 2785–2788.
Ko, Y.-K., Fisk, L. A., Geiss, J., Gloeckler, G. and Guhathakurta, M.: 1997, 'An empirical study of the electron temperature and heavy ion velocities in the south polar coronal hole', *Solar Phys.* **171**, 345–361.
Kohl, J., Noci, G., Antonucci, E., Tondello, G., Huber, M. C. E. *et al.*: 1997, 'First Results from the SOHO Ultraviolet Coronagraph Spectrometer', *Solar Phys.* **175**, 613–644.
Krieger, A. S., Timothy, A. F. and Roelof, E. C.: 1973, 'A Coronal Hole and its Identification as the Source of a High Velocity Solar Wind Stream', *Solar Phys.* **29**, 505–525.
Laming, J. M., Drake, J. J. and Widing, K. G.: 1996, 'Stellar Coronal Abundances IV: Evidence of the FIP Effect in the Corona of ϵ Eridani', *Astrophys. J.* **462**, 948–959.
Marsch, E., von Steiger, R. and Bochsler, P.: 1995, 'Element Fractionation by Diffusion in the Solar Chromosphere', *Astron. Astrophys.* **301**, 261–276.
McKenzie, J. F., Sukhorukova, G. V. and Axford, W. I.: 1998, 'Structure of a photoionization layer in the solar chromosphere', *Astron. Astrophys.* **332**, 367–373.
Meyer, J.-P., Drury, L. O. and Ellison, D. C.: 1997, 'Galactic Cosmic Rays from Supernova Remnants. I. A Cosmic-Ray Composition Controlled by Volatility and Mass-to-Charge Ratio', *Astrophys. J.* **487**, 182.
Neugebauer, M., Forsyth, R. J., Galvin, A. B., Harvey, K. L., Hoeksema, J. T., Lazarus, A. J., Lepping, R. P., Linker, J., Mikic, Z., Steinberg, J. T., von Steiger, R., Wang, Y.-M. and Wimmer-Schweingruber, R.: 1998, 'Spatial Structure of the Solar Wind and Comparisons with Solar Data and Models', *J. Geophys. Res.* **103**, 14,587–14,599.
Owocki, S. P. and Scudder, J. D.: 1983, 'The Effect of a Non-Maxwellian Electron Distribution on Oxygen and Iron Ionization Balances in the Solar Corona', *Astrophys. J.* **270**, 758–768.
Peter, H.: 1998, 'Element Separation in the Chromosphere', *Space Sci. Rev.*, this volume.

Raymond, J. C., Kohl, J., Noci, G., Antonucci, E., Tondello, G., Huber, M. C. E. et al.: 1997, 'Composition of Coronal Streamers from the SOHO Ultraviolet Coronagraph Spectrometer', *Solar Phys.* **175**, 645–665.

Raymond, J. C., Suleiman, R., Kohl, J. L. and Noci, G.: 1998, 'Elemental Abundances in Coronal Structures', *Space Sci. Rev.*, this volume.

Reames, D. V.: 1998, 'Solar Energetic Particles: Sampling Coronal Abundances', *Space Sci. Rev.*, this volume.

Rutten, R. J.: 1998, 'The Lower Solar Atmosphere', *Space Sci. Rev.*, this volume.

Schwabe, H.: 1838, 'Ueber die Flecken der Sonne', *Astr. Nachr.* **15**, 243–248.

Tagger, M., Falgarone, E. and Shukurov, A. M.: 1995, 'Ambipolar Filamentation of Turbulent Magnetic Fields', *Astron. Astrophys.* **299**, 940–946.

Tosi, M.: 1998, 'Galactic Evolution of D and ^{3}He', *Space Sci. Rev.* **84**, 207–218.

Vauclair, S.: 1996, 'Element Segregation in the Solar Chromosphere and the FIP Bias: The "Skimmer" Model', *Astron. Astrophys.* **308**, 228–232.

von Steiger, R.: 1988, *Modelle zur Fraktionierung der Häufigkeiten von Elementen und Isotopen in der solaren Chromosphäre*, Ph. D. thesis, University of Bern.

von Steiger, R.: 1996, 'Solar Wind Composition and Charge States', in D. Winterhalter, J. T. Gosling, S. R. Habbal, W. S. Kurth and M. Neugebauer (eds), *Solar Wind Eight*, Vol. 382 of *AIP Conference Proceedings*, AIP Press, Woodbury, NY, pp. 193–198.

von Steiger, R. and Geiss, J.: 1989, 'Supply of Fractionated Gases to the Corona', *Astron. Astrophys.* **225**, 222–238.

von Steiger, R., Geiss, J. and Gloeckler, G.: 1997, 'Composition of the Solar Wind', in J. R. Jokipii, C. P. Sonett and M. S. Giampapa (eds), *Cosmic Winds and the Heliosphere*, The University of Arizona Press, Tucson, pp. 581–616.

von Steiger, R., Wimmer Schweingruber, R. F., Geiss, J. and Gloeckler, G.: 1995, 'Abundance Variations in the Solar Wind', *Adv. Space Res.* **15**, (7)3–(7)12.

Widing, K. G. and Feldman, U.: 1992, 'Elemental Abundances and their Variations in the Upper Solar Atmosphere', in E. Marsch and R. Schwenn (eds), *Solar Wind Seven*, COSPAR Colloquia Series, Vol. 3, Pergamon Press, Goslar, Germany, pp. 405–410.

Wieler, R.: 1998, 'The Solar Noble Gas Record in Lunar Samples and Meteorites', *Space Sci. Rev.*, this volume.

Wilhelm, K. and Bodmer, R.: 1998, 'Solar EUV and UV Observations Above a Polar Coronal Hole', *Space Sci. Rev.*, this volume.

Wimmer-Schweingruber, R. F.: 1994, *Oxygen, Helium, and Hydrogen in the Solar Wind: SWICS/Ulysses Results*, Ph. D. thesis, University of Bern.

Wimmer-Schweingruber, R. F., von Steiger, R. and Paerli, R.: 1997, 'Solar Wind Stream Interfaces in Corotating Interaction Regions: SWICS/Ulysses Results', *J. Geophys. Res.* **102**, 17,407–17,417.

Wimmer-Schweingruber, R. F., von Steiger, R., Geiss, J., Gloeckler, G., Ipavich, F. M. and Wilken, B.: 1998a, 'O^{5+} in High Speed Solar Wind Streams: SWICS/Ulysses Results', *Space Sci. Rev.*, this volume.

Wimmer-Schweingruber, R. F., von Steiger, R. and Paerli, R.: 1998b, 'Solar Wind Stream Interfaces in Corotating Interaction Regions: SWICS/Ulysses Results II', *J. Geophys. Res.*, submitted.

Zurbuchen, T. H., Fisk, L. A., Gloeckler, G. and Schwadron, N. A.: 1998, 'Element and Isotopic Fractionaton in Closed Magnetic Structures', *Space Sci. Rev.*, this volume.

Address for correspondence: vsteiger@issi.unibe.ch

IS THE SUN A SUN-LIKE STAR?

B. GUSTAFSSON
Uppsala Astronomical Observatory
Box 515, S-751 20 Uppsala, Sweden
(Bengt.Gustafsson@astro.uu.se)

Abstract. Various observable properties of the Sun are compared with those of solar-type stars. It is concluded that the Sun, to a remarkable degree, is "solar-type". As regards its particular mass and age, and probably its non-binarity, "anthropic" explanations may seem in place. The possible tendency for the Sun, as compared with similar stars, to be somewhat rich in iron relative to other elements needs further exploration. This is also true concerning its presently small micro-variability amplitude.

1. Introduction

The suspicion that even faint stars are Suns is at least 300 years old, as is illustrated e.g. in the famous study on vacuum by Otto von Guericke *Experimenta nova...* from 1672, where stars are drawn as suns surrounding the solar system. When stellar parallaxes were finally measured towards the end of the 1830s, and stellar spectra a few decades later could verify the similarity between the spectrum of Capella or Procyon and the solar spectrum, the suspicion was proven to be true. It follows that the Sun is a star, but how ordinary? How much of an odd-ball is it? This question has a fundamental importance, for stellar physics as well as in a more fundamental sense, at least for humanity. The astrophysical significance lies in the fact that the number of individual stellar parameters needed to essentially completely characterize a star is thought to be relatively small (basically reflecting the fact that stars are systems in equilibrium in many respects). Thus, all the properties of a star are often thought to be functions of its initial mass, its age, its initial chemical composition as described by its over-all metallicity, and its initial angular momentum. If this is so, one would expect the Sun not to deviate significantly from stars having essentially solar fundamental parameters. However, one might conjecture that the formation of a solar system, or planets with life, could require a very special environment, such as an abnormal abundance of carbon, and that the Sun therefore could well be special. Our existence might require restrictions as regards our nearby star, and a proof that it is special or even abnormal could then have implications concerning the abundance of life in the Universe. In the following review we shall explore the extent to which the Sun is thought to be special. Indeed, we shall find it to be a very normal star, except for in some respects which might be related to our own existence.

2. Solar Mass and Solar Age

A glance at evolutionary tracks for solar-type stars (like those of VandenBerg 1985, Claret and Gimenez, 1992, or Schaller *et al.*, 1992) suggests that the mass of the Sun could hardly have been greater than about 1.3 solar masses, since a greater mass would lead to an increase of the solar luminosity by more than a factor of about 2 within 4.6 Gyears which exceeds the maximum temperature change on Earth that life can survive (i.e. avoiding the formation of cooling CO_2 clouds in the early evolution of the terrestrial atmosphere, or the loss of water through photolysis and hydrogen escape in the end, cf. Kasting *et al.*, 1993, Rampino and Caldeira, 1994). All stars less massive than that mass limit could, however, do. In view of the fact that the less massive stars are far more common than the stars with one solar mass, one may then wonder why the Sun happens to be so massive. One may conjecture that this is due to the radial extension Δr of the "life zone" around the star, i.e. the zone in which a planet with a circular orbit has to be located in order to catch the right amount of solar radiation to maintain its surface temperature in the interval desired for life. Δr roughly scales as $10^{-0.2 M_{bol}}$, and since the luminosity function $\psi(M)$ raises towards low luminosities less steeply than this decrease, one finds that the maximum probability of finding a planet in the life zone, $\psi(M) \times 10^{-0.2 M_{bol}}$, will be at maximum at the high luminosity end, now for the moment assuming that the distribution of absolute planetary distances from the central star is not too sensitively dependent on stellar mass. If, on the other hand, the probability of finding a planet in the life zone of a star is not proportional to the radial extension of the life zone but to its area or even its volume (cf. Huang, 1959), this maximum at the high-luminosity end gets even more peaked.

It was early pointed out by Cameron (1963), that the assumption of the independence of the distribution of planetary distances on stellar mass does not hold for the Solar system itself, taking the satellite systems into consideration. Instead a logarithmic distance distribution of planets and satellites was proposed by Cameron that would increase the estimated number of habitable planets around low mass stars considerably. Simple scaling arguments may suggest (cf. Kasting *et al.*, 1993) that the distances where terrestrial planets form roughly scale as the mass of the central star which would then admit the existence of a fair number of habitable planets around stars at least down to 0.2 solar masses. However, the numerical simulations of the formation of terrestrial planets around stars by Wetherill (1996) do not support this; they indicate distance distributions fairly insensitive to the stellar mass and with the peak around 1 AU, which would then suggest masses around 1 solar mass as most probable for the central star of a system with an inhabited terrestrial planet. Wetherill thus finds that model stars around 1 solar mass almost always get a planet of at least 1/3 Earth mass in their inhabitable zone, while the probability is reduced to 10% for 0.5 or 1.5 solar mass stars since their inhabitable zones are inside or outside the 1 AU planetary orbit. We also note, as was early pointed out by Dole (see Kasting *et al.*, 1993 for references and further discussion) that planets within the life zone around M stars (with mass less than 0.5 solar masses) will be close enough to the star to become tidally locked and thus get synchronous rotation, which may reduce the possibilities of life.

Our preliminary conclusion is that the special property of the Sun, its relatively high mass compared to most stars, seems to be related to our own existence – with a smaller mass there might not have been a planet with suitable temperature around. A star with higher mass on the other hand would have damaged life on its planets by now, even though – if such stars have planetary systems – the chance for planets at a suitable distance from the star in the beginning of its main-sequence phase may well be greater than for the Sun.

3. Chemical Composition and Solar Twins

Is the metallicity of the Sun typical for stars of solar mass and age? Fig. 14a of Edvardsson *et al.* (1993), showing the iron abundance [Fe/H] vs age for the 189 disk solar-type stars in their study, indicates that there is no general age-metallicity relation but a very considerable spread (of about 0.2 dex in [Fe/H]) and that the Sun is rather more metal rich (by about 0.2 dex) than most stars of its age. However, the sample of Edvardsson *et al.* was chosen to be biased towards low-metallicity disk stars. If the sample is corrected for the bias and then restricted to the stars with mean distances from the Galactic Centre between 8 and 9 kpc, and with ages between 4 and 6 Gyears, we find an average [Fe/H] = –0.09, with a standard deviation of 0.22 (the random error in the individual determinations being only about 0.05 dex or less). This may suggest that the Sun, although not being abnormal, is a bit iron-rich as compared with its similar contemporaries. This result does not reflect any obvious differences in methods of analysis, since it is based on strictly differential analyses: measurements of stellar and solar-disk (flux) spectra were made with the same instrument, and model atmospheres from the same grid were used, both for the Sun and the stars. One should note, however, that the spectroscopic comparison between stars and the Sun is tricky, not the least since photometry is also involved in the determination of effective temperatures by Edvardsson *et al.* (1993). Also, the correction for bias in the sample of stars is non-trivial. Therefore, further studies are needed before one might conclude that the Sun is indeed rather iron-rich. A comparison between the Sun and its neighbours, based on the surveys available in uvbyβ and/or Geneva photometry, as well as Hipparcos parallaxes and proper motions and Coravel radial velocities, would be of interest.

What can be said about the abundances of other elements relative to iron for the Sun, as compared with solar-type stars? In Table I some results from the study by Edvardsson *et al.* (1993) are given in the second and third column, while the fourth column gives results for the 9 solar-type stars with accurate abundances determined by Tomkin *et al.* (1997) (using very similar methods and on the same scale as those of Edvardsson *et al.* but with higher quality data). The results are

Table I

Mean abundance ratios with standard deviations for two sets of stars from Edvardsson et al. (1993) and the 9 stars in Tomkin et al. (1997)

Element	$3.6 \leq$ age ≤ 5.6 Gy $-0.2 \leq$ [Fe/H] ≤ 0.2 N = 22	$8.2 \leq R_m \leq 8.7$ kpc $-0.03 \leq$ [Fe/H] ≤ 0.02 N = 5	NaMgAl stars N = 9
[Fe/H]	-0.03 ± 0.11	0.00 ± 0.02	0.16 ± 0.07
[O/Fe]	-0.01 ± 0.06		0.02 ± 0.08
[Na/Fe]	0.05 ± 0.05	-0.01 ± 0.09	0.06 ± 0.04
[Mg/Fe]	0.08 ± 0.04	0.02 ± 0.03	-0.01 ± 0.07
[Al/Fe]	0.07 ± 0.05	0.02 ± 0.05	0.04 ± 0.03
[Si/Fe]	0.04 ± 0.04	0.02 ± 0.02	0.05 ± 0.02
[Ca/Fe]	0.00 ± 0.05	0.01 ± 0.03	-0.01 ± 0.02
[Ti/Fe]	0.04 ± 0.06	-0.02 ± 0.08	0.01 ± 0.02
[Ni/Fe]	0.03 ± 0.05	0.02 ± 0.07	0.03 ± 0.02
[Ba/Fe]	-0.02 ± 0.12	0.10 ± 0.14	-0.08 ± 0.03

presented in the usual notation [X/Fe] = $\log[(\epsilon(X)/\epsilon(Fe))/(\epsilon_\odot(X)/\epsilon_\odot(Fe))]$ where $\epsilon(X)$ denotes the absolute abundance of element X. From the second column of Table I we may trace a tendency for the Sun to be rich in iron relative to the lighter elements from Na to Si, as compared with other contemporary stars. This tendency may also be present, though smaller and hardly significant for the stars with solar mean distance from the Galactic Centre and with solar [Fe/H], as given in the third column. Tendencies of this character were also discussed by Edvardsson et al. (1993) and were partly ascribed to the presence in their sample of a somewhat peculiar group of stars, "NaMgAl stars", with some excess of those three elements as compared with iron. The spectra of 6 such stars were subsequently explored in detail by Tomkin et al. (1997) and compared to 3 "normal" stars in the Edvardsson et al. sample, but no significant difference was found, disproving the existance of the NaMgAl stars as a group with special abundance characteristics. From the last column of Table I one may conclude that there could still remain a tendency for the Sun to be somewhat poor in Na, Al and Si relative to Fe, as compared with stars of similar type. This possibility that the Sun is systematically more iron rich, as compared with the lighter elements, should be explored further; one should note, however, that a major interesting result of the Edvardsson et al. study is that the cosmic scatter in the ratio between the abundances of light elements and Fe for disk stars (excluding the most metal-poor ones) at a given [Fe/H] is very small, e.g. significantly below 0.05 dex for Mg and Si. Thus, if the solar Mg/Fe or Si/Fe ratios would depart from the stellar mean by as much as 0.05 dex, the Sun would be significantly off. We also note from our present study of carbon in solar-type stars (Gustafsson et al., 1998) that there may be a tendency for the C/Fe ratio of

Table II

Mean abundance ratios for suggested "solar twins" according to Cayrel de Strobel and Bentolila (1989), HD 44594, 76151 and 20630, and Friel et al. (1993), 16 Cyg A and B

Star	[Fe/H]	⟨[Al, Si, Ti/Fe]⟩	⟨[Na, Al, Si, Ti/Fe]⟩
HD44594	0.13 ± 0.06	0.04	
HD76151	0.07 ± 0.05	0.02	
HD20630	0.04 ± 0.06	0.00	
16 Cyg A	0.06 ± 0.04		0.06
16 Cyg B	0.02 ± 0.04		0.04

the Sun to be lower by about 0.05 dex as compared with similar stars in the solar neigbourhood.

In this connection it is also of interest to compare to the systematic studies of the composition of "solar twins" by Cayrel de Strobel and her collaborators. Results are given by Cayrel de Strobel and Bentolila (1989) for three solar analogue stars, HD 44594, 76151 and 20630, and by Friel et al. (1993) for the more evolved binary 16 Cyg A and B. The results are shown in Table II. It is seen that there may again be a small tendency for the stars to be somewhat richer in the light elements relative to iron than for the Sun. Further detailed comparisons should be carried out, e.g. between the Sun and the star 18 Sco which, according to the new study by Cayrel de Strobel and Friel (1997) based on Hipparcos parallaxes, should be very similar to the Sun in mass, age and overall metallicity.

We finally note that the FIP effect, with abundances of elements (Mg, Al, Si, Fe) with low First-Ionization Potential being enhanced relative to those with high potentials (like O, Ne and S) for the coronal lines as compared with lines formed in the photosphere, has also been traced in the EUV spectra of α Cen A and B (G2V and K0V)) by Drake et al. (1997). Even if the reason for this effect is still not well understood, it seems clear that also in this respect the Sun is at least not qualitatively different from the stars.

4. Abundances of Li and Be

The solar abundances of the lightest elements are of considerable interest from several points of view, one being that differences of mixing processes in the Sun, as compared with similar stars, could show up as differences in the degree to which the light elements have been depleted in the atmospheres. In the study by Pasquini et al. (1994) the solar Li abundance is compared with that of dwarf stars with similar effective temperatures. The range in ϵ(Li), whether at given T_{eff}, at given chromospheric activity indicator as an age measure or at given [Fe/H], is very considerable, with some stars having about 30 times greater Li abundance, and some an abundance below the solar. In a more recent study Pasquini *et al.* (1997) compared the Li abundances in solar type stars of three open clusters, M 67 (with solar metallicity and solar age), NGC 752 (with an age around 2 Gyr) and the Hyades (0.8 Gyr). They found a scatter of about a factor of 10 in Li abundances for M 67 stars at a given effective temperature, with maximum values not much less that those of the younger Hyades stars. This and theoretical models of rotation-induced mixing (cf. Charbonnel *et al.*, 1994 and references therein) suggest that the Li abundance may reflect the angular momentum history, or possibly the (extensive) mass loss history, of the star. From the comparisons of Pasquini *et al.* we anyhow conclude that the Sun may be typical of the more than 50% of solar-type stars with low Li. The absence of reliable Li abundance determinations for most of these stars makes it possible, however, that many of them have abundances far below that of the Sun.

With Li being depleted in the Sun but not totally extinguished, one might expect the Be abundance to be unaffected by interior processes, since Be "burns" at higher temperatures. Therefore, the standard photospheric solar abundance ($\log \epsilon(Be) = 1.0$), significantly below the meteoritic one ($\log \epsilon(Be) = 1.4$) is problematic and gives implications as regards the mixing processes occurring in the Sun. However, the photospheric estimate has recently been challenged by Balachandran and Bell (1998), allowing for the effect on the ultraviolet Be II doublet lines by an unknown and extensively discussed ultraviolet opacity. They estimate the strength of this opacity by matching ultraviolet and infrared OH lines, and find that an increase of the opacity by a factor of 1.6 is needed. This implies a Be abundance in agreement with the meteoritic value, i.e. no depletion would be needed. We note that Deliyannis *et al.* (1998) recently found a correlation between the Li abundance and the Be abundance for F-type dwarfs, which suggests a slow mixing process as responsible for the depletion of the light elements, at least for the F stars. At any rate, the solar Be line strengths are similar to those of solar-type stars in the temperature interval 5600 K – 6000 K (Stephens *et al.*, 1997).

5. Angular Momentum, Differential Rotation, Granulation and Turbulence

The angular momentum of solar-type stars may be determined from the measurement of rotational-broadened lines or from direct measurements of the rotational periods, e.g. as derived from the periodic variation of the CaII H and K line fluxes. Of particular significance here is the sample of 112 lower Main-Sequence stars monitored for 25 years at Mount Wilson. Using the first method Soderblom (1983) argued against earlier claims and concluded that "there is no evidence that the Sun rotates abnormally slowly". Soderblom (1985) found a nice relation between the ratio R'_{HK} of the flux in the H and K emission to the total bolometric flux and the Rossby number (the ratio between the period and the convective turn-over time).

The Sun fits this correlation very nicely. Adopting the relation $v_{rot} \propto t^{-1/2}$ (Skumanich, 1972; Soderblom, 1983), Soderblom (1985) next reconstructed the angular momentum evolution of the solar type stars with R'_{HK} observed and found an angular momentum of the Sun characteristic of old 1 solar-mass stars. More recently Baliunas *et al.* (1996) published direct period determinations for a number of late-type dwarfs in Mount Wilson survey. For the 20 stars with $0.60 \leq (B - V) \leq 0.70$ (the solar B–V being around 0.65) they found rotational periods ranging from 3 days to 41 days, with 6 stars in their sample having longer rotational periods than that of the Sun (25 days), while 9 stars have periods below 15 days. We note that the evolutionary rotating solar models of Pinsonneault *et al.* (1989) suggest that an order of magnitude difference in initial angular momentum only leads to small differences in the surface rotation speed on the main sequence after some 100 million years. This is supported by the fact that the stars in the Hyades cluster have similar rotation speeds while those in the younger Pleiades show a much larger scatter. Thus, the present solar rotation speed could be independent of the initial angular momentum.

The Sun is known to have differential rotation, with different rotational periods at different latitudes. Hall and Henry (1990) traced differential rotation in binary stars from light curves modulated by star spots. Donahue, Saar and Baliunas (1996) have traced similar phenomena in studying the variations ΔP of the periods in H and K emission in the Mount Wilson survey. These variations are then supposedly occurring as a result of active regions being located at different latitudes at different times. The authors find a relation between ΔP and the mean period, $\langle P \rangle$: $\Delta P \propto \langle P \rangle^{1.3 \pm 0.1}$, which seems independent of the stellar mass. The interpretation of this relation is certainly not obvious but it should be noted that the Sun is close to the line (cf. Donahue *et al.*, 1996, fig. 3), suggesting that the solar differential rotation is typical for stars of its mean rotational period. The systematic study of rotational properties of a more selective sample of stars with solar mass and composition could be rewarding.

Also as regards the dynamic properties at smaller scales than global rotation the Sun seems typical for solar-type stars. Thus, Gray (1982) found the scale factors of the spectral line bisectors, measuring the strength of granular motions, to be solar for G2V stars, while e.g. G0V and G5V stars depart significantly from solar values. We also note that Dravins and Nordlund (1990) in their detailed study of line asymmetries in solar-type stars found the minor differences in line profiles between the two G2 V stars α Cen A and β Hyi and the Sun to be consistent with the fundamental parameter differences as predicted by numerical simulations of convection.

Gray(1984) in his study on macroturbulence dispersion, ξ_t, which is a steep function of spectral type and probably describes effects on spectral lines of granular motions, found the solar value typical for that of a G2V star (cf. his fig. 3). The formula derived by Edvardsson *et al.* (1993) from observations of microturbulence parameters of solar-type dwarfs predicts a solar value for the integrated disk flux of this parameter of 1.15 km/s, which seems significantly smaller than, e.g., that derived from solar Ca I lines by Smith et al. (1986); the value of Edvardsson et al. is, however, based on gf values derived from the solar spectrum assuming a similarly low value of the parameter, and the result obtained then shows the consistency between the stars and the Sun independently of the absolute value.

6. Activity

As was mentioned above Soderblom (1985) found that the solar H and K emission was characteristic for relatively slowly rotating dwarfs. In fact, his survey indicated that the solar R'_{HK} is typical for a comparatively inactive majority of about 65% of the of F7–K2 dwarfs. This result was confirmed by Henry et al. (1996) in their more extensive survey of southern stars where the "inactive" solar-like group includes more than 60%; in addition to that there are more active stars but also a very inactive group, amounting to about 8% of the stars. The solar activity seems typical for its age and rotation, also when other activity criteria are explored. One example is the study of X-ray luminosities by Güdel et al. (1997) who found a relation between X-ray luminosity and rotation period for stars of solar type and the Sun to lie close to the relation.

The solar activity cycle also seems normal for solar-type stars. In their study of activity cycles based on the Mount Wilson survey of H and K emission, Baliunas et al. (1995) found cyclic behaviours, like that of the solar 11 years cycle, to be most frequent. They found the distribution function of activity periods for low-activity stars to peak around 10 years.

However, in another respect the Sun and its activity seems to depart from most, or at least a significant fraction of its similar neighbours. Lockwood et al. (1992, 1996, 1997) have found that many solar-type stars are variable on the 0.3% level or more in the b and y bands, while the variations of the solar irradiance are presently smaller than that. Stars with solar age and solar R'_{HK} may show typically 2–3 times greater amplitude in their b and y variations than the Sun. The flux in these relatively wide wavelength bands will be increased by faculae and decreased by star spots relative to the flux from inactive stellar disks; probably the first effect is most important. The question whether the Sun presently is in a low state of variability, maybe still a reminiscence of the Maunder minimum, or more permanently has a lower degree of activity, still remains to be settled. It may well be, however, that differences in aspect angle contribute to the different amplitude observed for different stars, as is suggested by the model calculations by Schatten (1993).

7. Binarity

In one more respect the Sun departs from the majority of solar-type stars: it is single. The frequency of binaries among solar-type stars is still not so well known;

typical studies suggest, however, that the fraction of single stars is significantly below 1/2. E.g., Duquennoy and Mayor (1991) found that only about 30% of the solar type dwarfs have no companion with a mass above 0.01 solar masses. (We note, in passing, that these authors derived a mass distribution for the companions that is close to that observed for single field stars.) So, even if the solitude of the Sun could well be just accidental, it may be tempting to try explanations in terms of the need for the existence of a stable planetary system for us to finally emerge. Obviously (proto-planetary?) dust disks may form around binaries, as was traced in the case of BD +31°643 by Kalas and Jewitt (1997). Bouvier *et al.* (1997) conclude from their study of Pleiades low-mass binaries that the existence of a companion does not significantly affect disk accretion and disk lifetimes. The question is, however, whether resulting planetary systems in binaries are stable enough. E.g., Holman and Weigert (1996) found from orbit integrations that distances between the components greater than 400 AU may be needed for the solar-system like planetary system to survive beyond a billion years around one of the components of a binary. Note, however, that Kasting *et al.* (1993) conclude that about 5% of the "external cases" (where the planet orbits the centre of mass of the binary), but as much as 50% of the "internal cases" (with the planet in an orbit around one of the stars) could support habitable planets in stable orbits.

8. Conclusions

The Sun seems indeed very normal, for its mass and age. That is the main conclusion of the present investigation. The mass and the age may perhaps most easily be explained in terms of our own existence; a higher mass would not have admitted a time interval long enough for complex life to develop, while a smaller mass could have decreased the chance for a planet to lie within the life-zone around the star. Also the fact that the Sun is single with no stellar companion may be explained in similar "anthropic" terms: the greater chance for an enduring stable environment with the formation of life like our own may bias the inhabited planetary systems towards those around single, instead of binary or multiple, stars. It remains to be proven whether other minor differences traced, e.g. in the iron richness or the micro-variability of the Sun as compared with other stars, are significant and, if so, whether they also are related in one way or another to our own existence.

With the main conclusion being the normality of the Sun one may ask what that further implies. One astrophysical conclusion may be that the Sun could not depart in a very individual way, because the stellar properties are just defined by a few parameters like the initial mass, chemical composition, angular momentum, and age. Another, alternative or additional conclusion which has more biological implications may be that life really does not need a very special environment to form.

Further, more detailed comparisons between the Sun and a well defined sample of solar-type stars seem well motivated.

Acknowledgements

Thanks are due to S. Solanki, K. Eriksson and N. Ryde for many valuable suggestions and comments on the manuscript.

References

Balachandran and Bell: 1998, *Nature* **392**, 791.
Baliunas, S.L., Donahue, R.A., Soon, W.H., Horne, J.H., and Frazer, J.: 1995, *ApJ* **438**, 269.
Baliunas, S.L., Sokoloff, D., and Soon, W.: 1996, *ApJL* **457**, L99.
Bouvier, J., Rigaut, F., and Nadeau, D.: 1997, *A&A* **323**, 139.
Cameron, A.G.W.: 1993, *Interstellar Communication*, W.A. Benjamin, Inc. New York, Amsterdam, p. 107.
Cayrel de Strobel, G., and Bentolila, C.: 1989, *A&A* **211**, 324.
Cayrel de Strobel, G., and Friel, E.: 1997, 2nd Lowell Obs. Workshop on *Solar Analogs*.
Charbonnel, C., Vauclair, S., Maeder, A., Meynet, G., and Schaller, G.: 1994, *A&A* **283**, 155.
Claret, A., and Gimenez, A.: 1992, *A&AS* **96**, 255.
Deliyannis, C.P., Boesgaard, A.M., Stephens, A., King, J.R., Vogt, S.S., and Keane, M.J.: 1998, *ApJ* **498**, L147.
Donahue, R.A., Saar, S.H., and Baliunas, S.L.: 1996, *ApJ* **466**, 384.
Drake, J., Laming, J.M., and Widing, K.G.: 1997, *ApJ* **478**, 403.
Dravins, D., and Nordlund, A.: 1990, *A&A* **228**, 203.
Duquennoy, A., and Mayor, M.: 1991, *A&A* **248**, 485.
Edvardsson. B., Andersen, J., Gustafsson, B., Lambert, D.L., Nissen, P.E., and Tomkin, J.: 1993, *A&A* **275**, 101.
Friel, E., Cayrel de Strobel, G., Chmielewski, Y., Spite. M., Lebre, A., and Bentolila, C.: 1993, *A&A* **274**, 825.
Gray, D. F.: 1982, *ApJ* **255**, 200.
Gray, D. F.: 1984, *ApJ* **281**, 71.
Güdel, M., Guinan, E.F., and Skinner, S.L.: 1997, *ApJ* **483**, 947.
Gustafsson, B., Edvardsson, B., Karlsson, T., Olsson, E., and Ryde, N.: 1998, in preparation.
Hall, D.S., and Henry, G.W.: 1990, Proceedings of the NATO Advanced Study Institute on *Active Close Binaries*, ed. C. Ibanoglu, Kluwer, Dordrecht, p. 287.
Henry, T.J., Soderblom, D.R., Donahue, R.A., and Baliunas, S.L.: 1996, *AJ* **111**, 439.
Holman, M., and Weigert, P.: 1996, *BAAS* **28**, 1212.
Huang, S.-S., *Am. Sci.* **47**, 397.
Kalas, P., and Jewitt, D.: 1997, *Nature* **386**, 52.
Kasting, J.F., Whitmire, D.P., and Reynolds, R.T.: 1993, *Icarus* **101**, 108.
Lockwood, G.W., Skiff, B.A., Baliunas, S.L., and Radick, R.R.: 1992, *Nature* **360**, 653.
Lockwood, G.W., and Skiff, B.A.: 1996, *BAAS* **189**, 8107.
Lockwood, G.W., Skiff, B.A., and Radick, R.R.: 1997, *ApJ* **485**, 789.
Pasquini. L., Liu, Q., and Pallavicini, R.: 1994, *A&A* **287**, 191.
Pasquini, L., Randich, S., and Pallavicini, R.: 1997, *A&A* **325**, 535.
Pinsonneault, M.H., Kawaler, D., Sofia, S., and Demarque, P.: 1994, *ApJ* **338**, 424.
Rampino, M.R., and Caldeira, K.: 1994, *Ann. Rev. Astr. Ap.* **32**, 83.
Schaller, G., Schaerer, D., Meynet, G., and Maeder, A.: 1992, *A&AS* **96**, 269.
Schatten, K.H.: 1993, *J. Geophys. Research* **98**, 18907.
Skumanich, A.: 1972, *ApJ* **171**, 565.
Soderblom, D.R.: 1983, *ApJS* **53**, 1.
Soderblom, D.R.: 1985, *AJ* **90**, 2103.
Stephens, A., Boesgaard, A.M., King, J.R., and Deliyannis, C.P.: 1997, *ApJ* **491**, 339.
Tomkin, J., Edvardsson, B., Lambert, D.L., and Gustafsson, B.: 1997, *A&A* **327**, 587.
VandenBerg, D.A.: 1985, *ApJS* **58**, 711.
Wetherill, G. W.: 1996, *Icarus* **119**, 219.

Author Index

Anderson-Huang, L. S., **203**

Berthomieu, G., 117
Blöcker, T., **105**
Bochsler, P., **291**, 355
Bodmer, R., 369
Bonnet, R. M., **1**

Christensen-Dalsgaard, J., **19**, 133
Cranmer, S. R., **339**

Däppen, W., **49**
Del Zanna, L., **347**
Dziembowski, W. A., **37**

Feldman, U., **227**
Fisk, L. A., 395
Freytag, B., 105

Gabriel, M., **113**
Gadun, A. S., 261
Galvin, A. B., 355
Geiss, J., **241**, 355, 385
Gliem, F., 355
Gloeckler, G., 355, 385, 395
Gough, D., **141**
Grünwaldt, H., 355
Grevesse, N., **161**
Gustafsson, B., **417**

Hénoux, J.-C., **215**
Hanslmeier, A., 261
Hefti, S., 355
Herwig, F., 105
Hilchenbach, M., 355
Holweger, H., 105
Hovestadt, D., 355

Iglesias, C. A., 61
Ipavich, F. M., 355, 385

Judge, P. G., **187**

Kallenbach, R., **355**
Kohl, J. L., 283, 339
Kucharek, H., 355

Leske, R. A., 377

Ludwig, H.-G., 105

Mason, H. E., 315
Mewaldt, R. A., 377
Morel, P., 117

Noci, G., 283, 339

Peter, H., 187, **253**
Ploner, S. R. O., **261**
Provost, J., **117**

Raymond, J., **283**
Reames, D. V., **327**
Rogers, F. J., **61**
Rutten, R. J., **269**

Sauval, A. J., 161
Schwadron, N. A., 395
Solanki, S. K., **175**, 261
Steffen, M., 105
Stone, E. C., 377
Suleiman, R., 283
Suzuki, Y., **91**

Turck-Chièze, S., **125**
Turcotte, S., **133**

Vauclair, S., **71**
Velli, M., 347
von Steiger, R., 347, 385, **405**

Wieler, R., **303**
Wilhelm, K., **369**
Wilken, B., 385
Williams, D. L., **377**
Wimmer-Schweingruber, R. F., **385**

Young, P. R., **315**

Zahn, J.-P., **79**
Zurbuchen, T. H., 395

List of Participants

L. S. Anderson, *Univ. of Toledo, Ohio, USA*, lsa@physics.utoledo.edu
M. Asplund, *Nordita, Copenhagen, Denmark*, martin@nordita.dk
P. Bochsler, *Univ. of Bern, Switzerland*, bochsler@phim.unibe.ch
R. Bonnet, *ESA HQ, Paris, France*, dbauer@hq.esa.fr
C. Charbonnel, *Lab. d'Astrophysique, Toulouse, France*, corinne@obs-mip.fr
J. Christensen-Dalsgaard, *Aarhus University, Denmark*, jcd@obs.aau.dk
W. Däppen, *University of Southern California, USA*, dappen@usc.edu
L. Del Zanna, *ISSI, Bern, Switzerland*, delzanna@issi.unibe.ch
W. Dziembowski, *Copernicus Astronomical Center, Warsaw, Poland*, wd@camk.edu.pl
U. Feldman, *Naval Research Lab., Washington DC, USA*, ufeldman@ssd2.nrl.navy.mil
C. Fröhlich, *PMO, Davos, Switzerland*, cfrohlich@obsun.pmodwrc.ch
M. Gabriel, *Univ. de Liège, Belgium*, gabriel@astro.ulg.ac.be
J. Geiss, *ISSI, Bern, Switzerland*, geiss@issi.unibe.ch
D. Gough, *Univ. of Cambridge, UK*, douglas@ast.cam.ac.uk
N. Grevesse, *Univ. de Liège, Belgium*, U2129RS@vm1.ulg.ac.be
B. Gustafsson, *Uppsala Astronomical Obs., Sweden*, bg@astro.uu.se
J.-C. Hénoux, *Obs. de Paris, France*, henoux@mesiob.obspm.fr
H. Holweger, *Univ. Kiel, Germany*, holweger@astrophysik.uni-kiel.de
M. C. E. Huber, *ESTEC, Netherlands*, mhuber@astro.estec.esa.nl
Ph. Judge, *HAO Boulder, USA*, judge@hao.ucar.edu
R. Kallenbach, *ISSI, Bern, Switzerland*, kallenbach@soho.unibe.ch
J. Kohl, *Harvard-Smithsonian, Cambridge MA, USA*, kohl@cfa.harvard.edu
R. Mewaldt, *Caltech, Pasadena, USA*, dick@citsrl.srl.caltech.edu
H. Peter, *HAO/NCAR, Boulder, USA*, hpeter@jabba.hao.ucar.edu
J. Provost, *Obs. Côte d'Azur, Nice, France*, provost@obs-nice.fr
J. Raymond, *Center for Astrophysics, Cambridge MA, USA*, jraymond@cfa.harvard.edu
D. Reames, *NASA/GSFC, Greenbelt MD, USA*, reames@lheavx.gsfc.nasa.gov
F. Rogers, *Lawrence Livermore Lab., USA*, opal@west.llnl.gov
R. Rutten, *Sterrekundig Instituut, Utrecht, Netherlands*, rutten@fys.ruu.nl
S. Solanki, *ETH Zürich, Switzerland*, solanki@astro.phys.ethz.ch
Y. Suzuki, *Univ. of Tokyo, Japan*, suzuki@icrr.u-tokyo.ac.jp
S. Turck-Chièze, *CE Saclay, France*, turck@discovery.saclay.cea.fr
S. Turcotte, *Teoretisk Astrofysik Center, Aarhus, Denmark*, turcotte@obs.aau.dk
S. Vauclair, *Obs. Midi-Pyrénées, Toulouse, France*, svcr@obs-mip.fr
R. von Steiger, *ISSI, Bern, Switzerland*, vsteiger@issi.unibe.ch
K. Widing, *Naval Research Lab., Washington DC, USA*, widing@ssd0.nrl.navy.mil
R. Wieler, *ETH Zürich, Switzerland*, wieler@erdw.ethz.ch
K. Wilhelm, *MPAE, Lindau, Germany*, wilhelm@solar.stanford.edu
R. F. Wimmer-Schweingruber, *Univ. of Bern, Switzerland*, wimmer@phim.unibe.ch
P. Young, *Univ. of Cambridge, UK*, p.r.young@damtp.cam.ac.uk
J.-P. Zahn, *Obs. de Meudon, France*, jean-paul.zahn@obspm.fr
T. Zurbuchen, *Univ. of Michigan, USA*, thomasz@engin.umich.edu

Space Science Series of ISSI

1. R. von Steiger, R. Lallement and M.A. Lee (eds.): *The Heliosphere in the Local Interstellar Medium.* 1996   ISBN 0-7923-4320-4
2. B. Hultqvist and M. Øieroset (eds.): *Transport Across the Boundaries of the Magnetosphere.* 1997   ISBN 0-7923-4788-9
3. L.A. Fisk, J.R. Jokipii, G.M. Simnett, R. von Steiger and K.-P. Wenzel (eds.): *Cosmic Rays in the Heliosphere.* 1998   ISBN 0-7923-5069-3
4. N. Prantzos, M. Tosi and R. von Steiger (eds.): *Primordial Nuclei and Their Galactic Evolution.* 1998   ISBN 0-7923-5114-2
5. C. Fröhlich, M.C.E. Huber, S.K. Solanki and R. von Steiger (eds.): *Solar Composition and its Evolution – From Core to Corona.* 1998   ISBN 0-7923-5496-6

Kluwer Academic Publishers – Dordrecht / Boston / London